Metal Matrix Composites

Edited by
Karl U. Kainer

Related Titles

Leyens, C., Peters, M. (eds.)

Titanium and Titanium Alloys
Fundamentals and Applications

2003
ISBN 3-527-30534-3

Krenkel, W., Naslain, R., Schneider, H. (eds.)

High Temperature Ceramic Matrix Composites

2001
ISBN 3-527-30320-0

Kainer, K. U. (ed.)

Magnesium Alloys and Technologies

2003
ISBN 3-527-30570-X

Metal Matrix Composites

Custom-made Materials for Automotive
and Aerospace Engineering

Edited by
Karl U. Kainer

WILEY-
VCH

WILEY-VCH Verlag GmbH & Co. KGaA

The Editor

Prof. K. U. Kainer
GKSS Forschungszentrum
Institut für Werkstoffforschung
Max Planck-Str. 1
21502 Geesthacht
Germany
karl.kainer@gkss.de

Original title:
K. U. Kainer (ed.)
Metallische Verbundwerkstoffe
Wiley-VCH, 2003

Translation:
Dr. Petra Maier

Cover illustration:
V8 light metal engine block,
by Josef Schmid, NAGEL Maschinen-
und Werkzeugfabrik GmbH

■ All books published by Wiley-VCH are carefully produced. Nevertheless, authors, editors, and publisher do not warrant the information contained in these books, including this book, to be free of errors. Readers are advised to keep in mind that statements, data, illustrations, procedural details or other items may inadvertently be inaccurate.

Library of Congress Card No.
applied for

British Library Cataloguing-in-Publication Data
A catalogue record for this book is available from the British Library.

Bibliographic information published by Die Deutsche Bibliothek
Die Deutsche Bibliothek lists this publication in the Deutsche Nationalbibliografie; detailed bibliographic data is available in the Internet at http://dnd.ddb.de.

© 2006 WILEY-VCH Verlag GmbH & Co. KGaA, Weinheim

All rights reserved (including those of translation into other languages). No part of this book may be reproduced in any form – by photoprinting, microfilm, or any other means – nor transmitted or translated into a machine language without written permission from the publishers. Registered names, trademarks, etc. used in this book, even when not specifically marked as such, are not to be considered unprotected by law.

Composition   Fotosatz Detzner, Speyer
Printing   Betz-Druck GmbH, Darmstadt
Bookbinding   J. Schäffer GmbH, Grünstadt

ISBN-13: 978-3-527-31360-0
ISBN-10: 3-527-31360-5

Preface

Metal matrix composites can no longer be excluded from daily life. The individual consumer is not aware of the variety of material systems and their application; in many cases they are even unknown. Examples are carbides for machining of materials in product engineering, noble metal composite systems for contacts in electronics and electro-technology, copper–graphite sliding contacts for generators and electric motors and multicompound systems for brake linings in high speed brakes. After the massive effort in recent years to develop metal matrix composites (MMCs) with light metal matrixes the successful application of these materials has taken place in traffic engineering, especially in automotive and transport technology. New applications are, for example, partially fiber-reinforced pistons and hybrid reinforced crank cases in passenger cars and truck engines, and particle-reinforced brake discs for light trucks, motorcycles, passenger cars and rail-mounted vehicles. A further application area of these materials is in civil and military air and space flight. These innovative materials are of great interest for modern material applications due to the possibility to develop MMCs with specific properties. Ongoing from this potential, the metal matrix composites meet the desired concepts of the design engineer, because they represent custom-made materials. This material group is becoming of interest for construction and functional materials, if the property profile of conventional materials does not meet the requirement of light-weight construction. The advantages of metal matrix composites are of use if a meaningful cost–performance relationship is possible during production of the components. Of special economic and ecological interest is the need for integration of processing residues, cycle scrap and waste products from these materials into the material cycle.

In the area of communication and power engineering, the replacement of the currently dominant structural materials with functional materials based on non-ferrous and noble metals is a development aim. For example, good electrical and thermal properties combined with high strength and wear resistance are demanded. Although the application of metal matrix composites in the area of contact materials has been known for years, it has been necessary to carry out further optimization for the use of new or modified material systems. Of special interest are materials which are able to dissipate the heat which develops during the use of electrical and electronic units (heat sinks). Additional requirements concerning the effect

Metal Matrix Composites. Custom-made Materials for Automotive and Aerospace Engineering.
Edited by Karl U. Kainer.
Copyright © 2006 WILEY-VCH Verlag GmbH & Co. KGaA, Weinheim
ISBN: 3-527-31360-5

of temperature on resistance are development challenges. In many cases improvements in wear resistance will be of benefit, leading to the development of multifunctional material systems.

This book gives an overview of the current state of research and development as well as a realistic introduction to materials in various application areas. Besides basic knowledge of metal matrix composites, the idea of appropriate material systems and production and processing methods, special importance is attached to the presentation of the potential of materials and their application possibilities. This summary arose from a further education seminar of the same name, presented by the German Society of Material Science, which has taken place regularly since 1990. Because of the overview character of the book it is addressed to engineers, scientists and technicians in the material development, production and design areas. As the editor I would like to thank all the authors for their efforts in providing appropriate articles. A special thanks to the publisher Wiley-VCH, represented by Dr. Jörn Ritterbusch, for the support grants and excellent supervision, especially necessary in the critical phases.

Geesthacht, December 2005

Prof. Dr.-Ing. habil. Karl Ulrich Kainer

Contents

Preface V

1	**Basics of Metal Matrix Composites** 1	
	Karl Ulrich Kainer	
1.1	Introduction 1	
1.2	Combination of Materials for Light Metal Matrix Composites 4	
1.2.1	Reinforcements 4	
1.2.2	Matrix Alloy Systems 6	
1.2.3	Production and Processing of Metal Matrix Composites 7	
1.3	Mechanism of Reinforcement 12	
1.3.1	Long Fiber Reinforcement 13	
1.3.2	Short Fiber Reinforcement 16	
1.3.3	Strengthening by Particles 20	
1.3.4	Young's Modulus 22	
1.3.5	Thermal Expansion Coefficient 24	
1.4	Interface Influence 26	
1.4.1	Basics of Wettability and Infiltration 27	
1.4.2	Objective of Adhesion 35	
1.5	Structure and Properties of Light Metal Composite Materials 40	
1.6	Possible Applications of Metal Matrix Composites 48	
1.7	Recycling 52	
	References 52	
2	**Particles, Fibers and Short Fibers for the Reinforcement of Metal Materials** 55	
	Hajo Dieringa and Karl Ulrich Kainer	
2.1	Introduction 55	
2.2	Particles 56	
2.2.1	Fibers 59	
2.3	Continuous Fibers 61	
2.3.1	Monofilaments 62	
2.3.2	Multifilament Fibers 65	
2.3.2.1	Carbon Fibers 65	

Metal Matrix Composites. Custom-made Materials for Automotive and Aerospace Engineering.
Edited by Karl U. Kainer.
Copyright © 2006 WILEY-VCH Verlag GmbH & Co. KGaA, Weinheim
ISBN: 3-527-31360-5

Contents

2.3.2.2 Oxide Ceramic Fibers 68
2.3.2.3 SiC Multifilament Fibers 70
2.3.2.4 Delivery Shapes of Multifilament Fibers 72
2.4 Short Fibers and Whiskers 73
References 75

3 Preforms for the Reinforcement of Light Metals – Manufacture, Applications and Potential 77
R. Buschmann
3.1 Introduction 77
3.2 Manufacturing Principle of Preforms 78
3.2.1 Short Fiber Preforms 78
3.2.2 Hybrid Preforms 85
3.3 Current Applications 88
3.3.1 Aluminum Diesel Piston with Fiber Reinforced Combustion Bowl 88
3.3.2 Aluminum Cylinder Heads with Fiber Reinforced Valve Bridges 90
3.3.3 Cylinder Liner Reinforcement in Aluminum Crankcases, Lokasil 90
3.3.4 Al-MMC Bearing Blocks for Crankshafts 92
3.3.5 Al-MMC Brake Disks 92
3.4 Summary and Outlook 93
References 94

4 Aluminum-matrix Composite Materials in Combustion Engines 95
E. Köhler and J. Niehues
4.1 Introduction 95
4.2 Cylinder Crankcase Design Concepts and Cylinder Surface Technology 96
4.2.1 ALUSIL® 98
4.2.2 Heterogeneous Concepts 99
4.2.3 Quasi-monolithic Concept 100
4.2.4 The LOKASIL® Concept [9] 101
4.3 Production of LOKASIL® Cylinder Crankcases 103
4.3.1 Introduction 103
4.3.2 Preform Manufacture 103
4.3.3 Casting Process 105
4.4 Summary and Outlook 108
References 108

5 Production of Composites or Bonding of Material by Thermal Coating Processes 111
B. Wielage, A. Wank, and J. Wilden
5.1 Introduction 111
5.2 Thermal Spraying 112
5.2.1 Spraying Additive Materials 113
5.2.2 Substrate Materials 115

5.2.3	Surface Preparation	115
5.2.4	Structure and Properties of Spray Coatings	116
5.2.5	Adhesion of Thermally Sprayed Coatings	119
5.2.6	Thermal Spraying Processes	120
5.2.6.1	Flame Spraying	120
5.2.6.1.1	Powder Flame Spraying	120
5.2.6.1.2	Plastic Flame Spraying	121
5.2.6.1.3	Wire/Rod Flame Spraying	121
5.2.6.2	Detonation Spraying	122
5.2.6.3	High Velocity Flame Spraying	123
5.2.6.4	Cold Gas Spraying	125
5.2.6.5	Arc Spraying	126
5.2.6.6	Plasma Spraying	127
5.2.6.6.1	DC Plasma Spraying	127
5.2.6.6.2	HF Plasma Spraying	128
5.2.7	New Applications	129
5.2.8	Quality Assurance	131
5.2.9	Environmental Aspects	132
5.3	Cladding	132
5.3.1	Coating Material	133
5.3.2	Substrate Materials	137
5.3.3	Cladding Processes	137
5.3.3.1	Autogenous Cladding	137
5.3.3.2	Open Arc Cladding (OA)	137
5.3.3.3	Underpowder Cladding (UP)	138
5.3.3.4	Resistance Electro Slag Welding (RES)	139
5.3.3.5	Metal Inert Gas Welding	140
5.3.3.6	Plasma MIG Cladding	141
5.3.3.7	Plasma Powder Transferred Arc Welding (PTA)	142
5.3.3.8	Plasma Hot Wire Cladding	143
5.4	Summary and Outlook	144
	References	145
6	**Machining Technology Aspects of Al-MMC**	**147**
	K. Weinert, M. Buschka, and M. Lange	
6.1	Introduction	147
6.2	Machining Problems, Cutting Material Selection and Surface Layer Influence	147
6.3	Processing of Components of Metal Matrix Composites	152
6.3.1	Materials, Cutting Materials and Process Parameters	153
6.3.2	Evaluation of Machinability	154
6.3.3	Turning of SiC-particle-reinforced Brake Drums	154
6.3.4	Boring of Si-particle- and Al_2O_3-fiber-reinforced Al Cylinder Surfaces	161
6.3.4.1	Pre-boring Operation	161

6.3.4.2	Finish Bore Processing 163
6.3.5	Drilling and Milling of TiB_2-particle-reinforced Extruded Profiles 166
6.3.5.1	Drilling 166
6.3.5.2	Milling 168
6.4	Summary 171
	References 171

7	**Mechanical Behavior and Fatigue Properties of Metal-Matrix Composites** 173
	H. Biermann and O. Hartmann
7.1	Introduction 173
7.2	Basics and State of Knowledge 174
7.2.1	Thermal Residual Stresses 174
7.2.2	Deformation Behavior of Metal-Matrix Composites 174
7.2.3	Determination of the Damage in Composites 176
7.2.4	Basic Elements and Terms of Fatigue 178
7.2.5	Fatigue Behavior of Composites 183
7.3	Experimental 185
7.3.1	Materials 185
7.3.2	Mechanical Tests 186
7.4	Results and Comparison of Different MMCs 186
7.4.1	Cyclic Deformation Behavior 186
7.4.2	Fatigue Life Behavior 188
7.4.3	Damage Evolution 192
7.5	Summary 193
	Acknowledgement 194
	References 194

8	**Interlayers in Metal Matrix Composites: Characterisation and Relevance for the Material Properties** 197
	J. Woltersdorf, A. Feldhoff, and E. Pippel
8.1	Summary 197
8.2	The Special Role of Interfaces and Interlayers 197
8.3	Experimental 198
8.4	Interlayer Optimisation in C/Mg–Al Composites by Selection of Reaction Partners 199
8.5	Interlayer Optimisation in C/Mg–Al Composites by Fiber Precoating 205
	Acknowledgements 211
	References 211

9	**Metallic Composite Materials for Cylinder Surfaces of Combustion Engines and Their Finishing by Honing** 215
	J. Schmid
9.1	Introduction 215

9.2	Composites Based on Light Metals	*215*
9.2.1	Manufacturing Possibilities	*215*
9.2.1.1	Casting of Over-eutectic Alloys	*215*
9.2.1.2	Infiltration	*216*
9.2.1.3	Sintering	*216*
9.2.1.4	Stirring of Hard Particles into the Melt	*216*
9.2.1.5	Spray Forming	*218*
9.2.1.6	Addition of Reactive Components into the Melt	*218*
9.2.1.7	Thermal Coating	*218*
9.2.1.8	Laser Alloying	*218*
9.2.2	Selection Criteria	*219*
9.2.2.1	Strength	*219*
9.2.2.2	Tribology	*220*
9.2.2.3	Flexibility	*220*
9.2.2.4	Design Criteria	*220*
9.2.2.5	Processing and Machining	*221*
9.2.2.6	Strength of the Material Composite	*221*
9.2.2.7	Heat Transmission Ability and Heat Expansion	*221*
9.2.3	Fine Processing	*221*
9.2.3.1	Processing before Honing	*222*
9.2.3.2	Honing Step 1	*222*
9.2.3.3	Honing Step 2	*223*
9.2.3.4	Honing Step 3	*224*
9.2.4	Marginal Conditions	*228*
9.2.4.1	Expanding systems	*228*
9.2.4.2	Cooling Lubricants	*228*
9.2.4.3	Cutting Speeds	*229*
9.2.5	Summary	*229*
9.3	Plasma Coatings	*230*
9.3.1	General	*230*
9.3.1.1	Coating Materials	*231*
9.3.1.1.1	Layer Characteristics and Tribological Properties	*231*
9.3.1.1.2	Application Potential	*232*
9.3.1.2	Comparison of Honing with other Processing Technologies	*232*
9.3.2	Definition of the Process Task	*234*
9.3.2.1	Process Adding, Geometry	*234*
9.3.2.2	Requirements of the Surface Processing	*234*
9.3.3	Results of Honing Tests	*236*
9.3.3.1	Investigation of Adhesion	*236*
9.3.3.1.1	Adhesion of the Plasma Coating at High Machining Rates (Rough Honing)	*236*
9.3.3.1.2	Adhesion Strength during Honing of Thin Coatings	*236*
9.3.3.2	Surface Qualities, Removal of Spalling	*237*
9.3.3.3	Reachable Form Accuracies	*239*
9.3.3.4	Processing of Metal–Ceramic Coatings	*239*

9.3.3.5	Cooling Lubricants	*241*
9.3.3.5.1	Honing of Pure Metallic Plasma Coatings	*241*
9.3.3.5.2	Metal Composites	*241*
9.3.4	Summary	*241*
	References	*242*

10 Powder Metallurgically Manufactured Metal Matrix Composites *243*
Norbert Hort and Karl Ulrich Kainer

10.1	Summary	*243*
10.2	Introduction	*243*
10.3	Source Materials	*245*
10.3.1	Metallic Powders	*245*
10.3.2	Ceramic Reinforcement Components	*248*
10.4	Manufacture of MMCs	*249*
10.4.1	Mechanical Alloying	*253*
10.4.2	*In situ* Composite Materials	*258*
10.4.3	Mixing	*259*
10.4.4	Consolidation	*260*
10.4.5	Spray Forming	*262*
10.4.6	Subsequent Processing	*262*
10.5	Materials	*263*
10.5.1	Magnesium-based MMCs	*263*
10.5.2	Aluminum-based MMCs	*264*
10.5.3	Titanium-based MMCs	*266*
10.5.4	Copper-based MMCs	*267*
10.5.5	Iron-based MMCs	*269*
10.5.6	Nickel-based MMCs	*271*
10.6	Summary and Outlook	*272*
	References	*272*

11 Spray Forming – An Alternative Manufacturing Technique for MMC Aluminum Alloys *277*
P. Krug, G. Sinha

11.1	Introduction	*277*
11.2	Spray Forming	*280*
11.3	Techniques	*281*
11.3.1	Rapid Solidification (RS) Technique	*281*
11.3.2	Spray Forming Technique	*282*
11.3.3	Melting Concept	*283*
11.3.4	Atomisation of the Metal Melt	*284*
11.3.5	Nozzle Unit	*284*
11.3.6	Primary Gas Nozzle	*285*
11.3.7	Secondary Gas Nozzle	*285*
11.3.8	Plant Safety	*287*
11.3.9	Re-injection of Overspray Powder	*288*

11.3.10	Functional Ways of Injecting for Re-injection *288*	
11.4	Materials *289*	
11.4.1	Spray Forming Products for Automotive Applications *289*	
11.4.2	Spray Forming MMC Materials *290*	
	Acknowledgement *293*	
	References *293*	
12	**Noble and Nonferrous Metal Matrix Composite Materials** *295*	
	C. Blawert	
12.1	Introduction *295*	
12.2	Layer Composite Materials *295*	
12.2.1	Contact- and Thermo-bimetals *296*	
12.2.2	Wear-protection Layers with Embedded Ceramic Particles *298*	
12.3	Particle Reinforced Composites *299*	
12.4	Infiltration Composites *302*	
12.5	Fiber Reinforced Composites *303*	
	References *306*	

Subject Index *309*

List of Contributors

H. Biermann
Institute for Materials Engineering
Technische Universität
Bergakademie Freiberg
Gustav-Zeuner-Str. 5
09599 Freiberg
Germany

C. Blawert
Centre of Magnesium Technology
GKSS Research Centre Geesthacht
GmbH
Max-Planck-Straße 1
21502 Geesthacht
Germany

M. Buschka
Faculty of Mechanical Engineering
Department of Machining Technology
University of Dortmund
Baroper Straße 301
44227 Dortmund
Germany

R. Buschmann
Thermal Ceramics de France
Route de Lauterbourg BP 148
67163 Wissembourg
France

H. Dieringa
Centre of Magnesium Technology
GKSS Research Centre Geesthacht
GmbH
Max-Planck-Straße 1
21502 Geesthacht
Germany

A. Feldhoff
Universität Hannover
Institut für Physikalische Chemie und
Elektrochemie
Callinstraße 3-3A
30167 Hannover
Germany

O. Hartmann
Robert Bosch GmbH
Robert-Bosch-Straße 40
96050 Bamberg
Germany

N. Hort
Centre of Magnesium Technology
GKSS Research Centre Geesthacht
GmbH
Max-Planck-Straße 1
21502 Geesthacht
Germany

List of Contributors

K. U. Kainer
Centre of Magnesium Technology
GKSS Research Centre Geesthacht GmbH
Max-Planck-Straße 1
21502 Geesthacht
Germany

E. Köhler
KS Aluminum Technology AG
Hafenstraße 25
74172 Neckarsulm
Germany

P. Krug
PEAK Werkstoff GmbH
Siebeneiker Straße 235
42553 Velbert
Germany

M. Lange
Faculty of Mechanical Engineering
Department of Machining Technology
University of Dortmund
Baroper Straße 301
44227 Dortmund
Germany

J. Niehues
KS Aluminum-Technology AG
Hafenstraße 25
74172 Neckarsulm
Germany

E. Pippel
Max-Planck-Institut für
Mikrostrukturphysik
Weinberg 2
06120 Halle
Germany

J. Schmidt
Head of R&D Department
NAGEL Maschinen- und
Werkzeugfabrik GmbH
Oberboihinger Straße 60
72622 Nuertingen
Germany

G. Sinha
PEAK Werkstoff GmbH
Siebeneiker Straße 235
42553 Velbert
Germany

A. Wank
Department of Composite Materials
Technical University of Chemnitz
Straße der Nationen 62
09107 Chemnitz
Germany

K. Weinert
Faculty of Mechanical Engineering
Department of Machining Technology
University of Dortmund
Baroper Straße 301
44227 Dortmund
Germany

B. Wielage
Department of Composite Materials
Technical University of Chemnitz
Straße der Nationen 62
09107 Chemnitz
Germany

J. Wilden
Department of Manufacturing
Engineering
Technical University of Ilmenau
Neuhaus 1
98693 Ilmenau
Germany

J. Woltersdorf
Max-Planck-Institut für
Mikrostrukturphysik
Weinberg 2
06120 Halle
Germany

1
Basics of Metal Matrix Composites

Karl Ulrich Kainer

1.1
Introduction

Metal composite materials have found application in many areas of daily life for quite some time. Often it is not realized that the application makes use of composite materials. These materials are produced *in situ* from the conventional production and processing of metals. Here, the Dalmatian sword with its meander structure, which results from welding two types of steel by repeated forging, can be mentioned. Materials like cast iron with graphite or steel with a high carbide content, as well as tungsten carbides, consisting of carbides and metallic binders, also belong to this group of composite materials. For many researchers the term metal matrix composites is often equated with the term light metal matrix composites (MMCs). Substantial progress in the development of light metal matrix composites has been achieved in recent decades, so that they could be introduced into the most important applications. In traffic engineering, especially in the automotive industry, MMCs have been used commercially in fiber reinforced pistons and aluminum crank cases with strengthened cylinder surfaces as well as particle-strengthened brake disks.

These innovative materials open up unlimited possibilities for modern material science and development; the characteristics of MMCs can be designed into the material, custom-made, dependent on the application. From this potential, metal matrix composites fulfill all the desired conceptions of the designer. This material group becomes interesting for use as constructional and functional materials, if the property profile of conventional materials either does not reach the increased standards of specific demands, or is the solution of the problem. However, the technology of MMCs is in competition with other modern material technologies, for example powder metallurgy. The advantages of the composite materials are only realized when there is a reasonable cost – performance relationship in the component production. The use of a composite material is obligatory if a special property profile can only be achieved by application of these materials.

The possibility of combining various material systems (metal – ceramic – non-metal) gives the opportunity for unlimited variation. The properties of these new materials are basically determined by the properties of their single components. Figure 1.1 shows the allocation of the composite materials into groups of various types of materials.

The reinforcement of metals can have many different objectives. The reinforcement of light metals opens up the possibility of application of these materials in areas where weight reduction has first priority. The precondition here is the improvement of the component properties. The development objectives for light metal composite materials are:

- Increase in yield strength and tensile strength at room temperature and above while maintaining the minimum ductility or rather toughness,
- Increase in creep resistance at higher temperatures compared to that of conventional alloys,
- Increase in fatigue strength, especially at higher temperatures,
- Improvement of thermal shock resistance,
- Improvement of corrosion resistance,
- Increase in Young's modulus,
- Reduction of thermal elongation.

To summarize, an improvement in the weight specific properties can result, offering the possibilities of extending the application area, substitution of common materials and optimisation of component properties. With functional materials there is another objective, the precondition of maintaining the appropriate function of the material. Objectives are for example:

- Increase in strength of conducting materials while maintaining the high conductivity,
- Improvement in low temperature creep resistance (reactionless materials),
- Improvement of burnout behavior (switching contact),
- Improvement of wear behavior (sliding contact),
- Increase in operating time of spot welding electrodes by reduction of burn outs,
- Production of layer composite materials for electronic components,
- Production of ductile composite superconductors,
- Production of magnetic materials with special properties.

For other applications different development objectives are given, which differ from those mentioned before. For example, in medical technology, mechanical properties, like extreme corrosion resistance and low degradation as well as biocompatibility are expected.

Although increasing development activities have led to system solutions using metal composite materials, the use of especially innovative systems, particularly in the area of light metals, has not been realised. The reason for this is insufficient process stability and reliability, combined with production and processing problems and inadequate economic efficiency. Application areas, like traffic engineering, are very cost orientated and conservative and the industry is not willing to pay additional costs for the use of such materials. For all these reasons metal matrix composites are only at the beginning of the evolution curve of modern materials, see Fig. 1.2.

1.1 Introduction

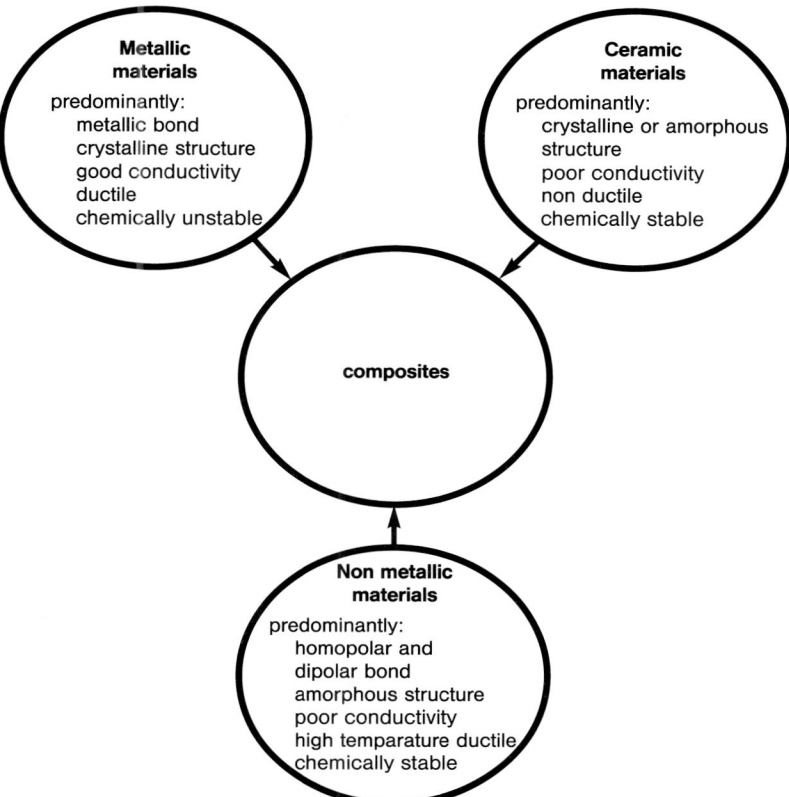

Fig. 1.1 Classification of the composite materials within the group of materials [1].

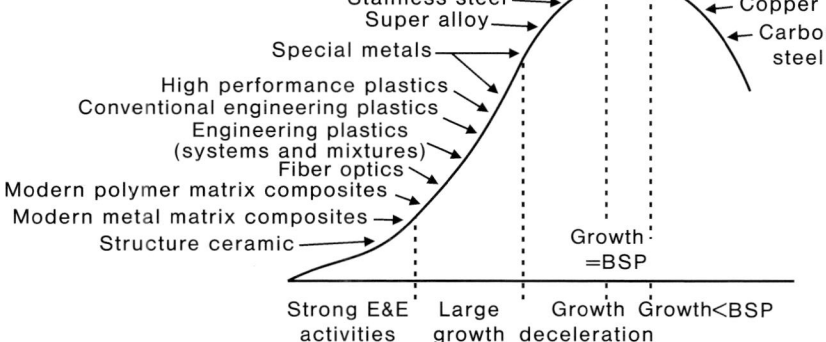

Fig. 1.2 Development curve of the market for modern materials [2].

Metal matrix composites can be classified in various ways. One classification is the consideration of type and contribution of reinforcement components in particle-, layer-, fiber- and penetration composite materials (see Fig. 1.3). Fiber composite materials can be further classified into continuous fiber composite materials (multi- and monofilament) and short fibers or, rather, whisker composite materials, see Fig. 1.4.

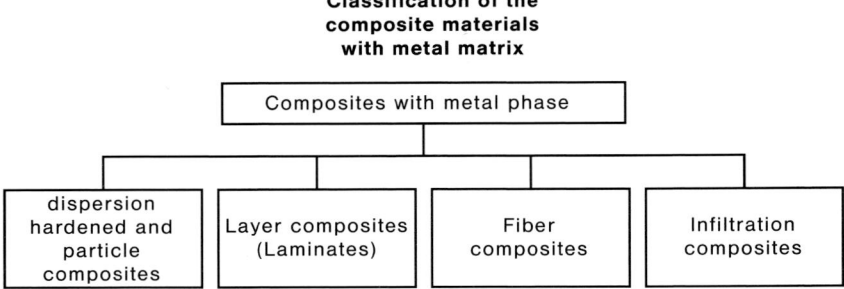

Fig. 1.3 Classification of composite materials with metal matrixes.

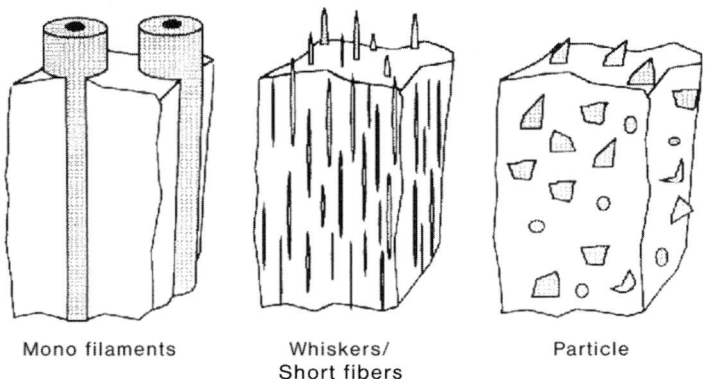

Fig. 1.4 Schematic presentation of three shapes of metal matrix composite materials [3].

1.2
Combination of Materials for Light Metal Matrix Composites

1.2.1
Reinforcements

Reinforcements for metal matrix composites have a manifold demand profile, which is determined by production and processing and by the matrix system of the composite material. The following demands are generally applicable [4]:

- low density,
- mechanical compatibility (a thermal expansion coefficient which is low but adapted to the matrix),
- chemical compatibility,
- thermal stability,
- high Young's modulus,
- high compression and tensile strength,
- good processability,
- economic efficiency.

These demands can be achieved only by using non-metal inorganic reinforcement components. For metal reinforcement ceramic particles or, rather, fibers or carbon fibers are often used. Due to the high density and the affinity to reaction with the matrix alloy the use of metallic fiber usual fails. Which components are finally used, depends on the selected matrix and on the demand profile of the intended application. In Refs. [4, 5] information about available particles, short fibers, whiskers and continuous fibers for the reinforcement of metals is given, including data of manufacturing, processing and properties. Representative examples are shown in Table 1.1. The production, processing and type of application of various reinforcements depends on the production technique for the composite materials, see Refs. [3, 7]. A combined application of various reinforcements is also possible (hybrid technique) [3, 8].

Every reinforcement has a typical profile, which is significant for the effect within the composite material and the resulting profile. Table 1.2 gives an overview of possible property profiles of various material groups. Figure 1.5 shows the specific strength and specific Young's modulus of quasi-isotropic fiber composite materials with various matrixes in comparison to monolithic metals. The group of discontinuous reinforced metals offers the best conditions for reaching development targets; the applied production technologies and reinforcement components, like short fibers, particle and whiskers, are cost effective and the production of units in large item numbers is possible. The relatively high isotropy of the properties in comparison to the long-fiber continuous reinforced light metals and the possibility

Tab. 1.1 Properties of typical discontinuous reinforcements for aluminium and magnesium reinforcements [6].

Reinforcement	Saffil (Al_2O_3)	SiC particle	Al_2O_3 particle
crystal structure	δ-Al_2O_3	hexagonal	hexagonal
density (g cm^{-3})	3.3	3.2	3.9
average diameter (µm)	3.0	variable	variable
length (µm)	ca. 150	–	–
Mohs hardness	7.0	9.7	9.0
strength (MPa)	2000	–	–
Young's Modulus (GPa)	300	200–300	380

Tab. 1.2 Property potential of different metal matrix composites, after [2].

MMC type	Properties Strength	Young's modulus	High temperature properties	Wear	Expansion coefficient	Costs
mineral wool: MMC	*	*	**	**	*	medium
discontinuous reinforced MMC	**	**	*	***	**	low
long fiber reinforced MMC: C fibers	**	**	**	*	***	high
other fibers	***	***	***	*	**	high

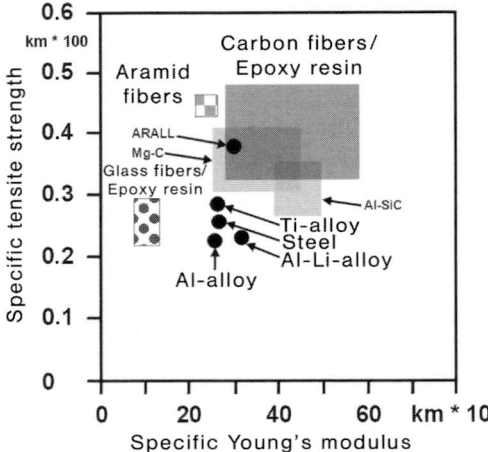

Fig. 1.5 Specific tensile strength and specific Young's modulus of different quasi-isotropic fiber composite materials in comparison to some metal alloys, after [2].

of processing of composites by forming and cutting production engineering are further advantages.

1.2.2
Matrix Alloy Systems

The selection of suitable matrix alloys is mainly determined by the intended application of the composite material. With the development of light metal composite materials that are mostly easy to process, conventional light metal alloys are applied as matrix materials. In the area of powder metallurgy special alloys can be applied due to the advantage of fast solidification during the powder production. Those systems are free from segregation problems that arise in conventional solidification. Also the application of systems with oversaturated or metastable structures is possible. Examples for matrix configurations are given in Refs. [7, 9–15]:

- conventional cast alloys
  - G-AlSi12CuMgNi
  - G-AlSi9Mg
  - G-AlSi7 (A356)
  - AZ91
  - AE42

- conventional wrought alloys
  - AlMgSiCu (6061)
  - AlCuSiMn (2014)
  - AlZnMgCu1.5 (7075)
  - TiAl6V4

- special alloys
  - Al–Cu–Mg–Ni–Fe-alloy (2618)
  - Al–Cu–Mg–Li-alloy (8090)
  - AZ91Ca

For functional materials non-alloyed or low-alloyed non-ferrous or noble metals are generally used. The reason for this is the demand for the retention of the high conductivity or ductility. A dispersion hardening to reach the required mechanical characteristics at room or higher temperatures is then an optimal solution.

1.2.3
Production and Processing of Metal Matrix Composites

Metal matrix composite materials can be produced by many different techniques. The focus of the selection of suitable process engineering is the desired kind, quantity and distribution of the reinforcement components (particles and fibers), the matrix alloy and the application. By altering the manufacturing method, the processing and the finishing, as well as by the form of the reinforcement components it is possible to obtain different characteristic profiles, although the same composition and amounts of the components are involved. The production of a suitable precursor material, the processing to a construction unit or a semi-finished material (profile) and the finishing treatment must be separated. For cost effective reasons prototypes, with dimensions close to the final product, and reforming procedures are used, which can minimize the mechanical finishing of the construction units.

In general the following product engineering types are possible:

- Melting metallurgical processes
  - infiltration of short fiber-, particle- or hybrid preforms by squeeze casting, vacuum infiltration or pressure infiltration [7, 13–15]
  - reaction infiltration of fiber- or particle preforms [16, 17]
  - processing of precursor material by stirring the particles in metallic melts, followed by sand casting, permanent mold casting or high pressure die casting [9, 10]

- Powder metallurgical processes
  - pressing and sintering and/or forging of powder mixtures and composite powders
  - extrusion or forging of metal-powder particle mixtures [11, 12]
  - extrusion or forging of spraying compatible precursor materials [7, 18, 19]
- Hot isostatic pressing of powder mixtures and fiber clutches
- Further processing of precursor material from the melting metallurgy by thixocasting or -forming, extrusion [20], forging, cold massive forming or super plastic forming
- Joining and welding of semi-manufactured products
- Finishing by machining techniques [21]
- Combined deformation of metal wires (group superconductors).

Melting metallurgy for the production of MMCs is at present of greater technical importance than powder metallurgy. It is more economical and has the advantage of being able to use well proven casting processes for the production of MMCs. Figure 1.6 shows schematically the possible methods of melting metallurgical production. For melting metallurgical processing of composite materials three procedures are mainly used [15]:

- compo-casting or melt stirring
- gas pressure infiltration
- squeeze casting or pressure casting.

Both the terms compo-casting and melt stirring are used for stirring particles into a light alloy melt. Figure 1.7 shows the schematic operational sequence of this procedure. The particles are often tend to form agglomerates, which can be only dissolved by intense stirring. However, here gas access into the melt must be absolutely avoided, since this could lead to unwanted porosities or reactions. Careful attention must be paid to the dispersion of the reinforcement components, so that the reactivity of the components used is coordinated with the temperature of the melt and the duration of stirring, since reactions with the melt can lead to the dissolution of the reinforcement components. Because of the lower surface to volume ratio of spherical particles, reactivity is usually less critical with stirred particle reinforcement than with fibers. The melt can be cast directly or processed with alternative procedures such as squeeze casting or thixocasting. Melt stirring is used by the Duralcan Company for the production of particle-strengthened aluminum alloys [9, 10]. At the Lanxide Company a similar process is used, with additional reactions between the reinforcement components and the molten matrix being purposefully promoted to obtain a qualitatively high-grade composite material [16]. In the reaction procedures of the Lanxide Company it may be desirable that the reinforcement component reacts completely with the melt to form the component *in situ*, which then transfers the actual reinforcement effect to the second phase in the MMC.

In gas pressure infiltration the melt infiltrates the preform with a gas applied from the outside. A gas that is inert with respect to the matrix is used. The melting

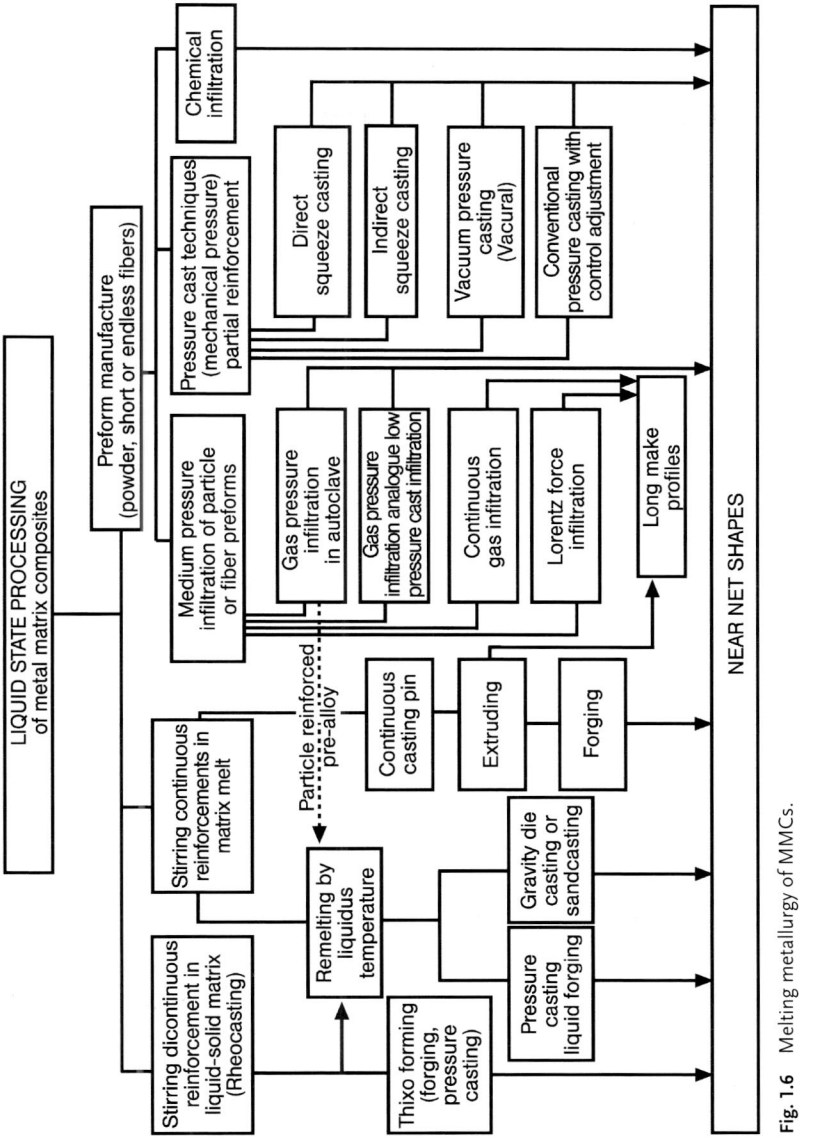

Fig. 1.6 Melting metallurgy of MMCs.

of the matrix and the infiltration take place in a suitable pressure vessel. There are two procedure variants of gas pressure infiltration: in the first variant the warmed up preform is dipped into the melt and then the gas pressure is applied to the surface of the melt, leading to infiltration. The infiltration pressure can thereby be coordinated with the wettability of the preforms, which depends, among other things, on the volume percentage of the reinforcement. The second variant of the

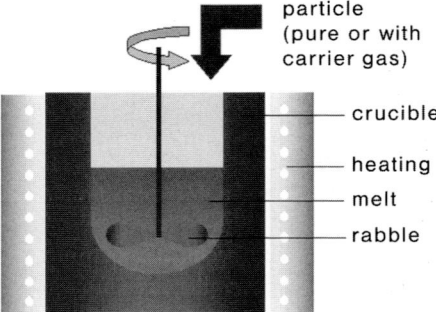

Fig. 1.7 Schematic operational sequence during melt stirring.

gas pressure infiltration procedure reverses the order: the molten bath is pressed to the preform by the applied gas pressure using a standpipe and thereupon infiltrates the bath (see Fig. 1.8). The advantage of this procedure is that there is no development of pores when completely dense parts are present. Since the reaction time is relatively short with these procedures, more reactive materials can be used than e.g. with the compo-casting. In gas pressure infiltration the response times are clearly longer than in squeeze casting, so that the materials must be carefully selected and coordinated, in order to be able to produce the appropriate composite material for the appropriate requirements.

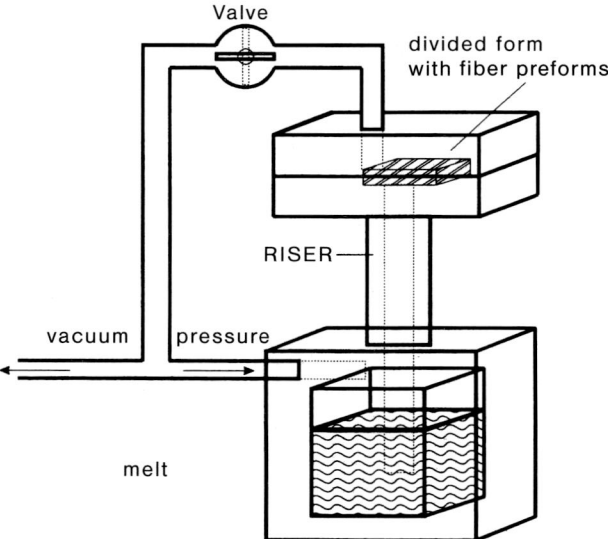

Fig. 1.8 Gas pressure infiltration technique.

Squeeze casting or pressure casting are the most common manufacturing variants for MMCs. After a slow mold filling the melt solidifies under very high pressure, which leads to a fine-grained structure. In comparison with die-casted parts the squeeze-casted parts do not contain gas inclusions, which permits thermal treatment of the produced parts. One can differentiate between direct and indirect squeeze casting (Fig. 1.9). With direct squeeze casting the pressure for the infiltration of the prefabricated preforms is applied directly to the melt. The die is thereby part of the mold, which simplifies the structure of the tools substantially. However, with the direct procedure there is a disadvantage in that the volume of the melt must be determined exactly, since no gate is present and thus the quantity of the melt determines the size of the cast construction unit. A further disadvantage is the appearance of oxidation products, formed in the cast part during dosage. In contrast, in indirect squeeze casting, where the melt is pressed into the form via a gate system, the residues will remain in this gate. The flow rate of the melt through a gate is, due to its larger diameter, substantially less than with die casting, which results in a less turbulent mold filling and gas admission to the melt by turbulences is avoided.

Both pressure casting processes make the production of composite materials possible, as prefabricated fiber or particle preforms are infiltrated with melt and solidify under pressure. A two-stage process is often used. In the first stage the melt is pressed into the form at low pressure and then at high pressure for the solidification phase. This prevents damage to the preform by too fast infiltration. The squeeze casting permits the use of relatively reactive materials, since the duration of the infiltration and thus the response time, are relatively short. A further advantage is the possibility to manufacture difficultly shaped construction units and to provide partial reinforcement, to strengthen those areas which are exposed to a higher stress during service.

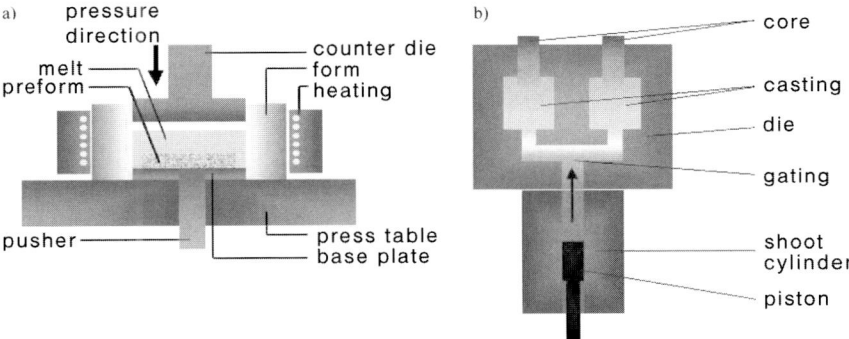

Fig. 1.9 Direct and indirect squeeze casting.

1.3
Mechanism of Reinforcement

The characteristics of metal matrix composite materials are determined by their microstructure and internal interfaces, which are affected by their production and thermal mechanical prehistory. The microstructure covers the structure of the matrix and the reinforced phase. The chemical composition, grain and/or sub-grain size, texture, precipitation behavior and lattice defects are of importance to the matrix. The second phase is characterised by its volume percentage, its kind, size, distribution and orientation. Local varying internal tension due to the different thermal expansion behavior of the two phases is an additional influencing factor.

With knowledge of the characteristics of the components, the volume percentages, the distribution and orientation it might be possible to estimate the characteristics of metallic composite materials. The approximations usually proceed from ideal conditions, i.e. optimal boundary surface formation, ideal distribution (very small number of contacts of the reinforcements among themselves) and no influence of the component on the matrix (comparable structures and precipitation behavior). However, in reality a strong interaction arises between the components involved, so that these models can only indicate the potential of a material. The different micro-, macro- and meso-scaled models proceed from different conditions and are differently developed. A representation of these models can be seen in Refs. [3, 23]. In the following, simple models are described, which facilitate our understanding of the effect of the individual components of the composite materials and their form and distribution on the characteristics of the composite.

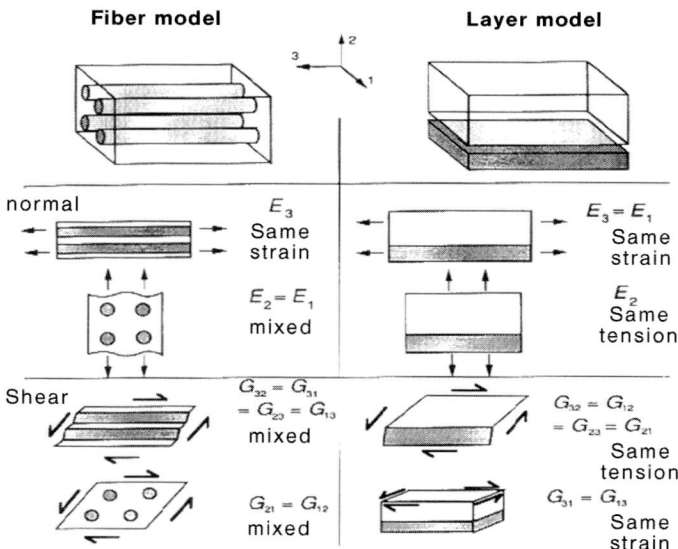

Fig. 1.10 Schematic presentation of elastic constants in composite materials.

Simply, we can consider a fiber and/or a plate model. Depending on the load direction, different elastic constants in the metallic composite material can result. Figure 1.10 illustrates the two different models and shows the resulting E and G-moduli as a function of the load type. On the basis of these simple considerations an estimate can be made of the attainable strength of the fiber reinforced composite material for the different forms of the fibers.

1.3.1
Long Fiber Reinforcement

For the optimal case of a single orientation in the direction of the stress, no fiber contact and optimal interface formation (Fig. 1.11), it is possible to use the linear mixture rule to calculate the strength of an ideal long fiber reinforced composite material with stress in the fiber orientation [23]:

$$\sigma_C = \Phi_F \cdot \sigma_F + (1 - \Phi_F) \cdot \sigma^*_M \tag{1}$$

where σ_C is the strength of the composite, Φ_F the fiber volume content, σ_F the fiber tensile strength and σ^*_M the matrix yield strength. From this basic correlation the critical fiber content $\Phi_{F,crit}$, which must be exceeded to reach an effective strengthening effect, can be determined. This specific value is important for the development of long fiber composites:

Limit of reinforcement:

$$\sigma_M = \Phi_{F,crit} \cdot \sigma_F + (1 - \Phi_{F,crit}) \cdot \sigma^*_M \tag{2}$$

Critical fiber content:

$$\Phi_{F,crit} = \frac{\sigma_M - \sigma^*_M}{\sigma_F - \sigma_M} \tag{3}$$

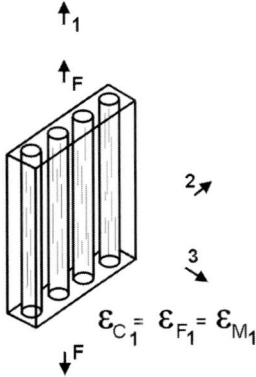

Fig. 1.11 Load of a unidirectional fiber composite layer with a force F in the fiber direction.

1 Basics of Metal Matrix Composites

Approximation of high fiber strength:

$$\Phi_{F,crit} = \frac{\sigma_M - \sigma^*_M}{\sigma_F} \tag{4}$$

Figure 1.12 shows the dependence of the tensile strength of unidirectional fiber composite materials on the fiber content. The basis is the use of a low strength ductile matrix and of high-strength fibers with high Young's modulus. For different matrix fiber combinations different behavior of the materials results. In Fig. 1.13

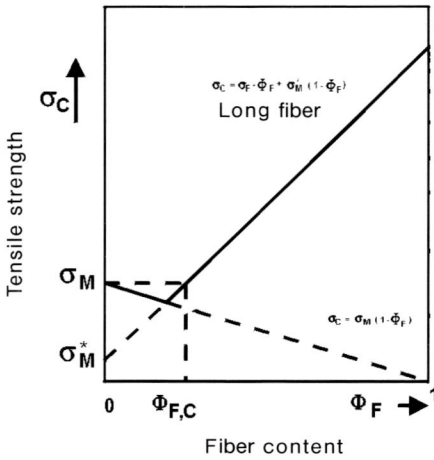

Fig. 1.12 Linear mixture rule for tensile strength of unidirectional fiber composite materials with a ductile matrix and high strength fibers [23].

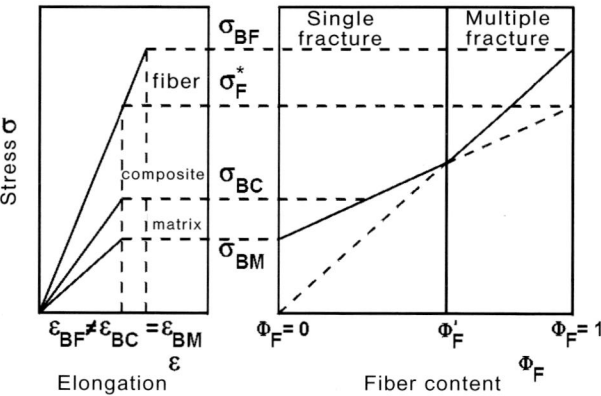

Fig. 1.13 Stress–strain behavior of a fiber composite material with a ductile matrix, in which elongation at fracture is higher than that of the fibers (σ_{BF}=tensile strength of the fiber, σ^*_F=effective fiber strength at the fracture of the composite material, σ_{BC}=strength of composite material, σ_{BM}=matrix strength, ε_{BF}= elongation at fracture of the fiber, ε_{BM}=elongation at fracture of the matrix, ε_{BC}= elongation at fracture of the composite material) [24].

1.3 Mechanism of Reinforcement

the stress–strain behavior of fiber composite materials with a ductile matrix, whose tensile strength is larger than of the fibers itself (according to Fig. 1.12) is shown. Above the critical fiber content $\Phi_{F,crit}$ the behavior is affected considerably by the fiber. On reaching the fiber strength a simple brittle failure develops and the composite material fails.

For composite materials with a brittle matrix, where no hardening arises and where the elongation to fracture is smaller than those of the fibers, the material fails on reaching the strength of the matrix below the critical fiber content (see Fig. 1.14). Above this critical parameter a higher number of fibers can carry more load and a larger reinforcement effect develops. In the case of a composite material with a ductile matrix and ductile fibers; where both exhibit hardening during the tensile test, the deformation behavior is, in principle, different (Fig. 1.15). The resulting

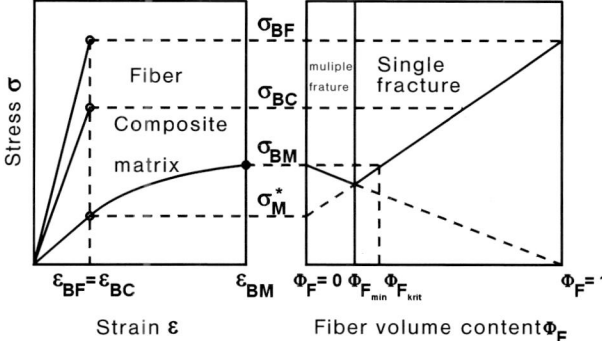

Fig. 1.14 Stress–strain behavior of a fiber composite material with a brittle matrix, which shows no strengthening behavior and whose elongation at fracture is smaller than that of the fibers (σ_{BF} = tensile strength of the fiber, σ^*_F = effective fiber strength at fracture of the composite material, σ_{BC} = strength of the composite material, σ_{BM} = matrix strength, ε_{BF} = elongation at fracture of the fiber, ε_{BM} = elongation at fracture of the matrix, ε_{BC} = elongation at fracture of composite material) [24].

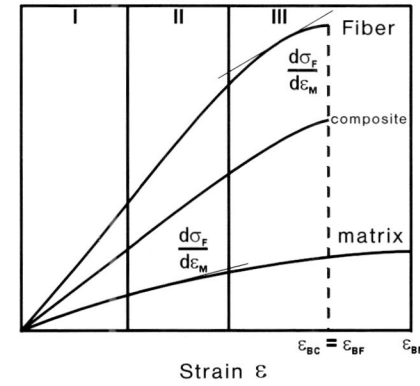

Fig. 1.15 Stress–strain behavior of fiber composite materials with a ductile matrix and fibers, both have strength in the tensile test (ε_{BF} = elongation at fracture of the fiber, ε_{BM} = elongation at fracture of the matrix, ε_{BC} = elongation at fracture of the composite material) [24].

stress–strain curve can be divided into three ranges: range I is characterized by the elastic behavior of both components by a Young's modulus in accord with the linear mixture rule. In range II only the matrix shows a strain hardening, the fiber is still elastically elongated. Here the composite material behaves as represented in Fig. 1.13. In range III both matrix and fiber show strain hardening behavior: the composite material fails after reaching the fiber strength.

1.3.2
Short Fiber Reinforcement

The effect of short fibers as reinforcement in metallic matrixes can be clarified with the help of a micromechanical model (shear lay model). The influence of the fiber length and the fiber orientation on the expected strength can be shown as a function of the fiber content and the fiber and matrix characteristics with the help of simple model calculations. The starting point is the mixture rule for the calculation of the strength of an ideal long-fiber-reinforced composite material with load in the fiber direction (Eq. (1) [23]). For short-fiber reinforcement the fiber length has to be considered [25]. During the loading of the composite materials, e.g. by tensions, the individual short fibers do not carry the full tension over their entire length. Only with over tension and predominantly shear stresses at the fiber/matrix interface will the load transfer partly to the fiber. Figure 1.16 shows the modeling of the load of a single fiber, which is embedded in a ductile matrix and stressed in the fiber direction.

The effective tension on the fiber in dependence on the fiber length can be calculated as follows:

$$\frac{d\sigma_F}{dx} \cdot dx \cdot r_F^2 \cdot \pi + 2\pi \cdot \tau_{FM} r_F \, dx = 0 \tag{5}$$

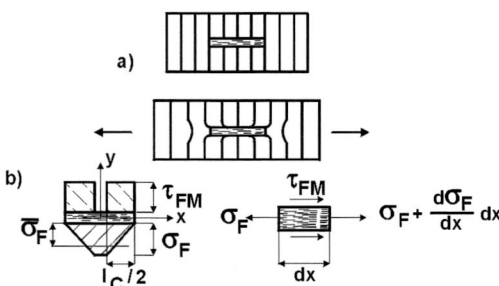

a) Strain field in the matrix
b) Shear strength at the interface fiber/matrix and tensile strength within the fiber

Fig. 1.16 Model of loading of a single fiber, embedded in a ductile matrix (after [23]): (a) Stress field in the matrix, (b) shear stress distribution at the interface fiber/matrix and tensile strength contribution in the fiber.

$$\sigma_F = \frac{2}{r_F} \cdot \tau_{FM} \cdot \left[\frac{2}{r_F} - x\right] \tag{6}$$

$$l_c = \frac{\sigma_F \cdot d_F}{2\,\tau_{FM}} \tag{7}$$

Where σ_F = fiber tension, r_F = fiber radius, d_F = fiber diameter, τ_{FM} = shear stress at the fiber/matrix interface. A critical fiber length l_c results, at which the fiber can be loaded to its maximum (Fig. 1.17).

The shear strength at the interface matrix/fiber is

$$\tau_{FM} = 0.5 \cdot \sigma_M^* \tag{8}$$

where σ_M^* = matrix yield point.

The effective fiber strength $\sigma_{F,eff}$ in dependence on the fiber length is

$$\sigma_{F,eff} = \eta \cdot \sigma_F \cdot \left(1 - \frac{l_c}{2 \cdot l_m}\right) \tag{9}$$

where η = fiber efficiency (deviation from optimum $0 < \eta < 1$) [28]; l_m = average fiber length.

According to Fig. 1.17 three cases, depending on the fiber length, can be distinguished [23–27]:

Fiber length $l_m > l_c$:

$$\sigma_C = \eta \cdot C \cdot \Phi_F \cdot \sigma_F \cdot \left(1 - d_F \cdot \frac{\sigma_F}{2 \cdot l_m \cdot \sigma_m^*}\right) \tag{10}$$

where C = orientation factor [26] (orientated C = 1, irregular C = 1/5, planar isotropic C = 3/8).

Fiber length $l_m = l_c$:

$$\sigma_C = \eta \cdot C \cdot 0.5 \cdot \Phi_F \cdot \sigma_F + (1 - \Phi_F) \cdot \sigma_M^* \tag{11}$$

Fiber length $l_m < l_c$:

$$\sigma_C = \eta \cdot C \cdot \sigma_M^* \cdot \frac{l_m}{2 \cdot d_F} + (1 - \Phi_F) \cdot \sigma_M^* \tag{12}$$

At a fiber length below the critical fiber length l_c the tensile strength of the fiber under load cannot be completely utilized. The reinforcement effect is lower [27]:

$$l_c = d_F \frac{\sigma_F - \sigma_M^*}{\sigma_M^*} \tag{13}$$

The models are based on idealised conditions: ideal adhesion between fiber and matrix and ideal adjustment and distribution of the long fibers or the arranged short fibers. Figure 1.18 shows schematically the influence of the length/thickness relationship of the fibers on the reinforcement effect under optimal conditions. By

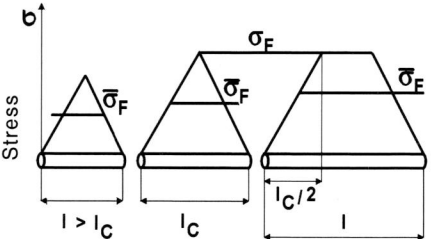

Fig. 1.17 Dependence of the effective fiber strength on the fiber length ($\sigma_F = \sigma_{F,\text{eff}}$), after [24].

increasing the fiber length the potential of long fibers (l/d 100) will be approached. For irregular or planar-isotropically arranged short fibers an optimal distribution is the basic condition for applicability. The result of an estimation is shown in Fig. 1.19. It represents the relationship of the strength of fiber-reinforced light metal alloys, calculated with Eqs. (1) and (10)–(12), to the strength of the non-reinforced matrix (reinforcement effect) as a function of the content of aligned fibers for different fiber length [29]. For the matrix characteristics the following mechanical properties at room temperature were used: tensile strength, 340 MPa; yield strength, 260 MPa. The aluminum oxide fiber Saffil (fiber tensile strength, 2000 MPa; diameter, 3 μm) was the fiber used. At small fiber contents a reduction in strength first occurs, up to a minimum fiber volume content, above this value the strength increases until a fiber content ϕ_1–ϕ_4 (depending on the fiber length) is reached, this is the strength of the nonreinforced matrix. Thereafter the reinforcement effect increases with increasing fiber content and length.

The presented calculations presuppose an orientation of the fibers in the stress direction, with an irregular arrangement there is a smaller reinforcement effect.

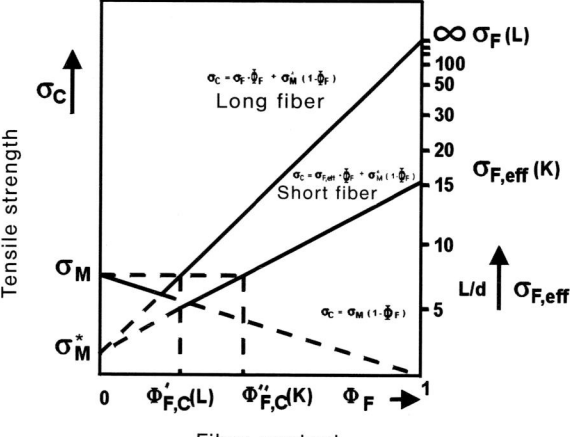

Fig. 1.18 Linear mixture rule of the tensile strength of unidirectional fiber composite materials, the right ordinate represents the effective fiber strength according to Eq. (9), after [23] and [27].

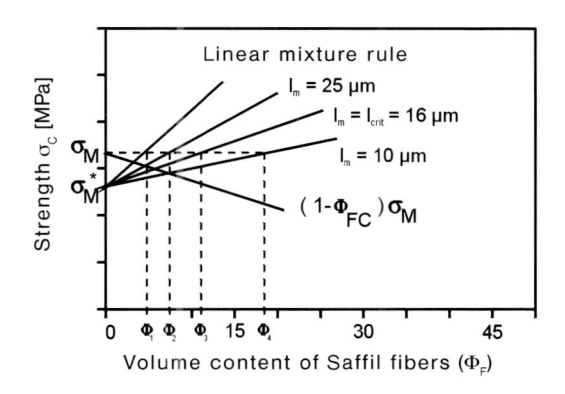

Fig. 1.19 Influence of the fiber length and volume content on the properties of magnesium composite materials (AZ91+Saffil-fibers) [28].

Figure 1.20 illustrates this using as an example magnesium alloy AZ91 strengthened by Saffil fibers. With increasing isotropy more fibers must be added in order to obtain a reinforcement effect. The amount of fibers required for the reinforcement effect is: for long fibers, a fiber content Φ_1 = 3.2 vol%; for aligned short fibers Φ_2=3.5 vol% and for planar-isotropically distributed fibers Φ_3= 12.5 vol%. This effect increases with increasing load temperature, as shown in Fig. 1.21. This figure gives the calculated strength of composite materials for two different yield

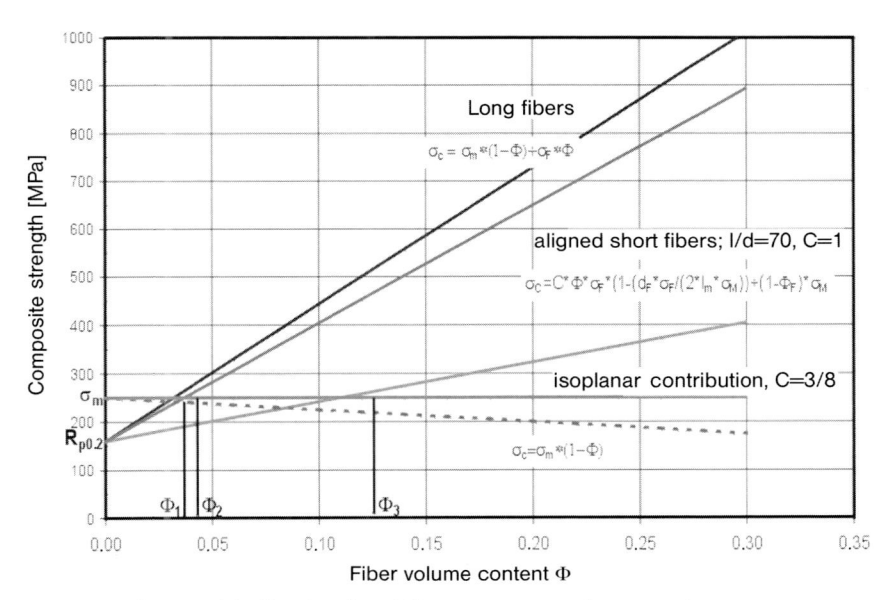

Fig. 1.20 Influence of the fiber length and fiber orientation on the composite material strength for the system magnesium alloy AZ91 (yield strength: 160 MPa, tensile strength 255 MPa) + C fiber (fiber strength 2500 MPa, fiber diameter 7 μm), schematic after Eqs. (1) and (10).

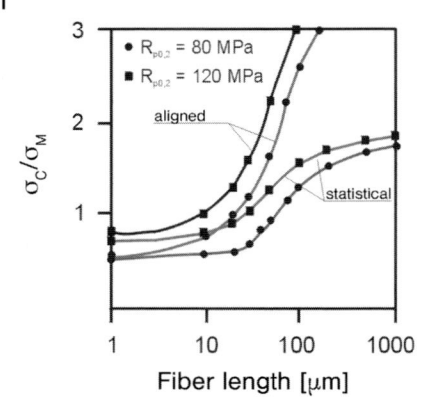

Fig. 1.21 Influence of the fiber length and fiber orientation on the reinforcement effect σ_C/σ_M for a composite material with 20 vol.% aluminum oxide fibers for different matrix yield strengths [29].

strengths of the nonreinforced matrix (80 MPa and 115 MPa) [29]. Although for the calculations only models with simplified boundary conditions were used, they show the objective for the production and processing of such composite materials. A goal is an optimal alignment of the fibers with the retention of a long fiber length.

1.3.3
Strengthening by Particles

The influence of ceramic particles on the strength properties of particle reinforced light metals can be described by using the following micromechanical model [30, 31]:

$$\Delta R_{p,C} = \Delta \sigma_\alpha + \sigma_{KG} + \Delta \sigma_{SKG} + \Delta \sigma_{KF} \tag{14}$$

where $\Delta R_{p,C}$ is the increase in tensile strength of aluminum materials by particle addition.

The influence of induced dislocations $\Delta \sigma_\alpha$ is given by:

$$\Delta \sigma_\alpha = \alpha \cdot G \cdot b \cdot \rho^{1/2} \tag{15}$$

with

$$\rho = 12 \Delta T \frac{\Delta C \Phi_p}{bd} \tag{16}$$

where $\Delta \sigma_\alpha$ is the yield strength contribution due to geometrical necessary dislocations and inner tension, α is a constant (values 0.5–1), G is the shear modulus, b the Burger's vector, ρ the dislocation density, ΔT the temperature difference, ΔC the difference in thermal expansion coefficient between matrix and particle, Φ_p the particle volume content and d the particle size.

The influence of the grain size $\Delta\sigma_{KG}$ is given by:

$$\Delta\sigma_{KG} = k_{Y1} D^{-1/2} \tag{17}$$

with

$$D = d\left(\frac{1-\Phi_p}{\Phi_p}\right)^{1/3} \tag{18}$$

where $\Delta\sigma_{KG}$ is the yield strength contribution from changes in grain size (for example recrystallization during thermomechanical treatment of composite materials, analogue Hall-Petch); k_{Y1} is a constant, D is the resulting grain size and Φ_p is the particle volume content.

The influence of the grain size $\Delta\sigma_{SKG}$ is given by:

$$\Delta\sigma_{SKG} = k_{Y2} \cdot D_s^{-1/2} \tag{19}$$

with

$$D_s = d\left(\frac{\pi d^2}{6\Phi_p}\right)^{1/2} \tag{20}$$

where $\Delta\sigma_{SKG}$ is the yield strength contribution due to changes in subgrain size (for example in a relaxation process during thermomechanical treatment of composite materials), k_{Y2} is a constant (typical value 0.05 MN m$^{-3/2}$), D_s is the resulting subgrain size and Φ_p is the particle volume content.

The yield point is usually measured as the yield strength with 0.2 % remaining elongation. A significant strain hardening occurs, which is dependent on the particle diameter and content.

The strain hardening contribution $\Delta\sigma_{KF}$ is given by

$$\Delta\sigma_{KF} = KG\,\Phi_p \left(\frac{2b}{d}\right)^{1/2} \cdot \varepsilon^{1/2} \tag{21}$$

where K is a constant, G the shear modulus, Φ_p the particle volume content, b the Burger's vector, d the particle diameter and ε the elongation.

According to whether the particle size or the particle content is the dominant effect, different characteristic tension contributions of the individual mechanisms to the technical yield strength $R_{P0.2}$ of the particle strengthened light metal alloys result. The example of a particle-strengthened composite material with two different particle diameters in Fig. 1.22 clarifies this in principle. Generally higher hardening contributions are made by smaller particle diameters than by coarser particles. For smaller particle diameters the work hardening and the grain size influence contributes the most to the increase in the yield strength. Figure 1.23 shows schematically the change in the substantial hardening contributions with increasing particle content for a constant particle diameter.

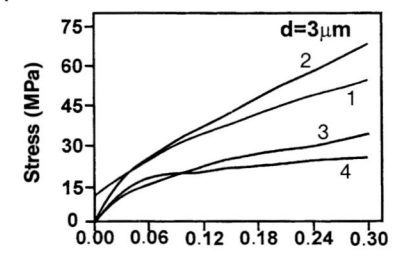

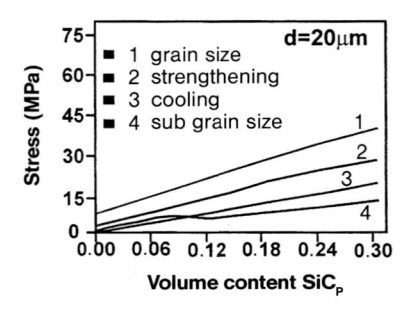

Fig. 1.22 Strain contribution of different mechanisms to the technical yield point calculated after the micromechanical model for aluminum alloys with SiC$_P$-addition, after [31].

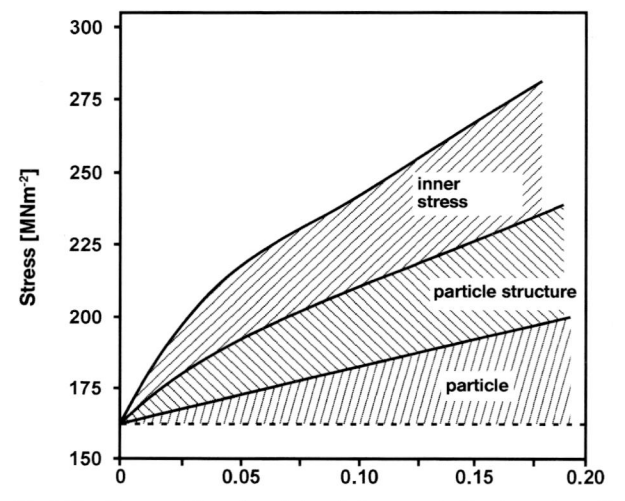

Fig. 1.23 Composition of particle reinforcement of various strengthening contributions (after [27]).

1.3.4
Young's Modulus

An objective in the development of light metal composite materials is to increase the modulus of elasticity (Young's modulus). Which potential arises here, can be estimated by the mixture rule, whereby the well-known border cases apply only to certain geometrical alignments of the components in the composite material. The universally used models are the following linear and inverse mixture rules [3]:

Linear mixture rule: Voigt-model (ROM)

$$E_C = \Phi_p E_p + (1 - \Phi_p) E_M \tag{22}$$

Inverse mixture rule: Reuss-model (IMR)

$$E_c = \left(\frac{\Phi_p}{E_p} + \frac{1-\Phi_p}{E_M}\right)^{-1} \qquad (23)$$

Where Φ_p is the volume content of particles or fibers, E_c the Young's modulus of the composite material, E_p the Young's modulus of the particle or fiber and E_M the Young's modulus of the matrix.

The Voigt model is only applicable for long-fiber-reinforced composite materials with a stress direction parallel to the fiber orientation, while the Reuss model applies to layer composite materials with a load perpendicular to the layers. An advancement of these models, which is also applicable for short fibers or particles, is the model by Tsai Halpin. By implementing an effective geometry factor, which can be determined from the structure of the composite materials as a function of the load direction, the geometry and the orientation of the reinforcement can be considered [32]:

$$E_c = \frac{E_M(1 + 2Sq\,\Phi_p)}{1 - q\,\Phi_p} \qquad (24)$$

with:

$$q = \frac{(E_p/E_M) - 1}{(E_p/E_M) + 2S} \qquad (25)$$

where S is the geometry factor of the fiber or particle $(1/d)$.

Figure 1.24 presents as an example the Young's moduli calculated using Eqs. (22)–(25) for SiC particle reinforced magnesium materials as a function of the particle content and for different geometry factors according to Eqs. (24) and (25).

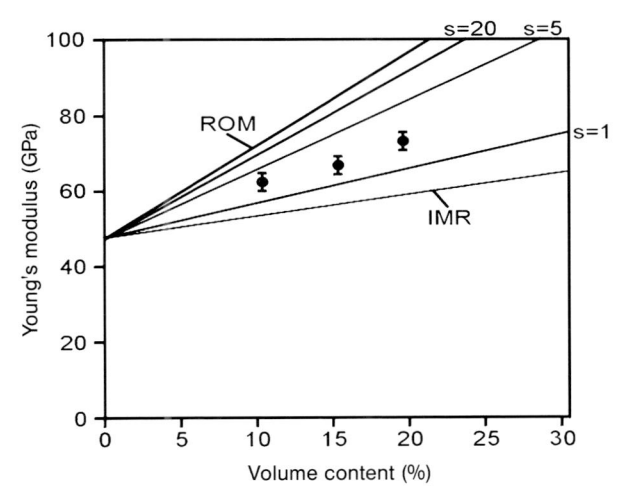

Fig. 1.24 Comparison of theoretically calculated Young's modulus values with the experimentally determined values for particle reinforced composite materials (ROM: linear mixture rule, IMR: inverse mixture rule [33].

Comparison of the measured Young's moduli is shown. A good agreement between calculated and experimental values can be seen; using a geometry factor $S=2$ for the applied SiC particles [33]. The basic condition for the application of such models is the presence of a composite material with an optimal structure, i.e. without pores, agglomerates of particles or nonreinforced areas.

1.3.5
Thermal Expansion Coefficient

Reinforcement of light metal alloys with ceramic fibers or particles entails a reduction in the thermal expansion coefficients. For this physical characteristic also, simple models are available to estimate the thermal expansion coefficients with the help of the characteristics of the individual components. The model of Schapery [34] was developed, in order to describe the influences on the thermal expansion coefficients:

$$\alpha_{3C} = \frac{E_F \alpha_F \Phi_F + E_M \alpha_M (1 - \Phi_F)}{E_C} \qquad (26)$$

where α_{3C} is the axial thermal expansion coefficient, α_F the thermal expansion coefficient of the fibers and α_M the thermal expansion coefficient of the matrix.

$$\alpha_{1C} = (1 + v_M) \alpha_M \Phi_M + (1 + v_F) \alpha_F \Phi_F - \alpha_{3C} v_{31C} \qquad (27)$$

$$v_{31C} = v_F \Phi_F + v_M (1 - \Phi_F) \qquad (28)$$

where α_{1C} is the transverse thermal expansion coefficient, v_F Poisson's ratio of the fibers and v_M Poisson's ratio of the matrix.

Since the Schapery model is conceived for the calculation of thermal expansion coefficients for aligned long fibers the model can only be used for short-fiber reinforced materials with restrictions. A basic condition is an alignment of the short fibers. The thermal prehistory of the materials, in order to be able to proceed from a uniform internal tensile state, also has to be considered. A representation of calculated and measured values for the thermal expansion coefficients of light metal alloy composite materials for the example of a magnesium alloy reinforced with aligned Al_2O_3-short fibers (Saffil) is shown in Fig. 1.25. Here the upper curve represents the calculated values for the transverse thermal expansion coefficients and the lower curve the calculated values for the axial coefficient. The lower limit curve was calculated with the help of the theoretical Young's modulus according to the linear mixture rule of Eq. (22). When using the experimentally determined values of the Young's modulus of the composite material, then the values marked by squares result. In this case the deviations from the optimal structure are considered to be in good agreement, for the Mg+15 vol% Al_2O_3, with the measured axial expansion coefficients. This agreement with the real measured Young's modulus of the composite materials can be found also when using particles as reinforcement [35].

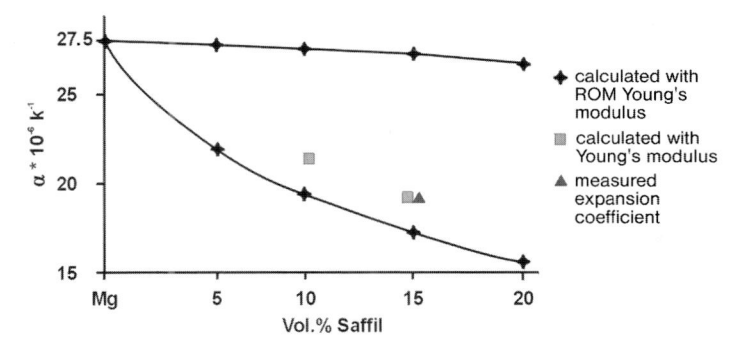

Fig. 1.25 Change in the thermal expansion coefficient with increasing fiber content (model after Shapery [34]) [33].

The thermal expansion coefficient is determined by the thermal prehistory of the composite materials, which results from the production and the application. Essentially the internal strain exercises influence. Figure 1.26 shows the temperature dependence of the thermal expansion coefficients of the monolithic magnesium alloy QE22 and the composite material QE22+20 vol% Saffil fibers for different orientations of the fibers. With the monolithic materials the expansion coefficient increases with increasing temperature. The same applies to the composite material with a fibers oriented perpendicular to the level of the planar-isotropic distribution of the fibers (90°). Since the fibers there are not optimally effective a lower reduction in the expansion develops. With increasing temperature the difference between the reinforced and the nonreinforced matrixes becomes less. In the case of an orientation parallel to the fiber level (0°) a stronger reduction effect results,

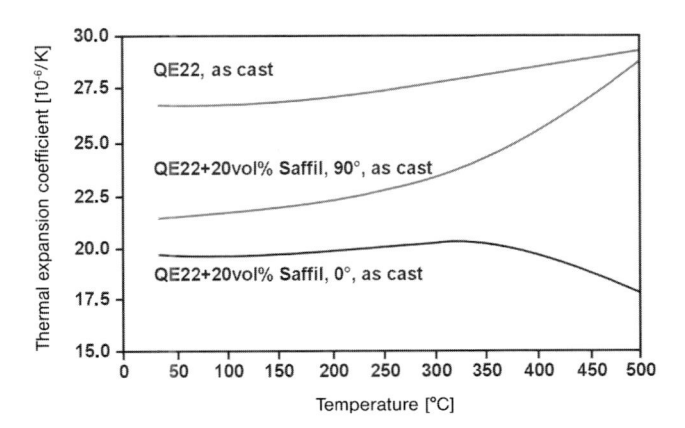

Fig. 1.26 Dependence of the thermal expansion coefficient of magnesium composite materials on the temperature and fiber orientation in comparison to the nonreinforced matrix [36].

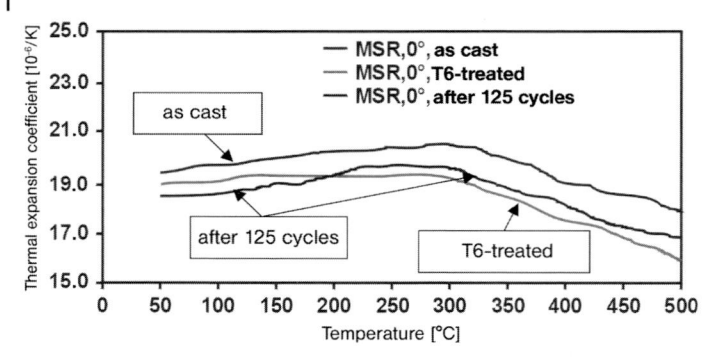

Fig. 1.27 Influence of the thermal prehistory on the expansion behavior of aluminum oxide fiber reinforced magnesium composite materials [36].

which increases with increasing temperature. The influence of the thermal prehistory on the thermal expansion coefficients is represented in Fig. 1.27 for the example of the magnesium alloy MSR + vol% Saffil. For the cast condition a comparable process results, as shown in Fig. 1.24. After a T6-heat treatment the curve shifts to higher values, particularly in the temperature range above the ageing temperature (204°C). After a thermal cyclic load a reduction in the internal strain appears and results in further increase in the values.

1.4
Interface Influence

Compared with monolithic materials the microstructure and the interfaces of metal matrix composite materials cannot be considered in isolation, they are mutually related. Chemical interactions and reactions between the matrix and the reinforcement component determine the interface adhesion, modify the characteristics of the composite components and affect the mechanical characteristics significantly.

In high temperature use of MMCs the microstructure has to remain stable for long service periods. Thermal stability and failure is determined by changes in the microstructure and at the interfaces, e.g. reaction and precipitation processes. Thermal stress of MMCs can take place both isothermally and cyclically. The effects show differences. During a cyclic load of monolithic materials, especially at high temperature gradients and cycle speeds, a high probability of failure by thermal fatigue is to be expected, e.g. short-fiber reinforced aluminum alloys possess good thermal shock stability.

The formation of the interface between the matrix and the reinforcing phase has a substantial influence on the production and characteristics of the metallic composite materials. The adhesion between both phases is usually determined by the interaction between them. During the production of the molten matrix e.g. by infiltration, wettability becomes significant.

1.4.1
Basics of Wettability and Infiltration

Basically the wettability of reinforcement with a metal melt can be shown by the edge angle adjustment of a molten droplet on a solid base as the degree of wettability according to Young:

$$\gamma_{SA} - \gamma_{LS} = \gamma_{LA} \cdot \cos \Theta \qquad (29)$$

where γ_{LA} is the surface energy of the liquid phase, γ_{SA} the surface energy of the solid phase, γ_{LS} the interface energy between the liquid and solid phases and Θ is the edge angle.

Figure 1.28 shows the edge angle adjustment of a molten droplet on a solid base for different values of the interface energy. At an angle of $>\pi/2$ a nonwettable system is described and for an angle limit of $<\pi/2$ a wettable system. With decreasing angle the wettability improves. In Table 1.3 the surface and interface stresses of selected metal – ceramic systems at different temperatures are summarized. Of special relevance is the system Al/SiC, since it is the basis for the melting metallurgy of particle reinforced aluminum composite materials.

As the contact develops, for example at the beginning of an infiltration, adhesion occurs. The adhesion work W_A for separation is [41]:

$$W_A = \gamma_{SA} - \gamma_{LA} = \gamma_{LS} \qquad (30)$$

$$W_A = \gamma_{LA} \cdot (1 + \cos \Theta) \qquad (31)$$

Tab. 1.3 Surface and interface strains of selected metal–ceramic systems at different temperatures.

Alloy, Ceramic, Systems	Temperature (K)	γ_{la} (mJ m^{-2})	γ_{lsa} (mJ m^{-2})	γ_{ls} (mJ m^{-2})	Ref.
Al	953	1050	–	–	37
Mg	943	560	–	–	37
Al$_2$O$_3$	0	–	930	–	38
MgO	0	–	1150	–	38
Cu/Al$_2$O$_3$	1370	1308	1485	2541	39
	1450	1292	1422	2284	39
Ni/Al$_2$O$_3$	1843	1751	1114	2204	39
	2003	1676	988	1598	39
Al/SiC	973	851	2469	2949	40
	1073	840	2414	2773	40
	1173	830	2350	2684	40

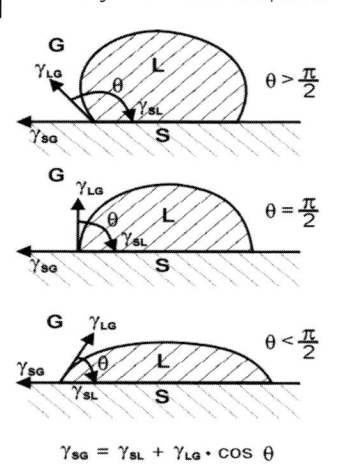

Fig. 1.28 Edge angle adjustment of a melt drop on a solid base for various values of the interface energy (after Young).

$$\gamma_{SG} = \gamma_{SL} + \gamma_{LG} \cdot \cos \theta$$

In the case of immersion the interface between the solid and the atmosphere disappears, while the interface between the solid and the liquid forms. The immersing work W_I is:

$$W_I = \gamma_{LS} - \gamma_{SA} \tag{32}$$

In the case of spreading the liquid is spread out on a solid surface. During this procedure the solid surface is reduced as well as a new liquid surface being formed and hence a new solid/liquid interface is formed. The spreading work W_S is:

$$W_S = \gamma_{SA} - \gamma_{LS} - \gamma_{LA} \tag{33}$$

The wetting procedure is kinetic and is dependent on time and temperature. Therefore, the kinetics can be affected by the temperature. Figure 1.29 shows, as

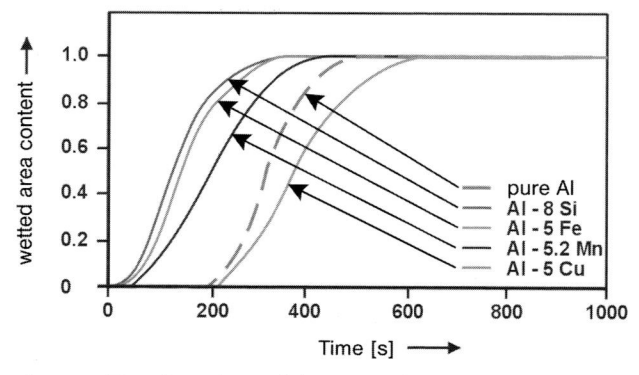

Fig. 1.29 Time dependence of the wetting degree (unit fraction) of SiC plates by aluminum melts at different alloy additions [42].

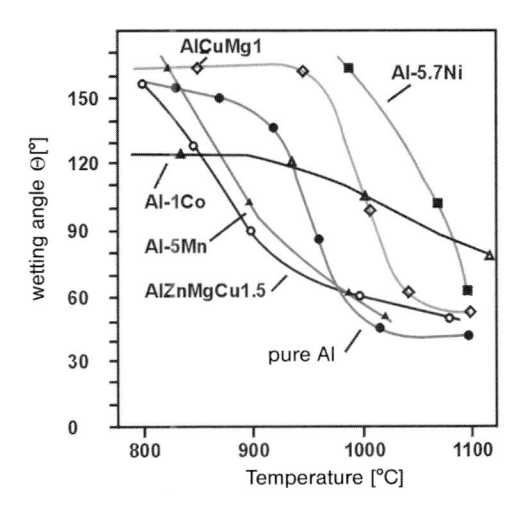

Fig. 1.30 Temperature dependence of the wetting angle of aluminum drops on a SiC plate [43].

an example of the time dependence of a wetting procedure, the wetting degree (surface fraction) of SiC plates by aluminum alloy melts of different composition. In Fig. 1.30 the temperature dependence of the wetting angle of one aluminum alloy droplet on a SiC plate is represented. Both figures show the further possibility of the influence of variation of the composition in the appropriate material system. The alloying elements act by changing the surface tension of the melt or by reaction with the reinforcement. On the one hand the composition of the matrix, or rather reinforcement, is modifiable and on the other hand there exists the possibility of purposeful influence by applying coatings on the intensifying phase. The role of a reaction at the interface is important, because from it a new system can result and the interface energies can be changed substantially, thus altering the wetting angle.

In Eq. (29) the change induced by the reaction, for example, of an oxide reinforcement Me_1O with a matrix alloy part Me_2, has to be taken into consideration [44]:

$$\gamma_{LS} - \gamma_{SA} = (\gamma_{LS} - \gamma_{SA})_0 - \Delta\gamma_r - \Delta G_r \tag{34}$$

$$Me_2 + Me_1O \leftrightarrow Me_2O + Me_1 \tag{35}$$

$$\gamma_{LA} \cdot \cos\Theta = (\gamma_{SA} - \gamma_{LS})_0 - \Delta\gamma_r - \Delta G_r \tag{36}$$

where $(\gamma_{ls} - \gamma_{sa})_0$ is the wettability without reaction, $\Delta\gamma_r$ the interface tension from the reaction of newly formed interfaces and ΔG_r the given free energy at the triple line solid/liquid/atmosphere (reaction energy).

Figure 1.31 shows the influence of pressure-free infiltration through different reactions by using different reactive binder systems and fiber contents such as Mg/Al_2O_3 fibers. With a very reactive SiO_2-containing binder premature infiltration happens at lower temperatures than with an Al_2O_3 binder [17].

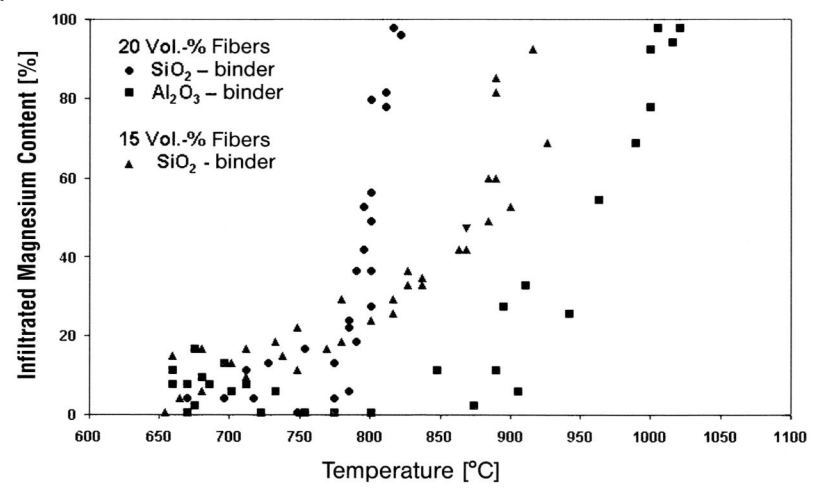

Fig. 1.31 Influence of pressure free infiltration by different reactions under use of different reactive fiber contents on the example of pure Mg/Al$_2$O$_3$ fibers, Dissertation Fritze and [17].

An additional possible influence exists through the change in the surrounding atmosphere or rather the atmosphere in the preform. It is possible, for example, that a preform before infiltration is flushed with gas, which can lead to a change in the oxygen partial pressure. Figure 1.32 shows the dependence of the wetting angle in the system Al$_2$O$_3$/pure aluminum on the temperature for two oxygen partial pressures [40]. For high partial pressures of oxygen high wetting angles occur at low temperatures. Only when starting from temperatures above 1150 K does the value for the wetting angle decrease to that for a low oxygen partial pressure. However, technically this influence is not relevant, because the atmosphere can be changed only with difficulty. An exception is the infiltration by the production of a vacuum (gas pressure infiltration); in this case gas pressures can be modified.

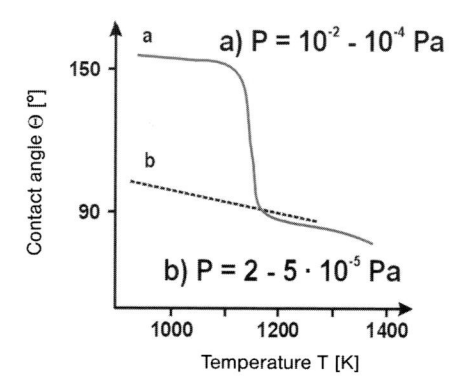

Fig. 1.32 Schematic presentation of the change in the wetting angle with change in the oxygen particle pressure [44].

1.4 Interface Influence

Wetting for the actual infiltration procedure is of substantial importance. This is shown in a simple schematic representation in Fig. 1.33. In the case of good wetting (small edge angle) a capillary effect occurs (Fig. 1.33a). At large edge angles this procedure is inhibited (Fig. 1.33b). Additionally this can occur in technical processes by a reaction between the melt and the surrounding atmosphere. Then, for example, an oxide film forms, as in the case of magnesium alloys, which affects the wetting behavior by formation of a new interface between the reinforcement and the melt, as clarified in Fig. 1.34. The statements made above apply only to considerations close to the equilibrium. The influence of the wetting on the infiltration during technically relevant processes is thus less, if applied pressure on the melt, or rather the flow rate of the melt in the perform, determines the kinetics of the wetting, for example in the production of a wetted system by high pressure in the melt. However, the wetting nevertheless still has an influence on the adhesion of the components in the composite, which will later be described in more detail.

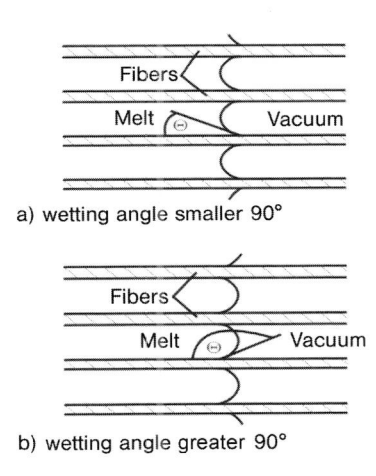

Fig. 1.33 Schematic presentation of an ideal melt infiltration of fiber preforms [45].

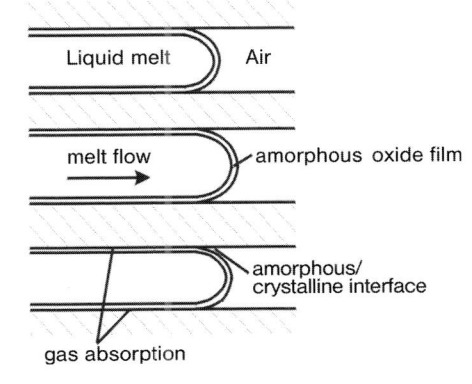

Fig. 1.34 Schematic presentation of the infiltration process of an aluminum oxide preform with molten aluminum [45].

The actual infiltration process for the production of metallic composite materials consists of several indexing steps: the formation of a contact between the melt and the reinforcement at the surface of a fiber or particle preform, the infiltration with the melt flow through the preform and the solidification procedure. At the beginning of the infiltration a minimum pressure must usually be developed, so that infiltration can follow. Usually a pressure-free spontaneous infiltration is not the rule, and is only possible with thin preforms with reactive systems and with long process times. The resulting pressure as driving force for the infiltration [46] is

$$\Delta P = P_0 - P_a - \Delta P_\gamma \tag{37}$$

where ΔP is the resulting pressure, the driving force for the reaction, P_0 the pressure in the melt on entering the preform (see Fig. 1.35), P_a the pressure in the melt at the infiltration front (see Fig. 1.36), ΔP_γ the pressure decrease in the melt at the infiltration front due to surface influences (effect of wettability).

When $P_0 = P_a$ a minimum infiltration pressure ΔP_μ can be defined:

$$\Delta P_\mu = \Delta P_\gamma = S_f (\gamma_{LS} - \gamma_{SA}) \tag{38}$$

where S_f is the surface interface per unit area.

Without external applied pressure the effect of induced infiltration by the capillary force can be presented as following [47]:

$$P_\gamma = \frac{2 \gamma_{LA} \cdot \cos \Phi}{r} \tag{39}$$

where r is the radius of the capillary.

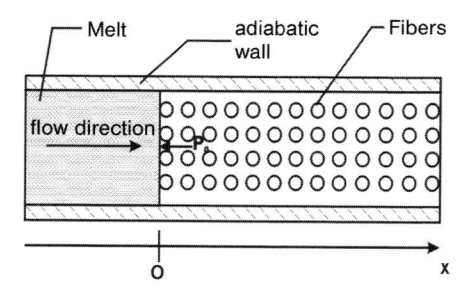

Fig. 1.35 Schematic presentation of an adiabatic, unidirectional infiltration, starting condition [46].

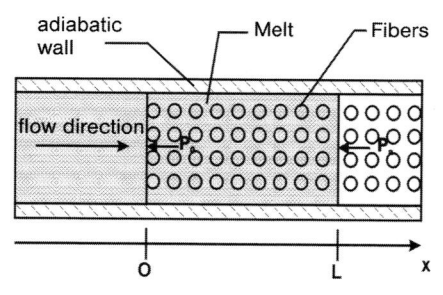

Fig. 1.36 Schematic presentation of an adiabatic, unidirectional infiltration [46].

1.4 Interface Influence

Using the hydrostatic pressure

$$S_f = \frac{6 V_f}{d_f (1 - V_f)} \qquad (40)$$

results in the rising height:

$$h_s = \frac{2 \gamma_{LA} \cdot \cos \Phi}{\rho \cdot g \cdot h} \qquad (41)$$

where h_s is the rising height, g the gravitational constant and ρ the density.

As a pressure-free infiltration a preform consisting of many capillaries can be imagined and thus the influence of wettability and structure parameter (surface and pores – or rather capillary diameter) can be seen when introducing the following structure parameter in Eq. (38) [47]:

For a spherical particle:

$$S_f = \frac{6 V_f}{d_f (1 - V_f)} \qquad (42)$$

For a long fiber bunch and short fiber preform:

$$S_f = \frac{4 V_f}{d_f (1 - V_f)} \qquad (43)$$

Where V_f is the fiber or particle content and d_f is the fiber or particle diameter.

Table 1.4 shows the change in the specific surface with increasing fiber portion in the Al_2O_3-preform of Saffil fibers [48]. The specific surface also influences the permeability of a preform. This characteristic is important for the even supply of the preform with the melt and affects the necessary pressure for the infiltration. This is shown in Fig. 1.37 for Saffil preforms with water infiltration [48, 49]. It is noticeable that from fiber contents of 20 vol% the permeability is significantly reduced. In this context the viscosity of a melt also has an important influence. By variation of the temperature and composition optimisation of the infiltration process is controllable. Fig. 1.38 gives information on the change in the viscosity of magnesium and aluminum melts as a function of the temperature for unalloyed systems.

The previous considerations provide only an explanation of the processes and influencing variables on the wetting and infiltration, which are relevant in material systems, for example for stirring particles into melts or infiltrating. For both processes further procedures are of course relevant. Examples are the solidification

Tab. 1.4 Specific surface of Al_2O_3 preforms, after [17] and [48].

Fiber volume content of Al_2O_3 preforms [vol.%]	10	20	24	25
Specific surface: $S_f = 10^6$ fiber surfaces (m²)/ pore volume (m³)	1.26	3.41	4.39	4.58

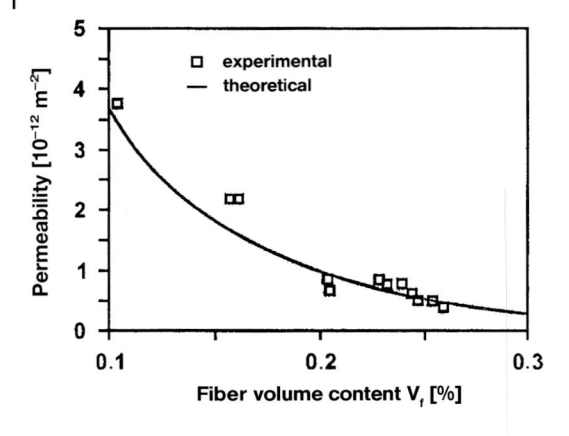

Fig. 1.37 Comparison of permeability of preforms for running water, after Mortensen and calculations of Sangini and Acrivos [48, 49].

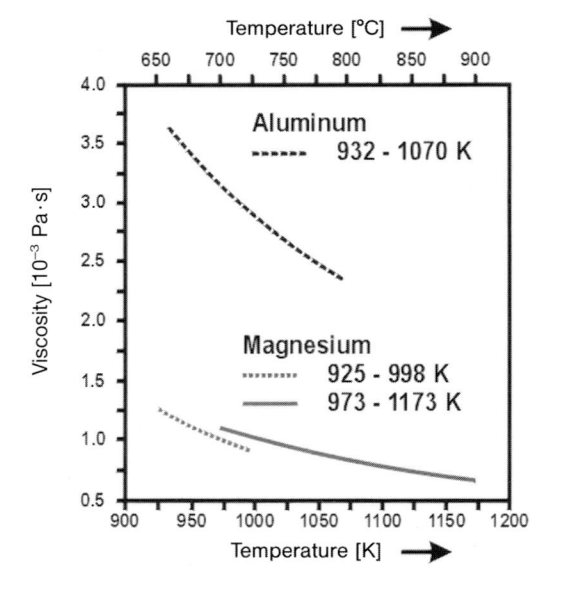

Fig. 1.38 Temperature dependence of the viscosity of magnesium and aluminum melts [50].

procedures of the melts. They overlay with the abovementioned procedures. For poor wetting of particles, for example in the production of particle-strengthened light alloys, a segregation or liquidization of the particles can take place. During the infiltration, solidification procedures can affect the permeability and prevent the complete treatment of the preform. Equation (38) (see Fig. 1.35 and 1.36) assumes a constant heat balance and no partial solidification. In reality there is heat dissipation over the tool and thus directed solidification occurs. Also the given free solidification heat has a substantial influence. The solidification worsens the permeability and influences the flow conditions in the preform. In reality a feed and a solidification procedure take place during the infiltration as a result of the directed heat

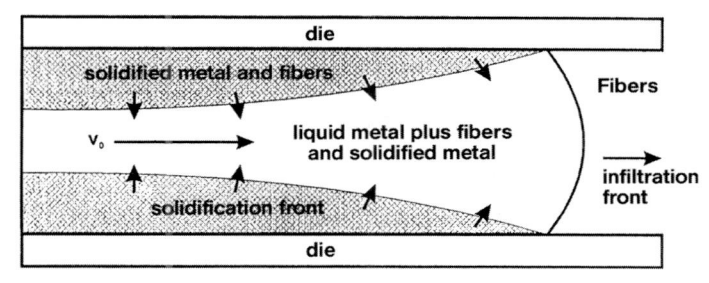

Fig. 1.39 Schematic presentation of solidification with external heat loss during a unidirectional infiltration [47].

dissipation, as represented in Fig. 1.39. The heat dissipation in the system preform/liquid or solidified melt is essentially determined by the thermal characteristics of the components (specific heat, thermal conductivity). Thus the reinforcements possess essentially higher specific heat values and smaller heat conductivities (exceptions are carbon fibers). In Table 1.5 these characteristic values are summarized for C and Al_2O_3 fibers. In the case of the applied alloys it is to be noticed that magnesium melts possess a smaller heat capacity than aluminum and therefore are processed at higher temperatures or the preforms have to be at higher temperatures than with aluminum alloys.

Tab. 1.5 Comparison of physical data of C fibers and aluminum oxide fibers (Saffil).

Fiber	Specific heat ($J\ m^{-3}\ K^{-1}$)	Coefficient of thermal conductivity ($W\ m^{-1}\ K^{-1}$)
Carbon P100	1.988×10^6	520
Carbon T300	1.124×10^6	20.1
Saffil	2.31×10^6	0

1.4.2
Objective of Adhesion

The interaction between wetting and adhesion has already been briefly mentioned in Section 1.1. Figure 1.40 describes this connection using the example of the adhesive strength of a solidified aluminum melt dropped onto a substrate as a function of the wetting angle, determined by the droplet shear test. For small edge angles high adhesive strength values with a failure by shearing result. At larger angles the adhesive strength decreases and the failure only occurs under tension. In systems with good wettability reactions play a substantial role. The adhesion in composite systems can be improved by reaction. However, in some cases the reactions can become too distinctive, so that they result in damage to the reinforcement, e.g. reduction of the tensile strength of fibers. Thus the reinforcement po-

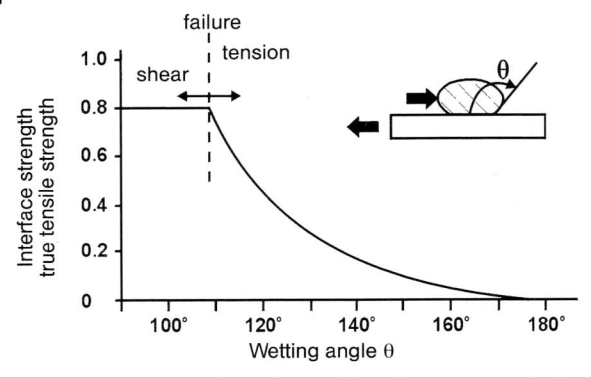

Fig. 1.40 Edge angle dependence of the adhesion strength of solidified Al melt drops [51].

tential is reduced. Later, brittle reaction products or pores can develop, which can again decrease the adhesion. Fot the example of the system Ni and Al_2O_3 in Fig. 1.41 it is clear that an optimum must be sought. With a proceeding reaction bonding is improved and the fiber strength decreases. In the case of poor binding the interface fails and the fiber failure dominates with increasing binding. In Fig. 1.42 the formation of an interface between an Al_2O_3 fiber and a magnesium alloy is represented for two conditions. In the cast condition in Fig. 1.42b only sporadic discontinuous MgO particles occur. The fibers are negligibly damaged and possess their full reinforcement effect. After a long-term annealing treatment the reaction products have grown and the fibers are damaged, the strength of the composite material decreases (see Fig. 42a) [53]. In an example of a thermal treatment of the composite material system magnesium alloy AZ91/Al_2O_3 fiber/(Saffil) the connection between reaction layer thickness and strength properties can be clarified. An untreated composite material of this system has a tensile strength of 220 MPa [54].

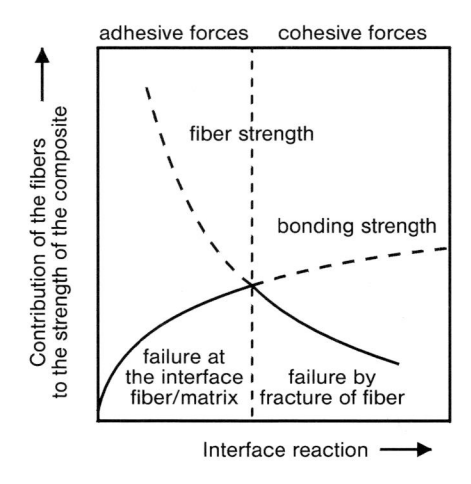

Fig. 1.41 Dependence on the reaction layer thickness of the shear strength of the interface between Ni and Al_2O_3 [52].

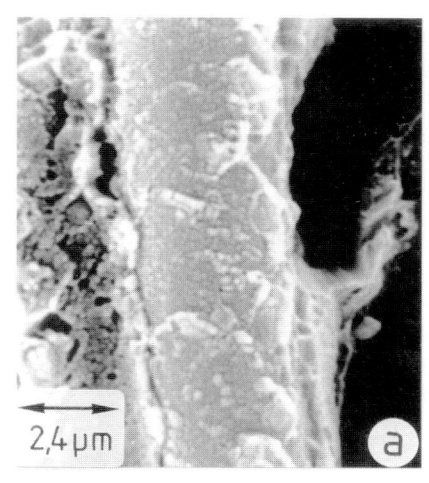

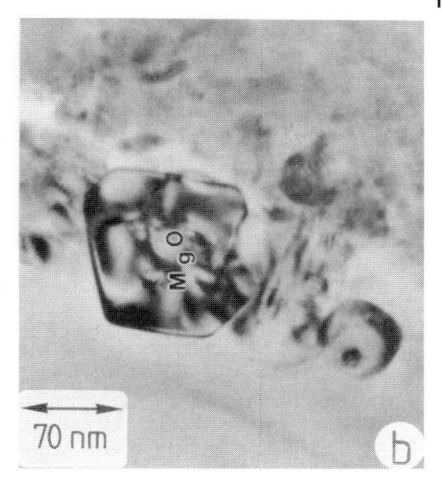

Fig. 1.42 Reaction products at the interface Mg alloy/Al$_2$O$_3$ fiber [53]:
(a) SEM image, long term loading 350 °C, 250 h; (b) TEM image, as-cast condition.

With increasing reaction layer thickness the tensile strength decreases to more than 50% of the output strength (Fig. 1.43). With support of thermodynamic calculations the risk of this damage can be estimated. Also the influence of the reaction by layer systems on fibers [2, 22] or by modification of the alloy composition [55] is calculable and thus predictable.

The formation of the interface has, as discussed, a crucial influence on the behavior of the metallic composite materials. The influence of the elastic constants and the mechanical properties on the failure is substantial. As an example the change in the crack growth behavior in fiber composite materials is represented schematically in Fig. 1.44. In the case of weak binding (Fig. 1.44a) the crack moves along the fiber, the interface delaminates and the stress leads to the fracture of the

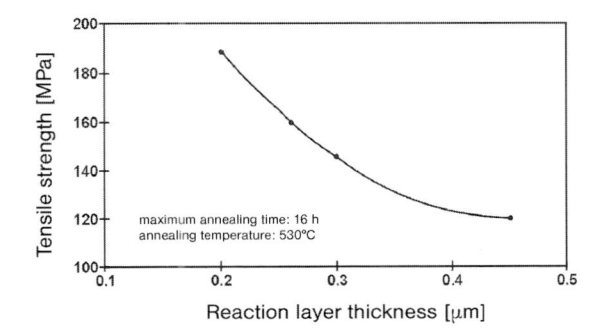

Fig. 1.43 Transverse pull strength of the fiber composite material AZ91/20 vol% Al$_2$O$_3$ fibers in dependence on the reaction layer thickness as a function of annealing time at 530 °C [54].

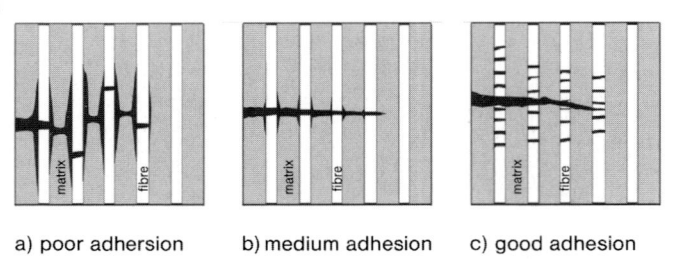

a) poor adhersion   b) medium adhesion   c) good adhesion

Fig. 1.44 Schematic presentation of the fiber/matrix dependence of the adhesion of a crack run [56, 57].

fibers successively. A classical "fiber pull out" develops. In the fracture image, for example the tensile test sample of a titanium composite material with SiC fibers (Fig. 1.45a) or aluminum alloy with C fibers (Fig. 1.46a), pulled out fibers are sporadically visible. For the case of very good adhesion of the matrix on the fiber no delamination (Fig. 1.44c) occurs. The crack opens up due to the tensile stress and the matrix deforms, due to the good adhesion the fiber is fully loaded and malfunctions. During further load the matrix continues to deform above the fiber fracture area also, thus the fiber is further loaded above and below the separation and malfunctions in further fragments. Macroscopically a brittle failure without pulling out of fibers (Fig. 1.45b, 1.46c,d) develops.

Depending on the interface formation transitions at very small delamination and fiber pull out also result (Fig. 1.44b and 1.46b). The adhesion for the tensile strength perpendicular to the fiber alignment (transverse pull strength) is of substantial importance, see Fig. 1.47 and 1.48. With very poor adhesion the fibers or particles work like pores and the strength is less than for the nonstrengthened matrix (Fig. 1.47a). In the case of very good binding a failure occurs in the matrix (Fig. 1.47c and 1.48a,b) or by disruption of the fiber (Fig. 1.47d and 1.48c). The strength of the composite material is comparable to the nonstrengthened matrix. At an average adhesion a mixed fracture occurs (Fig. 1.47b).

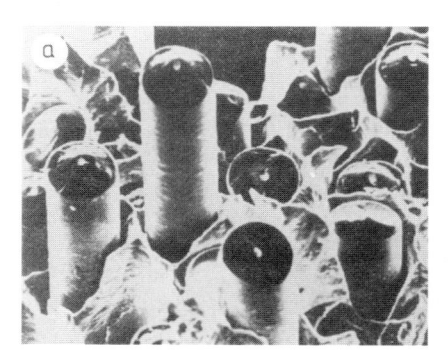

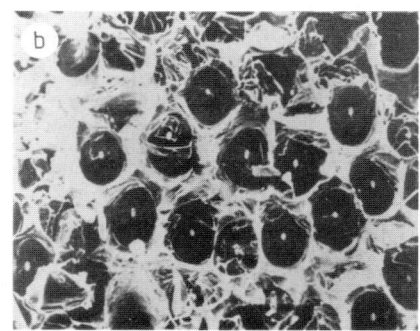

Fig. 1.45 Fracture surface in a monofilament composite material [58]: (a) low interface shear strength; (b) high interface shear strength.

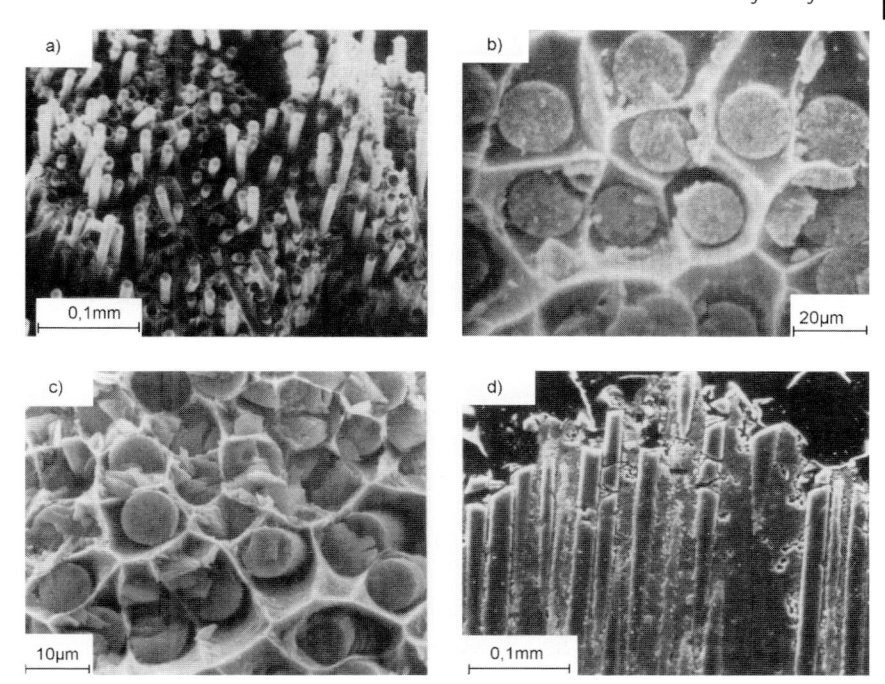

Fig. 1.46 Fracture surfaces in aluminum composite materials after tensile strain vertical to the fiber orientation [57]: (a) fiber/matrix- delamination (C/Al, weak adhesion); (b) shearing of fibers and dimple formation of a deformed matrix on the fibers (Al$_2$O$_3$/Al-2.5Li, medium adhesion); (c) Fracture in the matrix (SiC/Al good adhesion); (d) fracture run in multiple broken fibers (SiC/Al good adhesion).

poor adhesion a)
Fiber-matrix debonding

medium adhesion b)
Fiber crack near the interface and on fiber by matrix

good adhesion remains
Fracture within the matrix c)

good adhesion d)
splicing of the fiber

Fiber   Matrix

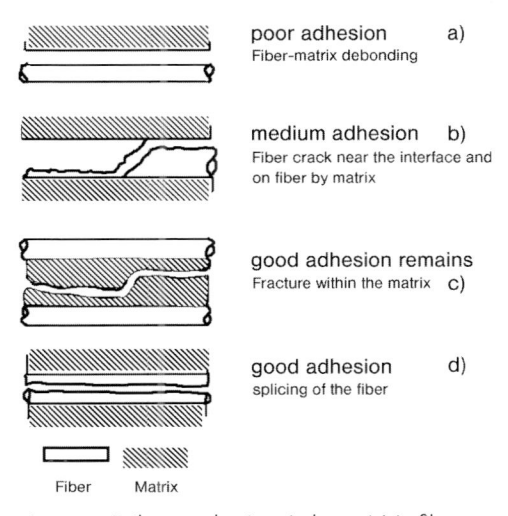

Fig. 1.47 Failure mechanism (schematic) in fiber composite materials for loading vertical to the fiber orientation [57].

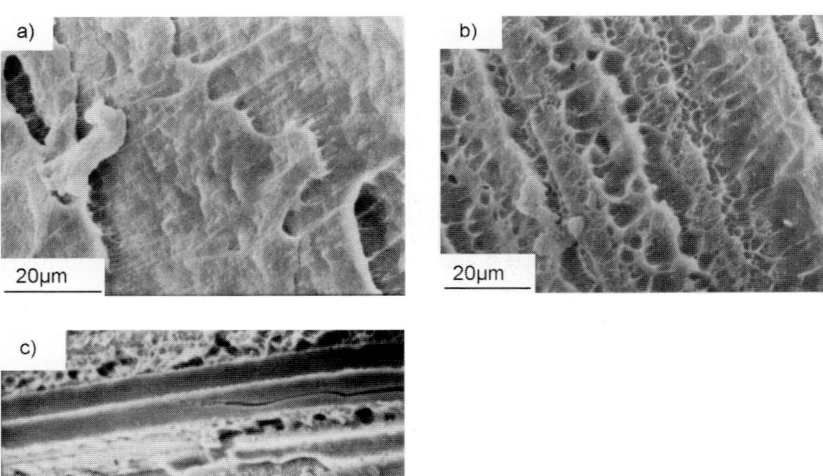

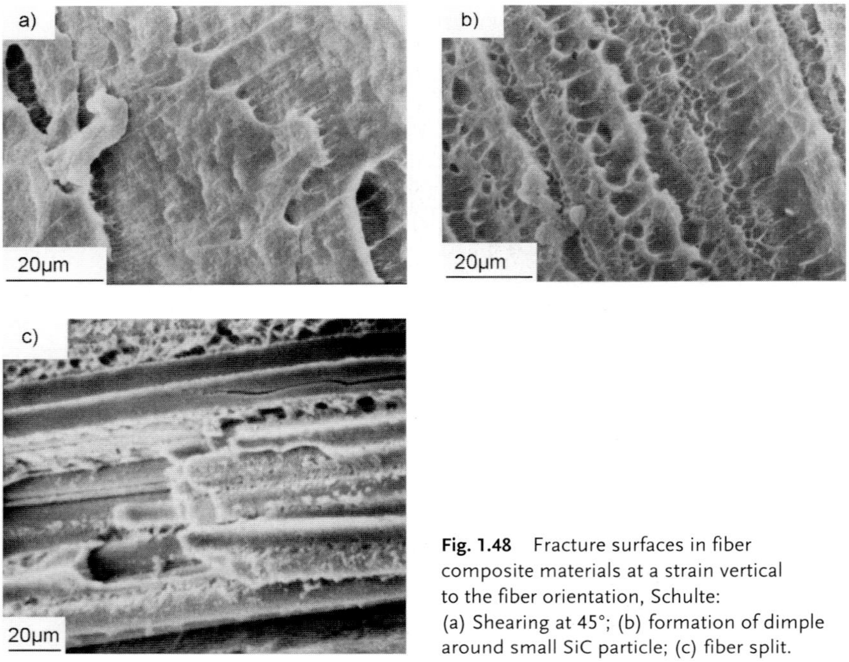

Fig. 1.48 Fracture surfaces in fiber composite materials at a strain vertical to the fiber orientation, Schulte: (a) Shearing at 45°; (b) formation of dimple around small SiC particle; (c) fiber split.

1.5
Structure and Properties of Light Metal Composite Materials

The structure of the composite materials is determined by the type and form of the reinforcement components, whose distribution and orientation are affected by the manufacturing processes. For composite materials, which are reinforced with long fibers, extreme differences result with different fibers. For multi-filament-strengthened composite materials (Fig. 1.49) the fiber/fiber contacts and nonreinforced areas are recognizable as a result of the infiltration of fiber bunch preforms. Structure defects, like fiber/fiber contacts, pores and nonreinforced areas are visible, which have a substantial influence on the composite characteristics. Figure 1.50 shows the optimal structure of a SiC monofilament/Ti composite material. With

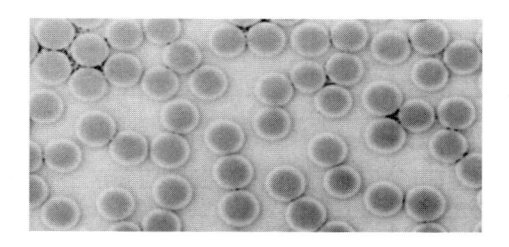

Fig. 1.49 Structure of a unidirectional endless fiber reinforced aluminum composite material (transverse grinding) [59]: matrix: AA 1085, 52 vol.% 15 µm Altex-fiber (Al_2O_3).

1.5 Structure and Properties of Light Metal Composite Materials

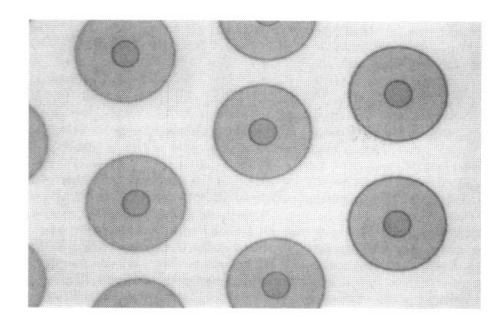

Fig. 1.50 Structure of a titan matrix composite material of SiC monofilaments [60].

monofilament-reinforced materials and with wire composite superconductors (Fig. 1.51 and 1.52) the uniformity of the fiber arrangement which result from the production process is remarkable. In Table 1.6 the material properties of different light alloy composite materials with continuous fibers are shown.

Figure 1.53 shows typical structure images of short-fiber reinforced light alloys. With short-fiber reinforced composite materials a planar-isotropic distribution of

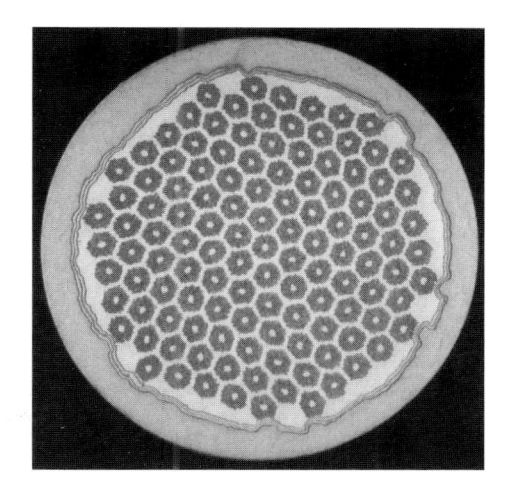

Fig. 1.51 Composite superconductor type Vacryflux NS 13 000 Ta: 13000 Nb filaments in CuSn and 35% stabilization material 30% Cu+5% Ta) in a shell [61].

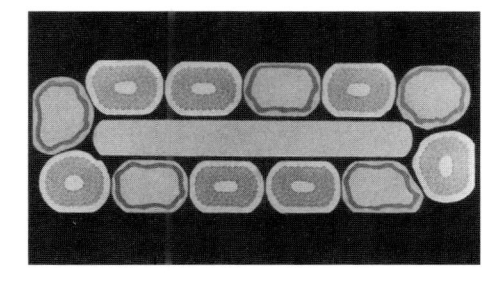

Fig. 1.52 Composite superconductor cable consisting of 7 superconductor cables and 5 stabilisation cables [61].

Tab. 1.6 Selected properties of typical long fiber reinforced light metal composites.

Material		Fiber content (%)	Density (g cm^{-3})	Tensile strength (MPa)	Young's modulus (GPa)	Ref.
System	Orientation					
Monofilaments						
B/Al	0°	50	2.65	1500	210	22
B/Al	90°	50	2.65	140	150	22
SiC/TiAl6V4	0°	35	3.86	1750	300	17, 5
SiC/TiAl6V4	90°	35	3.86	410		20, 2
Multifilaments						
SiC/Al	0°	50	2.84	259	310	21, 4
SiC/Al	90°	50	2.84	105		19, 3
Al$_2$O$_3$/Al–Li	0°	60	3.45	690	262	16, 9
Al$_2$O$_3$/Al–Li	90°	60	3.45	172–207	152	21, 4
C/Mg-Leg	0°	38	1.8	510		16, 6
C/Al	0°	30	2.45	690	160	6, 4
SiC/Al	Al+55–70% SiC		2.94		226	7, 2
MCX-736™	Al+55–70% SiC		2.96		225	7, 3

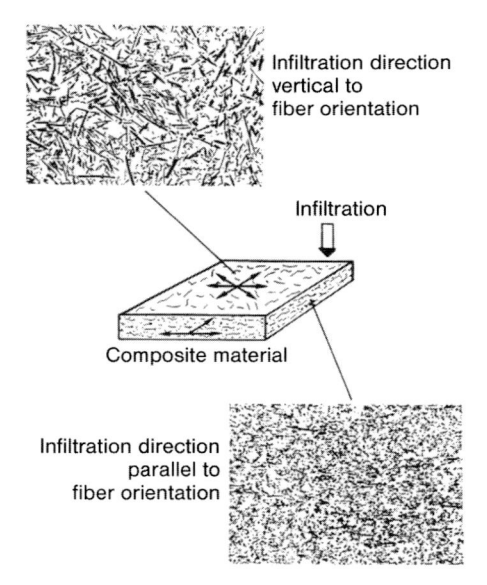

Fig. 1.53 Structure of formation of short fiber reinforced light metal composite materials [62].

the short fibers develops, due to the fiber molded padding production. The pressure-supported sedimentation technology leads to a layered structure. The infiltration direction is generally perpendicular to these layers. A reinforcement of light metal cast alloys by short fibers does not lead exclusively to an increase in strength, e.g. at room temperature, as the objective. It leads to a strength increase with increasing fiber content, as the example of AlSi12CuMgNi with a fiber content of 20 vol.% (Fig. 1.54) shows. However, the achievable effect is not economically justified in practice. The improvement in the properties, particularly at higher temperature where a doubling of the strength occurs (Fig. 1.54) and the strength properties under alternating flexural stress at 300 °C (Fig. 1.55), makes the material interesting for applications such as pistons or for reinforced cylinder surfaces in engines. A dramatic increase in the temperature alternating resistance at the same application temperature is attainable, see Fig. 1.56.

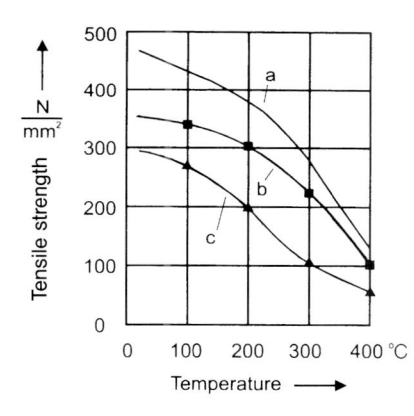

Fig. 1.54 Comparison of temperature dependence of the tensile strength of nonreinforced and reinforced piston alloy AlSi12CuMg (KS 1275) [13]: (a) KS 1275 with 20 vol.% SiC whisker; (b) KS 1275 with 20 vol.% Al_2O_3 fibers; (c) KS 1275 nonreinforced.

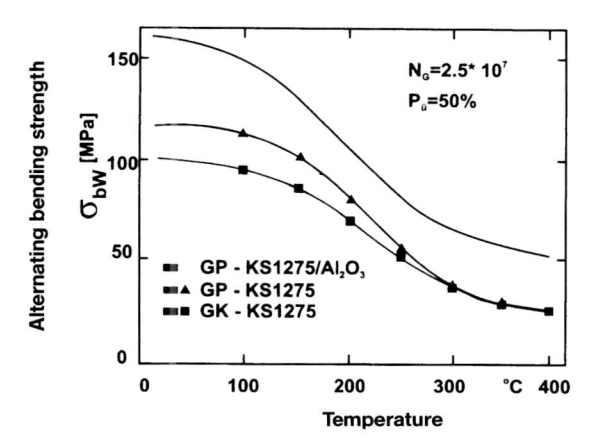

Fig. 1.55 Change in the alternating bending strength of nonreinforced and reinforced piston alloy (20 vol.% Al_2O_3 fibers) AlSi12CuMgNi (KS1275), with increasing temperature (GK=mold casting, GP=die casting) [14].

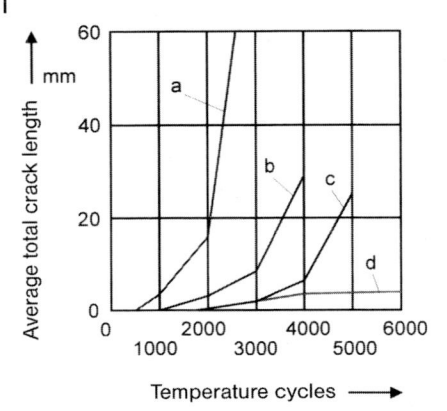

Fig. 1.56 Temperature shock resistance of the fiber reinforced piston alloy AlSi12CuMgNi (KS1275) for different fiber contents for a temperature of 350 °C [13]: (a) Nonreinforced, (b) 12 vol.% Al_2O_3 short fibers, (c) 17.5 vol.% Al_2O_3 short fibers, (c) 20 vol.% Al_2O_3 short fibers.

The cast particle-reinforced light alloys show typical particle distributions depending on the processing methods. Gravity die cast materials show nonreinforced areas due to the solidification conditions (Fig. 1.57a); while with pressure die cast materials the distribution of the particles is more optimal (Fig. 1.57b). Even better results are reached after the extrusion of feed material (Fig. 1.57c). In powder metallurgically manufactured composite materials (Fig. 1.57d) the extremely homogeneous distribution of the particles is noticeable after the extrusion

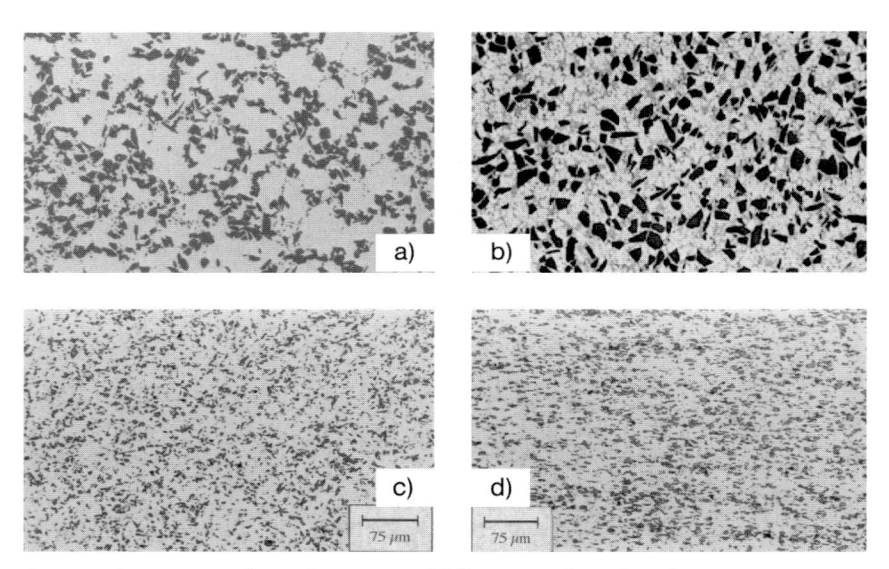

Fig. 1.57 Arrangement of typical structures of different particle reinforced light metal composite materials: (a) SiC-particle reinforced Al (mold cast [9]), (b) SiC-particle reinforced Al (die cast [10]), (c) SiC-particle reinforced Al (extruded powder mixture [11]), (d) SiC- particle reinforced Al (cast and extruded).

of powder mixtures. The possibility of combining particles and fibers to form a hybrid-reinforced composite material with the different effects of both reinforcement components is shown in Fig. 1.58.

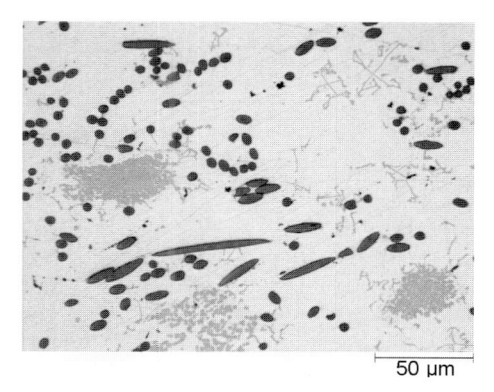

Fig. 1.58 Structure of formation of hybrid reinforced light metal composite materials with C short fibers and Mg$_2$Si particles [63].

With particle addition to light metals like aluminum, the hardness, the Young's modulus, the yield strength, the tensile strength and the wear resistance increase and the thermal expansion coefficient decreases. The order of magnitude of the improvement of these characteristics depends on the particle content and the selected manufacturing process. In Tables 1.7 and 1.8 characteristics of different particle-reinforced aluminum alloys are presented. In melting metallurgically manufactured materials by mixing in particles (Table 1.7) the upper limit of the particle addition is approx. 20 vol%. This limit is technically justified since a maximum tensile strength of over 500 MPa and a Young's modulus of 100 GPa are attainable with this particle content. Higher particle contents are made possible by reaction infiltration procedures, however, the materials then take on a more ceramic character becoming susceptible to brittle failure and, during tensile stress, a premature failure without plastic deformation takes place. However, the low thermal expansion despite the metallic character of these materials is outstanding.

For spray formed materials (Table 1.8) the limit for the particle content is approximately 13–15 vol%. However, the utilisation of special alloy systems, e.g. with lithium addition, can nevertheless lead to high specific characteristics. In powder metallurgical materials processed by extrusion from powder mixtures the particle content can be increased to over 40 vol%. Along with the fine-grained structure of the matrix, very high strength (up to 760 MPa), very high Young's modulus (up to 125 GPa) and low expansion coefficients (approximately 17×10^{-6} K^{-1}) are attainable. Unfortunately also the elongation at fracture and fracture toughness are worsened, however, the values are better than those of cast materials.

Tab. 1.7 Selected properties of typical aluminum cast composites, processed by ingot-, die cast or reaction infiltration, Manufacturers instruction after [11, 12, 16]. (T6 = solution annealed and artificially aged; T5 = artificially aged; *after ASTM G-77: Cast iron 0.066 mm^3; **CTE = thermal expansion coefficient, (a) after ASTM E-399 and B-645; (b) after ASTM E-23).

Material		Yield stress (MPa)	Tensile strength (MPa)	Ultimate strain (%)	Young's modulus (GPa)	(a) Fracture toughness (b) Impact strength	Wear* Volume decrease (mm^3)	Thermal conductivity 22 °C (cal cm^{-1} s^{-1} K^{-1})	CTE** 50–100 °C (10^{-6} K^{-1})
Name	Composition								
Gravity die casting (Chill casting)						(a) (MPa m$^{1/2}$)			
A356-T6	AlSi7g	200	276	6.0	75.2		0.18	0.360	21.4
F3S.10S-T6	AlSi9Mg10SiC	303	338	1.2	86.9	17.4			20.7
F3S.20S-T6	AlSi9Mg20SiC	338	359	0.4	98.6	17.4	0.02	0.442	17.5
F3K.10S-T6	AlSi110CuMgNi10SiC	359	372	0.3	87.6	15.9			20.2
F3K.20S-T6	AlSi110CuMgNi20SiC	372	372	0.0	101			0.346	17.8
Die casting						(b) (J)			
A390		241	283	3.5	71.0	1.4	0.18	0.360	21.4
F3D.10S-T5	AlSi110CuMnNi10SiC	331	372	1.2	93.8	1.4		0.296	19.3
F3D.20S-T5	AlSi110CuMnNi20SiC	400	400	0.0	113.8	0.7	0.018	0.344	16.9
F3N.10S-T5	AlSi110CuMnMg10SiC	317	352	0.5	91.0	1.4		0.384	21.4
F3N.20S-T5	AlSi110CuMnMg20SiC	338	365	0.3	108.2	0.7	0.018	0.401	16.6
Reaction infiltration		Flexural strength (MPa)		Density (g cm^{-3})		(a) (MPa m$^{1/2}$)			
MCX-693™	Al+55–70% SiC	300	2.98	255	9.0		0.430	6.4	
MCX-724™	Al+55–70% SiC	350	2.94	226	9.4		0.394	7.2	
MCX-736™	Al+55–70% SiC	330	2.96	225	9.5		0.382	7.3	

1.5 Structure and Properties of Light Metal Composite Materials

Tab. 1.8 Properties of aluminum wrought alloy composites. Manufacturers instruction after [11, 12, 18–20]. (T6 = solution annealed and artificially aged; *after ASTM G-77: Cast iron 0.066 mm³; **CTE = thermal expansion coefficient).

Material		Yield stress (MPa)	Tensile strength (MPa)	Ultimate strain (%)	Young's modulus (GPa)	Fracture toughness (MPa m$^{1/2}$) ASTM E-399	Wear* volume decrease (mm³)	Thermal conductivity 22°C (cal cm^{-1} s^{-1} K^{-1})	CTE** 50–100°C (10^{-6} K^{-1})
Name	Composition								
Cast prematerial (extruded or forged)									
6061-T6	AlMg1SiCu	355	375	13	75	30	10	0.408	23.4
6061-T6	+ 10% Al$_2$O$_3$	335	385	7	83	24	0.04	0.384	20.9
6061-T6	+ 15% Al$_2$O$_3$	340	385	5	88	22	0.02	0.336	19.8
6061-T6	+ 20% Al$_2$O$_3$	365	405	3	95	21	0.015		
Powder metallurgically processed prematerial (extruded)									
6061-T6	AlMg1SiCu	276	310	15	69.0				23.0
6061-T6	+ 20% SiC	397	448	4.1	103.4				15.3
6061-T6	+ 30% SiC	407	496	3.0	120.7				13.8
7090-T6		586	627	10.0	73.8				
7090-T6	+ 30% SiC	676	759	1.2	124.1				
6092-T6	AlMg1Cu1Si17.5SiC	448	510	8.0	103.0				
6092-T6	AlMg1Cu1Si25SiC	530	565	4.0	117.0				
Spray formed material (extruded)									
6061-T6	+ 15% Al$_2$O$_3$	317	359	5	87.6				
2618-T6	+ 13% SiC	333	450		89.0				19.0
8090-T6	AlCuMgLi	480	550		79.5				22.9
8090-T6	+ 12% SiC	486	529		100.1				19.3

1.6
Possible Applications of Metal Matrix Composites

Light alloy composite materials have, in automotive engineering, a high application potential in the engine area (oscillating construction units: valve train, piston rod, piston and piston pin; covers: cylinder head, crankshaft main bearing; engine block: part-strengthened cylinder blocks), see Table 1.9. An example of the successful use of aluminum composite materials within this range is the partially short-fiber reinforced aluminum alloy piston in Fig. 1.59, in which the recess range is strengthened by Al_2O_3 short fibers. Comparable construction unit characteristics are attainable only with the application of powder metallurgical aluminum alloys or when using heavy iron pistons. The reason for the application of composite materials is, as already described, the improved high temperature properties. Potential applications are in the area of undercarriages, e.g. transverse control arms and particle-strengthened brake disks, which can be also applied in the area of rail-mounted vehicles, e.g. for undergrounds and railway (ICE), see Fig. 1.60. In the

Tab. 1.9 Applications of metal composites.

I.	Drive shaft for people and light load motor vehicles (Fig. 1.61) [65]:	
	Material:	$AlMg1SiCu$ + 20 vol.% Al_2O_3P
	Processing:	extrusion form cast feed material
	Development aims:	– high dynamic stability, high Young's modulus (95 GPa)
		– low density (2.95 g cm^{-3})
		– high fatigue strength (120 MPa for $n = 5 \times 10^7$, $R = -1$, RT)
		– sufficient toughness (21.5 MPa m$^{1/2}$)
		– substitution of steels
II.	Vented passenger car brake disk (Fig. 1.62) [65]:	
	Material:	G-$AlSi12Mg$ + 20 vol.% SiC_P
	Processing:	sand- or gravity die casting
	Development aims:	– high wear resistance (better than conventional cast iron brake discs)
		– low heat conductivity (factor 4 higher than cast iron)
		– substitution of iron materials
III.	Longitudinal bracing beam (Stringer) for planes (Fig. 1.63) [66]:	
	Material:	$AlCu4Mg2Zr$ + 15 vol.% SiC_P
	Processing:	extrusion and forging of casted feed material
	Development aims:	– high dynamic stability. high Young's modulus (100 GPa)
		– low density (2.8 g cm^{-3})
		– high strength (R_m = 540 MPa. $R_{p0.2}$ = 413 MPa. RT)
		– high fatigue strength (240 MPa for n = 5×10^7, $R = -1$, RT)
		– sufficient toughness (19.9 MPa m$^{1/2}$)
IV.	Disk brake calliper for passenger cars (Fig. 1.64) [67]:	
	Material:	Aluminium alloy with Nextel ceramic fibre 610
	Weight reduction:	55% compared to cast iron.

1.6 Possible Applications of Metal Matrix Composites | 49

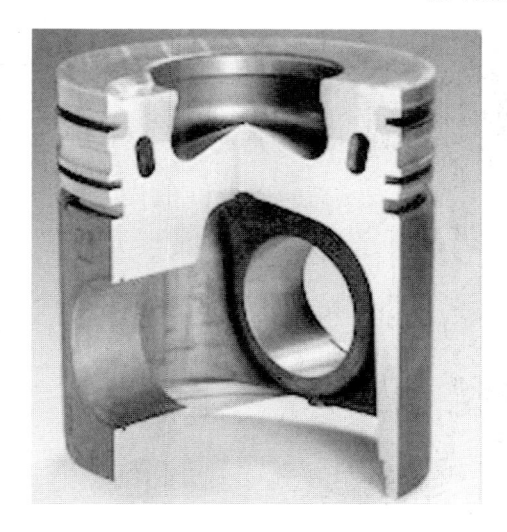

Fig. 1.59 Partial short fiber reinforced light metal diesel pistons [13, 14].

following some potential construction units made out of aluminum matrix composite materials with data concerning materials, processing and development targets are presented.

Fig. 1.60 Cast brake disk particle of reinforced aluminum for the ICE 2 [64].

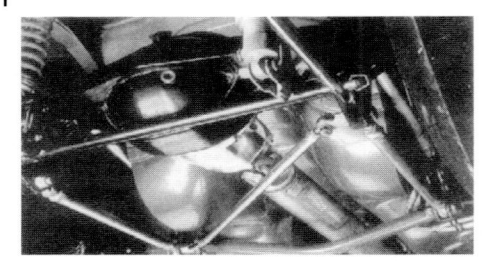

Fig. 1.61 Drive shaft particle of reinforced aluminum for passenger cars of [65].

Fig. 1.62 Vented passenger car brake disk of particle reinforced aluminum [65].

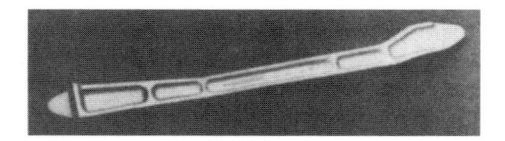

Fig. 1.63 Longitudinal bracing beam (Stringer) of particle reinforced aluminum [66].

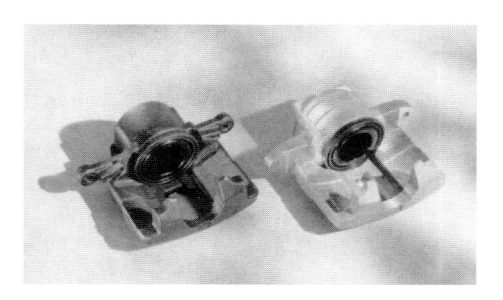

Fig. 1.64 Disk brake calliper for passenger cars of conventional cast iron (left) and an aluminum matrix composite material (AMC) with Nextel® ceramic fiber 610 [67].

In the aviation industry the high specific strength, the high Young's modulus, the small thermal expansion coefficient, the temperature resistance and the high conductivity of the strengthened light alloys are of interest compared with polymer materials, e.g. for reinforcements, axle tubes, rotors, housing covers and structures for electronic devices. A compilation of potential and realized applications of most different metal matrix composites can be seen in Table 1.10.

1.6 Possible Applications of Metal Matrix Composites

Tab. 1.10 Potential and realistic technical applications of metal matrix composites.

Application	Required properties	Material system	Processing technique
Automotive and heavy goods vehicle			
Bracing systems, piston rods, frames, piston, piston pins, valve spring cap, brake discs, disc brake calliper, brake pads, cardan shaft	High specific strength and stiffness, temperature resistance, low thermal expansion coefficient, wear resistance, thermal conductivity	Al-SiC, Al-Al$_2$O$_3$, Mg-SiC, Mg-Al$_2$O$_3$, discontinuous reinforcements	Fusion infiltration, extrusion, forging, gravity die casting, die casting, squeeze-casting
Accumulator plate	High stiffness, creep resistance	PbC, Pb-Al$_2$O$_3$	Fusion infiltration
Military and civil air travel			
Axle tubes, reinforcements, blade- and gear box casing, fan and compressor blades	High specific strength and stiffness, temperature resistance, impact strength, fatigue resistance	Al-B, Al-SiC, Al-C, Ti-SiC, Al-Al$_2$O$_3$, Mg-Al$_2$O$_3$, Mg-C continuous and discontinuous reinforcements	Fusion infiltration, hot pressing, diffusion welding and soldering, extrusion, squeeze-casting
Turbine blades	High specific strength and stiffness, temperature resistance, impact strength, fatigue resistance	W super alloys, z. B. Ni$_3$Al, Ni-Ni$_3$Nb	Fusion infiltration, aligned solidification near net-shaped components
Aerospace industry			
Frames, reinforcements, aerials, joining elements	High specific strength and stiffness, temperature resistance, low thermal expansion coefficient, thermal conductivity	Al-SiC, Al-B, Mg-C, Al-C, Al-Al$_2$O$_3$, continuous and discontinuous reinforcements	Fusion infiltration, extrusion, diffusion welding and joining (spacial structures)
Energy techniques (electrical components and conducting materials)			
Carbon brushes	High electrical and thermal conductivity, wear resistance	Cu-C	Fusion infiltration, powder metallurgy
Electrical contacts	High electrical conductivity, temperature and corrosion resistance, burn-up resistance	Cu-C, Ag-Al$_2$O$_3$, Ag-C, Ag-SnO$_2$, Ag-Ni	Fusion infiltration, powder metallurgy, extrusion, pressing
Super conductor	Superconducting, mechanical strength, ductility	Cu-Nb, Cu-Nb$_3$Sn, Cu-YBaCO	Extrusion, powder metallurgy, coating technologies
Other applications			
Spot welding electrodes	Burn-up resistance	Cu-W	Powder metallurgy, infiltration
Bearings	Load carrying capacity, wear resistance	Pb-C, Brass-Teflon	Powder metallurgy, infiltration

1.7 Recycling

Of special economic and ecological interest for newly developed materials is the necessity for recirculation of arrears, cycle scrap and other material from these composites into the material cycle. Since ceramic materials usually occur in the form of particles, short fibers or continuous fibers for the reinforcement of metallic materials, a material separation of the components with the goal being the reuse of the matrix alloy and the reinforcement is almost impossible. However, with conventional melting treatments in remelting factories the matrix alloy can be recycled without problems.

In melting or powder metallurgically manufactured discontinuous short fiber or particle-strengthened light alloys reuse of the composite materials from cycle scrap or splinters can be possible under certain conditions. Particularly for particle-strengthened aluminum cast alloys the use of cycle scrap is possible, however splinters on direct remelting cause difficulties due contamination problems. Ref. [69] gives an overview of different recycling concepts for light alloy composite materials in relation to the alloy composition, the kind of reinforcement and the production and processing prehistory.

References

1 G. Ondracek, *Werkstoffkunde: Leitfaden für Studium und Praxis*, Expert-Verlag, Würzburg (**1994**).
2 *TechTrends*, International Reports on Advanced Technologies: Metal Matrix Composites: Technology and Industrial Application, Innovation 128, Paris (**1990**).
3 T. W. Clyne, P. J. Withers, *An Introduction to Metal Matrix Composites*, Cambridge University Press, Cambridge (**1993**).
4 K.U.Kainer, *Keramische Partikel, Fasern und Kurzfasern für eine Verstärkung von metallischen Werkstoffen. Metallische Verbundwerkstoffe*, K.U. Kainer (Ed.), DGM Informationsgesellschaft, Oberursel (**1994**), pp. 43–64.
5 H. Dieringa, K. U. Kainer, this report
6 K. U. Kainer, *Werkstoffkundliche und technologische Aspekte bei der Entwicklung verstärkter Aluminumlegierungen für den Einsatz in der Verkehrstechnik*, Newsletter TU Clausthal, Issue 82 (**1997**), pp. 36–44.
7 K. U. Kainer (Ed.), *Metallische Verbundwerkstoffe*, DGM Informationsgesellschaft, Oberursel (**1994**).
8 J. Schröder, K. U. Kainer, Magnesium Base Hybrid Composites Prepared by Liquid Infiltration, *Mater. Sci. Eng.*, A **1991**, *135*, 33–36.
9 *DURALCAN Composites for Gravity Castings*, Duralcan USA, San Diego (**1992**).
10 *DURALCAN Composites for High-Pressure Die Castings*, Duralcan USA, San Diego (**1992**).
11 C. W. Brown, W.Harrigan, J. F. Dolowy, Proc. Verbundwerk 90, Demat, Frankfurt (**1990**), pp. 20.1–20.15.
12 *Manufacturers of Discontinuously Reinforced Aluminum (DRA)*, DWA Composite Specialities, Inc., Chatsworth USA (**1995**).
13 W.Henning, E. Köhler, *Maschinenmarkt* **1995**, *101*, 50–55.
14 S. Mielke, N. Seitz, Grosche, *Int. Conf. on Metal Matrix Composites*, The Institute of Metals, London (**1987**), pp. 4/1–4/3.

15 H. P. Degischer, Schmelzmetallurgische Herstellung von Metallmatrix-Verbundwerkstoffen, in *Metallische Verbundwerkstoffe*, K.U.Kainer (Ed.), DGM Informationsgesellschaft, Oberursel (**1994**), pp.139–168.
16 *Lanxide Electronic Components*, Lanxide Electronic Components, Inc., Newark USA (**1995**).
17 C. Fritze, Infiltration keramischer Faserformkörper mit Hilfe des Verfahrens des selbstgenerierenden Vakuums, Dissertation TU Clausthal (**1997**).
18 A. G. Leatham, A. Ogilvy, L. Elias, *Proc. Int. Conf. P/M in Aerospace, Defence and Demanding Applications*, MPIF, Princeton, USA (**1993**), pp. 165–175.
19 Cospray Ltd. Banbury, U.K., **1992**.
20 *Keramal Aluminum-Verbundwerkstoffe*, Aluminum Ranshofen Ges.m.b.H., Ranshofen, Österreich (**1992**).
21 F. Koopmann, Kontrolle Heft 1/2 (**1996**), pp. 40–44.
22 K. K. Chawla, *Composite Materials: Science and Engineering*, Springer-Verlag, New York (**1998**).
23 D. L. McDanels, R. W. Jech, J. W. Weeton, Analysis of Stress-Strain Behavior of Tungsten-Fiber-Reinforced Copper Composites, *Trans. Metall. Soc. AIME* **1965**, *223*, 636–642.
24 J. Schlichting, G. Elssner, K. M. Grünthaler, *Verbundwerkstoffe, Grundlagen und Anwendung*, Expert-Verlag, Renningen (**1978**).
25 A. Kelly, *Strong Solids*, Oxford University Press, London (**1973**).
26 C. M. Friend, The effect of matrix properties on reinforcement in short alumina fiber-aluminum metal matrix composites, *J. Mater. Sci.* **1987**, *22*, 3005–3010.
27 A. Kelly, G. J. Davies, The Principles of the Fiber Reinforcement of Metals, *Metall. Rev.* **1965**, *10*, 1–78.
28 G. Ibe, Grundlagen der Verstärkung von Metallmatrix-Verbundwerkstoffen, in *Metallische Verbundwerkstoffe*, K.U.Kainer (Ed.), DGM Informationsgesellschaft, Oberursel (**1994**), pp.1–41.
29 K. U. Kainer, Strangpressen von kurzfaserverstärkten Magnesium-Verbundwerkstoffen, *Umformtechnik* **1993**, *27*, 116–121.
30 F. J. Humphreys, Deformation and annealing mechanisms in discontinuously reinforced metal-matrix composites, *Proc. 9th Risø Int. Symp. on Mechanical and Physical Behavior of Metallic and Ceramic Composites*, S. I. Anderson, H. Lilholt, O. B. Pederson (Eds.), Risø National Laboratory, Roskilde (**1988**), pp. 51–74.
31 F. J. Humphreys, A. Basu, M. R. Djazeb, The microstructure and strength of particulate metal-matrix composites, *Proc. 12th Risø Int. Symp. on Materials Science, Metal-Matrix Composites – Processing, Microstructure and Properties*, N. Hansen et al. (Eds), Risø National Laboratory, Roskilde (**1991**), pp. 51–66.
32 J. C. Halpin, S. W. Tsai: Air Force Materials Laboratory (**1967**), AFML-TR-67-423.
33 F. Moll, K. U. Kainer, Properties of Particle Reinforced Magnesium Alloys in Correlation with Different Particle Shapes, *Proc. Int. Conf. Composite Materials 11*, Vol. III, pp. 511–519.
34 R. A. Schapery, *J. Comp. Mater.* **1968**, *23*, 380–404.
35 K. U. Kainer, U. Roos, B. L. Mordike, Platet-Reinforced Magnesium Alloys, *Proc. 1st Slovene-German Seminar on Joint Projects in Materials Science and Technology*, D. Kolar, D. Suvorov (Eds.), Forschungszentrum Jülich (**1995**), pp. 219–224.
36 C. Köhler, Thermische Beständigkeit von Kurzfaser-verstärkten Magnesium-Al$_2$O$_3$-Verbundwerkstoffen, Dissertation TU Clausthal (**1994**).
37 F. Delanny, L. Froyen , A.Deruyttiere, *J. Mater. Sci.* **1987**, *22*, 1–16.
38 J. Haag, *Bedeutung der Benetzung für die Herstellung von Verbundwerkstoffen unter Weltraumbedingungen – Größen, Einflüsse und Methoden-*, BMBF-Research Report W 81-021 (**1980**).
39 U. Angelopoulos, U. Jauch, P. Nikolopoulos, *Mat.-wiss. u. Werkstofftechnik* **1988**, *19*, 168–172.
40 S. Y. Oh, J. A. Cornie, K. C. Russel, Particulate Welding and Metal: Ceramic Interface Phenomena, Ceramic Engineering and Science, *Proc. 11th Annual Conf. On Composites and Advanced Materials* (**1987**).

41 K. C. Russel, S.Y. Oh , A. Figueredo, MRS Bull. **1991**, *16*, 46–52.
42 T. Choh, T. Oki, *Mater. Sci. Technol.* **1987**, *3*, 378.
43 R. Warren, C.-H. Andersson, *Composites* **1984**, *15*, 101.
44 A. Mortensen, I. Jin, *Int. Met. Rev.* **1992**, *37*, 101–128
45 G. A. Chadwick, *Mater. Sci. Eng., A* **1991**, *135*, 23–28.
46 A. Mortensen, L.J. Masur, J. A. Cornie, M. C. Flemings, *Metall. Trans. A* **1989**, *20*, 2535–2547.
47 A. Mortensen, J. A. Cornie, *Metall. Trans. A* **1987**, *18*, 1160–1163.
48 A. Mortensen, T. Wong , *Metall. Trans. A* **1990**, *21*, 2257–2263.
49 A.S. Sangini, A. Acrivos, *J. Multiphase Flow*, **1982**, *8*, 193–206.
50 R. P. Cchabra, D. K. Sheth, *Z. Metallkde.* **1990**, *81*, 264– 271.
51 L. J. Ebert, P. K. Wright: Mechanical Aspects of the Interface, in *Interfaces in Metal Matrix Composites*, A. G. Metcalfe (Ed.), Academic Press, New York **(1974)**, p. 31.
52 R. F. Tressler, Interfaces in Oxide Reinforced Metals, in *Interfaces in Metal Matrix Composites*, A. G. Metcalfe (Ed.), Academic Press, New York (1974), p. 285.
53 K. U. Kainer, Herstellung und Eigenschaften von faserverstärkten Magnesiumverbundwerkstoffen, in: DGM Informationsgesellschaft, K. U. Kainer (Ed.), Oberursel **(1994)**, pp. 219–244.
54 I. Gräf, K. U. Kainer, Einfluß der Wärmebehandlungen bei Al$_2$O$_3$-kurzfaserverstärktem Magnesium, *Prakt. Metall.* **1993**, *30*, 540–557.
55 K. U. Kainer, Alloying Effects on the Properties of Alumina-Magnesium-Composites, in *Metal Matrix Composites – Processing Microstructure and Properties*, N. Hansen et al. (Ed.), Risø National Laboratory, Roskilde **(1991)**, pp. 429–434.
56 E. Fitzer, G. Jacobsen, G. Kempe in *Verbundwerkstoffe*, G. Ondracek (Ed.) DGM Oberursel **(1980)**, p. 432.
57 Schulte in *Metal Matrix Composites – Processing Microstructure and Properties*, N. Hansen et al. (Ed.), Risø National Laboratory, Roskilde **(1991)**, pp. 429–434.
58 J. M. Wolla, *Proc. Int. Conf. ISTFA 87: Advanced Materials*, Los Angeles **(1987)**, p. 55
59 J. Janczak et al., *Grenzflächenuntersuchungen an endlosfaserverstärkten Aluminummatrix Verbundwerkstoffen für die Raumfahrttechnik*, Oberflächen Werkstoffe **(1995)** Heft 5.
60 http//www.mmc-asses.tuwien.ac.at/data/mfrm/tisic.htm#top3.
61 *Vaccumschmelze: Superconductors*, Corporate publications SL 021 **(1987)**
62 K. U. Kainer, B. L. Mordike, *Metall* **1990**, *44*, 438–443.
63 H. Dieringa, T. Benzler, K.U. Kainer, Microstructure, creep and dilatometric behavior of reinforced magnesium matrix composites, in *Proc. of ICCM13*, Peking, **2001**, p. 485.
64 F. Koopmann, *Kontrolle* Issue 1/2 **(1996)**, pp. 40–44.
65 P. J. Uggowitzer, O. Beffort, Aluminumverbundwerkstoffe für den Einsatz in Transport und Verkehr. Ergebnisse der Werkstofforschung, Volume 6, M. O. Speidel, P. J. Uggowitzer (Eds.), Verlag "Thubal-Kain", ETH-Zürich **(1994)**, 13–37.
66 C. Carre, V. Barbaux, J. Tschofen, *Proc. Int. Conf. on PM-Aerospace Materials*, MPR Publishing Services Ltd, London **(1991)**, pp. 36-1–36-12.
67 3M Metal Matrix Composites, *Firmenschrift* 98-0000-0488-1(51.5)ii **(2001)**.
68 K. U. Kainer, Konzepte zum Recycling von Metallmatrix-Verbundwerkstoffen, in *Recycling von Verbundwerkstoffen und Werkstoffverbunden*, G. Leonhardt, B. Wielage (Eds.), DGM Informationsgesellschaft, Frankfurt **(1997)**, pp. 39–44.

2
Particles, Fibers and Short Fibers for the Reinforcement of Metal Materials

Hajo Dieringa and Karl Ulrich Kainer

2.1
Introduction

The availability as well as the demand for reinforcing compounds for metal matrix composites is very extensive. Their selection depends on the condition of the matrix, the type of processing of the composite material and the demands on the material (temperature, corrosion, stress etc.). The following requirements apply in general:

- low density
- mechanical compatibility (a thermal expansion coefficient which is low but adapted to the matrix)
- chemical compatibility, which leads to optimal adhesion between matrix and reinforcement, but does not lead to corrosion problems
- thermal stability
- high Young's modulus
- high compression and tensile strength
- good processability
- low cost.

These demands can be almost exclusively fulfilled by nonmetal inorganic reinforcement components. Ceramic particles, or rather fibers or carbon fibers, are used for metal reinforcement. An application area of metal fibers is that of functional materials (for example for contacts, conductors and superconductors). However, their application in the structural area mainly fails because of the high density. Organic fibers cannot be employed because of their low Young's modulus, processing problems, poor thermal stability and poor compatibility. However, there have been attempts to use organic reinforcement components in metallic materials.

Considering economic criteria, the use of discontinuous reinforcement, like particles or short fibers, appears most favorable. Significant improvements in the property profile only become possible by using high-performance fibers. The

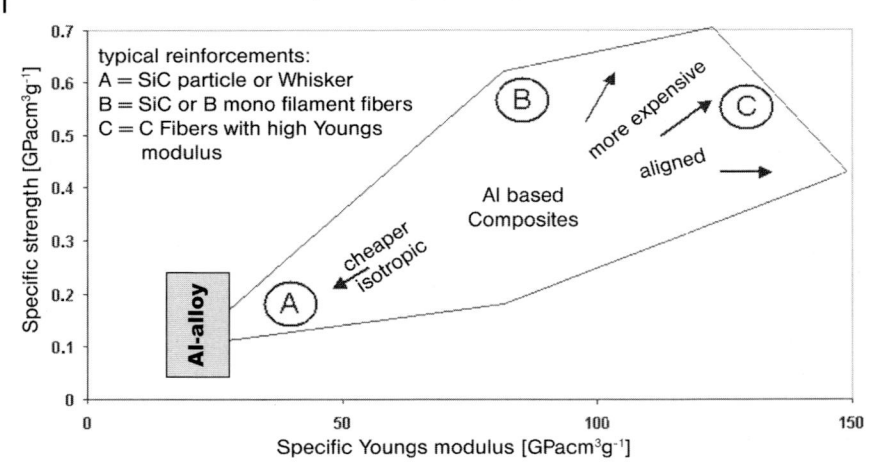

Fig. 2.1 Area of specific strength and specific Young's modulus of aluminum-composite materials with various reinforcements [1].

achievable properties of metal matrix composites using the same matrix alloying systems are dependent on, amongst others, the contribution, orientation and type of reinforcement compound. Figure 2.1 shows as an example the possible variation in the specific strength and specific Young's modulus of aluminum matrix composite materials with various reinforcements. With the employment of discontinuous reinforcements, the specific properties are only slightly improved. The advantages exist in the isotropic condition of the material and their low production costs. Fiber matrix composites, with aligned monofilament fibers with a high Young's modulus or carbide fibers, have high specific strength and Young's modulus, but have the disadvantage of anisotropic properties as well as high fiber and production costs.

2.2
Particles

For particle-reinforcement of light metals, hard particle material is used for reasons of economy; the conventional areas of use are in the ceramics and the grinding and polishing industries. A large number of different oxides, carbides, nitrides and borides are suitable for reinforcement; an overview is given in Table 2.1. The following have proved technically and economically interesting: silicon and boron carbides, aluminum oxide, aluminum and boron nitrides and titanium boride. A summary of the properties of these hard materials is given in Table 2.2.

The production of hard particle material usually takes place through endothermic reaction of oxides with the elements. Silicon carbide is produced from a mix of

Tab. 2.1 Potential particle- or platelet-shaped ceramic compounds for metal reinforcement.

Metal-basis	Carbide	Nitride	Boride	Oxide
boron	B_4C	BN	–	–
tantalum	TaC	–	–	–
zirconium	ZrC	ZrN	ZrB_2	ZrO_2
hafnium	HfC	HfN	–	HfO_2
aluminum	–	AlN	–	Al_2O_3
silicon	SiC	Si_3N_4	–	–
titanium	TiC	TiN	TiB_2	–
chromium	CrC	CrN	CrB	Cr_2O_3
molybdenum	Mo_2C, MoC	Mo_2N, MoN	Mo_2B, MB	–
tungsten	W_2C, WC	W_2N, WN	W_2B, WB	–
thorium	–	–	–	ThO_2

Tab. 2.2 Properties of various particles for reinforcement of metals [4–6]

Type of Particle	SiC	Al_2O_3	AlN	B_4C	TiB_2	TiC	BN
Type of Crystal	hex.	hex.	hex.	rhomb.	hex.	cub.	hex.
Melting point [°C]	2300	2050	2300	2450	2900	3140	3000
Young's modulus [GPa]	480	410	350	450	370	320	90
Density [g cm^{-3}]	3.21	3.9	3.25	2.52	4.5	4.93	2.25
Heat conductivity [W m^{-1} K^{-1}]	59	25	10	29	27	29	25
Mohs-hardness	9.7	6.5		9.5			1.0–2.0
Thermal coefficient of expansion [10^{-6} K^{-1}]	4.7–5.0	8.3	6.0	5.0–6.0	7.4	7.4	3.8
Producer	Wacker Ceramics Kempten, Electro Abrasive, H. C. Starck	Wacker Ceramics Kempten	H. C. Starck	Wacker Ceramics Kempten, Electro Abrasive, H. C. Starck	H. C. Starck	H. C. Starck	Wacker Ceramics Kempten, H. C. Starck

quartz sand (SiO_2) and coke (C), which is reacted in a resistance furnace at a temperature of 2000 °C [3]:

$$625.1 \text{ kJ} + SiO_2 + 3C \rightarrow SiC + 2CO$$

Boron nitride is produced from the elements by high temperature synthesis. Boron carbide is produced by heating boron or diboron oxide with coal at 2500 °C. Aluminum oxide is usually produced from bauxite by the Bayer technique and then melted and purified in an electric arc furnace [7]. Aluminum nitride is synthesized at 1600 °C and 100 bar from the elements. The carbides and the solidified oxides are broken up and ground. The classification (sizing) into different grain classes takes place by processes like sieving or sedimentation in order to obtain the favored particle size contribution. The notation for the grain classification is in line with FEPA regulation or Mesh notation [8, 9]. Examples are shown in Tables 2.3 and 2.4.

The geometries of the particles are manifold. It is possible to produce spherical, block-shaped, plate-shaped or needle-shaped geometries. However, due to the production process the particles are mostly irregular in shape, having sharp tips and edges. Figure 2.2a shows an example of a SiC particle in block-shaped geometry, Fig. 2.2b shows a SiC particle in the spherical form. The geometrical shape can have an unfavorable effect on the composite material. Therefore, attempts have been made to produce by suitable techniques, regular shaped particles in platelet (Fig. 2.2c,d) or rod forms, which have a positive influence on the mechanical properties of particle reinforced light metals [2]. Rod-shaped particles act, at an appropriate length–thickness ratio, like short fibers.

Tab. 2.3 F.E.P.A. codes for the grain size of SiC particles.

F.E.P.A.-Code	Main gain portion (>50%) [µm]
F 100	106 – 150
F 120	90 – 125
F 150	63 – 106
F 180	53 – 90
F 220	45 – 75
F 230	50 – 56
F 240	42.5 – 46.5
F 280	35.0 – 38.0
F 320	27.7 – 30.7
F 360	21.3 – 24.3
F 400	16.3 – 18.3
F 500	11.8 – 13.8
F 600	8.3 – 10.3
F 800	5.5 – 7.5
F 1000	3.7 – 5.3
F 1200	2.5 – 3.5

Tab. 2.4 Mesh notation with USA sieving-equivalent.

Mesh notation	USA sieving equivalent (µm)
42	355
48	300
60	250
65	212
80	180
100	150
115	125
150	106
170	90
200	75
250	63
270	53
325	45
400	38

Fig. 2.2 (a) SiC, particle in block shaped geometry [10]; (b) SiC, particle in spherical geometry [10]; (c) SiC, particle in platelet-shaped geometry [11]; (d) SiC, particle in platelet-shaped geometry [11].

2.2.1
Fibers

The application of thin fibers for the reinforcement of metallic materials is, amongst other reasons, related to their high strength. The high strength of these thin fibers is explained by the fiber paradox, which is expressed as follows: Fibrous material has a considerably high strength; the thinner the fibers, the greater the strength. The reason for this phenomenon is the contribution of defects, which are significantly responsible for the reduction in strength. The following example describes the phenomenon:

The defect density in a solid ρ_D and the volume which contains exactly one defect, V_D, are related by:

$$V_D = \rho_D^{-1} \qquad (1)$$

The average defect distance l_{D-D} corresponds graphically to the edge line of a cube with exactly one defect:

$$l_{D-D} = V_D^{1/3} = \rho_D^{-1/3} \qquad (2)$$

For an assumed defect density of 1000 cm^{-3} an average defect distance of 1 mm results from Eq. (2), see Fig. 2.3. At a fiber volume, V_F, where

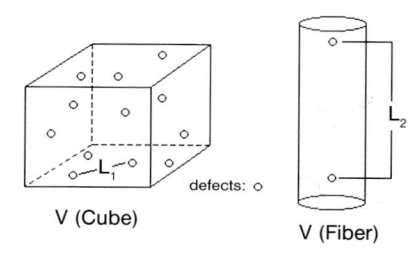

$$V_F = d^2 l\, \pi/4 \qquad (3)$$

Fig. 2.3 Defect distance in a cube and a fiber volume at similar defect density

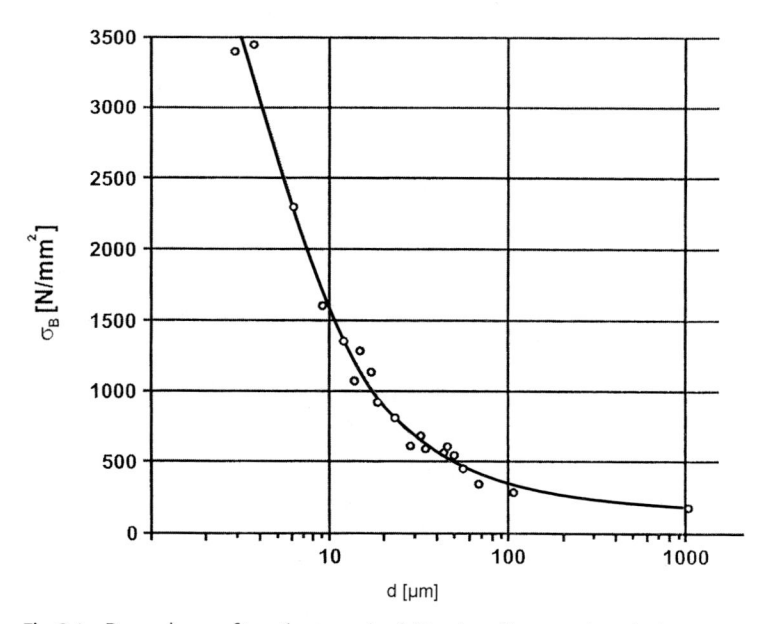

Fig. 2.4 Dependence of tensile strength of thin glass fibers on their thickness [12].

(the volume of a cylinder) one can calculate at similar defect density (1000 cm^{-3}) an average defect distance of approximately 12.7 m. This is the average defect distance in comparison to a cube 1.27×10^4 times larger (Fig. 2.3). Because the fiber diameter is present as the inverse square in the formula for the average defect density, the defect distance again increases substantially by application of thinner fibers, see Fig. 2.4. This effect can be used in both short fiber reinforced and long fiber reinforced composite materials.

2.3 Continuous Fibers

The potential of inorganic and organic long fibers (continuous fibers) for reinforcement of metals is manifold. The classification of the fibers into monofilament (single fibers with $d = 100$–150 µm) and multifilament (fiber strands with 500–1000 single fibers with $d = 6$–20 µm) results from the thickness of the fibers. The different chemical compositions, production processes and delivery shapes offer a large bandwidth of properties, see the overview of the tensile strength and Young's modulus of different fibers in Fig. 2.5.

If the properties are related to the density, one can classify fibers into different groups. Figure 2.6 shows the classification regarding the specific strength and specific Young's modulus. Oxidic fibers have average specific strength and Young's modulus, far above those of the metals. Multifilament fibers, based on SiC or SiO_2 have high specific strength at comparable specific Young's modulus.

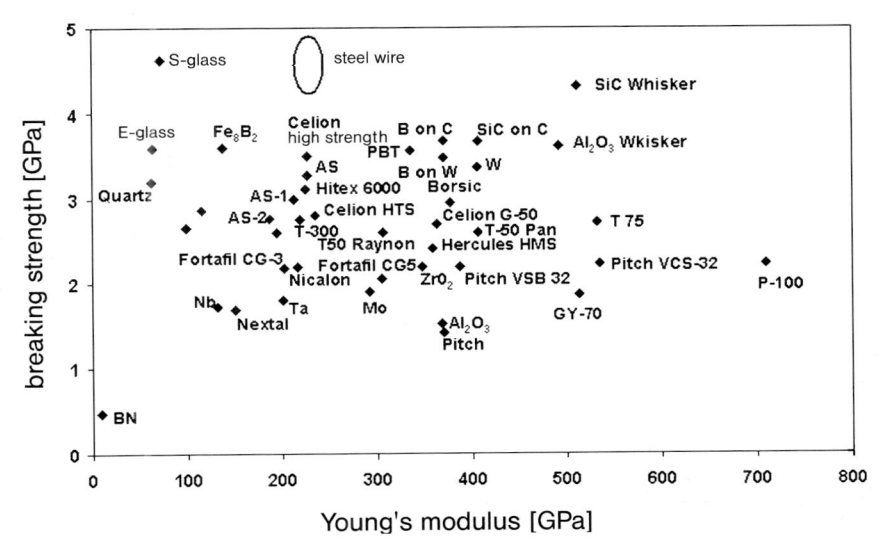

Fig. 2.5 Breaking strength and Young's modulus of some fibers [13].

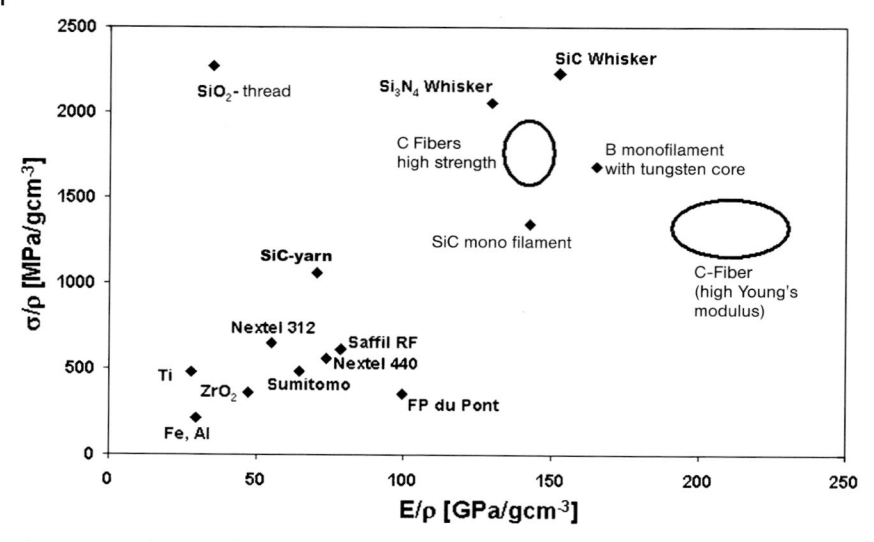

Fig. 2.6 Specific strength σ/ρ and specific Young's modulus E/ρ of different reinforcements and metals [14].

2.3.1
Monofilaments

The production of monofilament fibers takes place in general by chemical vapor deposition (CVD) [15]. Figure 2.7 shows an example of the production scheme for SiC monofilaments. The fiber substrate, typically carbon or tungsten, is guided through a reactor through which passes a stream of hydrogen and silane (hydro silicon) gas. The deposition process takes place at a temperature of 1000–1300 °C according to the following reaction [16]:

$$CH_3SiCl_3(g) \rightarrow SiC(s) + 3HCl(g)$$

The SiC fiber consists mainly of microcrystalline β-SiC in stoichiometric composition. By suitable changes in process control, a composition in the border area can be reached, which has special functions. First, pyrolytic carbon is deposited on the surface to heal surface defects in order to achieve a higher strength of the fibers. Then, the deposition of SiC in stoichiometric composition follows, to complete the task of increasing the moistening behavior in a metal matrix [17].

Due to their low density and high strength, boron fibers are an ideal strengthening component. However, their high cost reduces their application to a niche market. The production of boron monofilaments takes place by deposition from a boron trichloride–hydrogen gas mixture according to the following equation [16]:

$$2BCl_3(g) + 3H_2(g) \rightarrow 2B(s) + 6HCl(g)$$

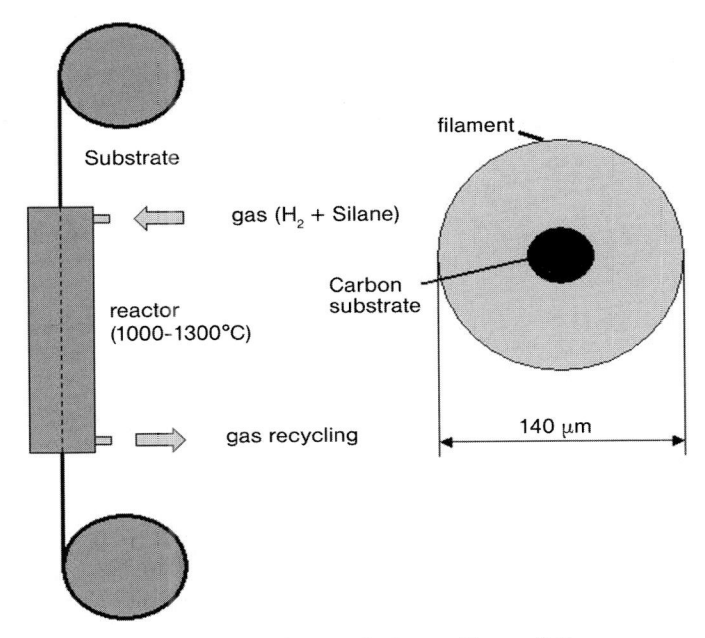

Fig. 2.7 Scheme for the production of SiC monofilament [15].

In this case tungsten fiber with a thickness of 12.5 μm is used exclusively as a substrate, and is guided continuously through the reaction chamber. The structure of the monofilaments, especially their surface morphology, can be varied by predetermined process techniques. This has a major influence on both the mechanical properties of the fibers and their adhesion to the metal matrix.

Monofilaments are characterized by very high strength and Young's modulus at a density below 3 g cm^{-3}. The property profiles of typical, commercially available monofilaments are presented in Table 2.5. It has been shown, by statistically checked measurements, that the strength of the monofilaments has to be described with a Weibull distribution. Figure 2.8 shows, as an example, the tensile strength of SiC monofilaments. Hence, information on the average strength of a

Tab. 2.5 Properties of three SiC fibers, manufacturer: Specialty Materials Inc.

Product name	SCS-6™	SCS-9™	SCS-Ultra™
Cross-section	round	round	round
Diameter (μm)	140	78	140
Tensile strength (MPa)	3450	3450	5865
Young's modulus (GPa)	380	307	415
Density (gcm^{-3})	3.0	2.8	3.0
CTE [10^{-6} K^{-1}]	4.1	4.3	4.1

2 Particles, Fibers and Short Fibers for the Reinforcement of Metal Materials

fiber is insufficient, the minimum strength σ_0 and the Weibull modulus should also be stated [18].

Thermal stress of monofilaments in air or protective gas atmospheres occurs resulting in a decrease in strength with increasing temperature, depending on the composition of the fibers and the atmosphere. Figure 2.9 shows the change in strength with increasing temperature for SiC and boron monofilaments. SiC fibers

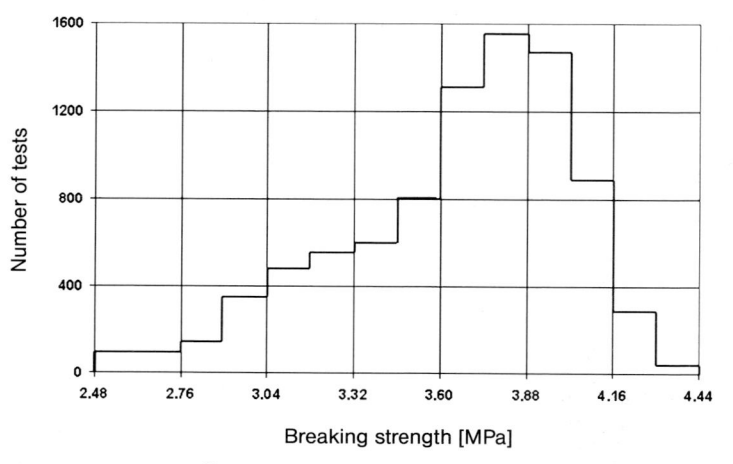

Fig. 2.8 Histogram of breaking strength tests on SiC monofilaments [16].

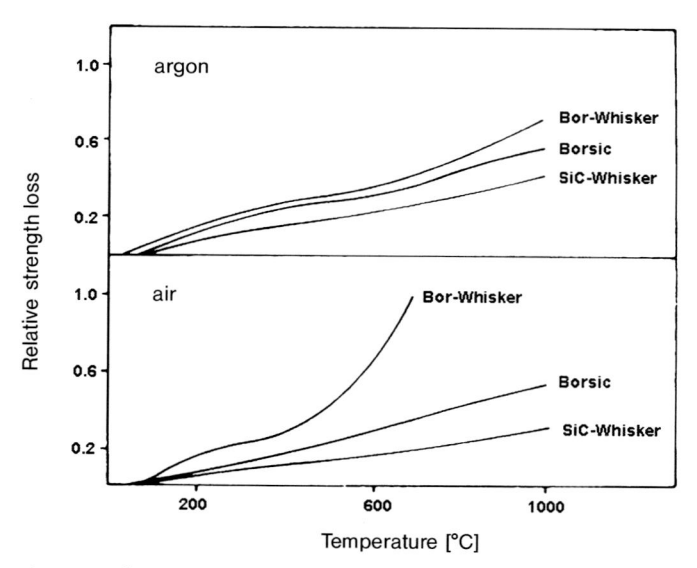

Fig. 2.9 Relative strength loss of different types of fibers (monofilaments) after treatment for 9 min in air or argon [16].

show sufficient stability at higher temperatures, whereas, untreated boron fibers show a decrease in strength in both argon and air. An appropriate coating, for example with SiC (Borsic fibers), can reduce the damage to the fiber.

2.3.2
Multifilament Fibers

Multifilament fibers consist of a multiplicity of single fibers. Usually fiber strands consist of 500, 1000, 2000 or 3000 filaments. The fiber diameter is usually less than 20 µm. There is a large variation possible in the chemical composition and fiber shape, including carbon fibers, carbide fibers and oxide fibers.

2.3.2.1 Carbon Fibers
Carbon fibers are the most developed fiber group. The reason for this is their excellent property profile:

- low density
- high strength
- high Young's modulus
- high stability to molten mass in various metal systems
- possible large variation in property profiles
- low coefficient of thermal expansion
- good thermal and electrical conductivity
- high availability
- cost effectiveness

For the production of C fibers two techniques are available, which can be distinguished by their use of different feed materials [19]. Figure 2.10 shows the schematic representation of the production processes. In the first technique polyacrylonitrile (PAN) precursors are used, which are manufactured by spinning. The final product here is the PAN fiber. In the second technique, spun tar and pitch fibers are used.

Before carbonation of the precursor fibers a suitable orientation of the molecules is necessary to reach the subsequent desired property profile for the C fibers. In the case of the pitch fibers, this takes place automatically due to the spinning process. However, the PAN precursor has to be stretched. Both precursor types are fixed by careful oxidation. The fibers are then carbonized at temperatures above 800 °C. Stabilization annealing at temperatures between 1000 and 3000 °C allows adjustment of the property profile of the fibers, see Fig. 2.11. Up to 1600 °C high-strength fibers are obtained, higher treatment temperatures give fibers with a high Young's modulus.

The elastic constants for the layer-wise graphite lattice amount to 1060 GPa in the direction of the layer level and 36.5 GPa perpendicular to the level. This means, that the better the alignment of the microstructure of the fiber in the direction of these levels, the higher the Young's modulus [19, 21]. The tensile strength of the C

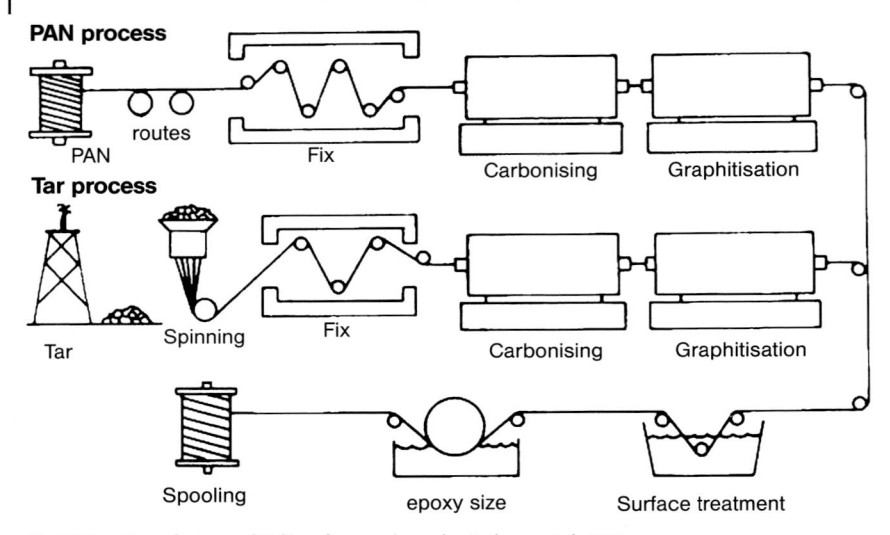

Fig. 2.10 Manufacture of C fiber from polyacrylonitrile or pitch [19].

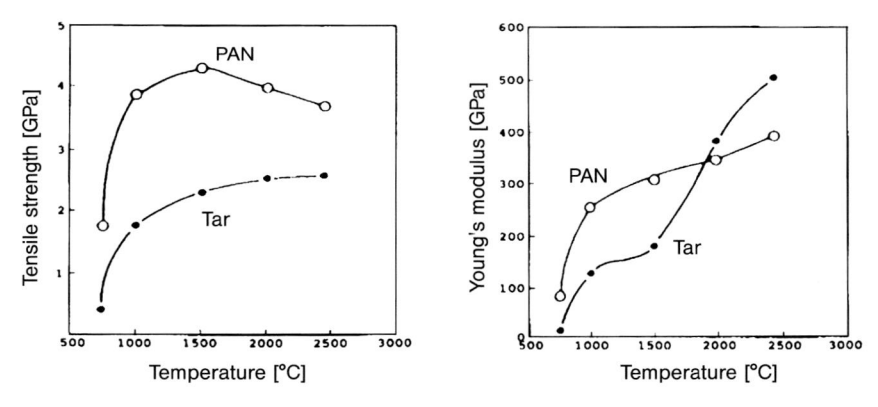

Fig. 2.11 Temperature dependence of the tensile strength and Young's modulus of fibers made from PAN or pitch [20].

fibers is mostly determined by the radial structure. Fibers with low Young's modulus and high strength have, in general, a radial structure, whereas fibers with high Young's modulus and low strength have shell-shaped structures. Crosssections of C fibers with different structures are shown in Fig. 2.12a–c. A summary of the properties of carbon fibers can be found in Table 2.6 (PAN fibers) and Table 2.7 (pitch fibers).

Fig. 2.12 (a) Round tar-based fibers [22]; (b) rectangular-shaped PAN-based fibers; (c) cross-shaped PAN-based fiber [22].

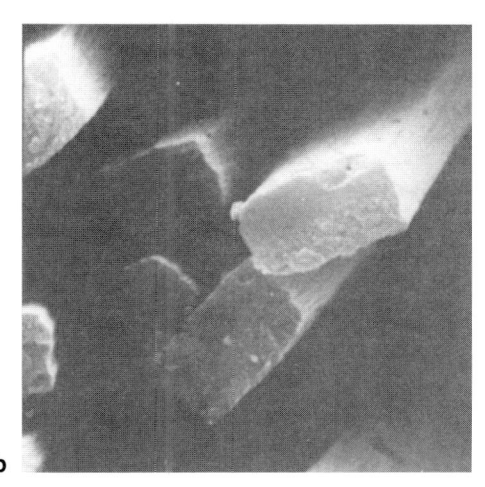

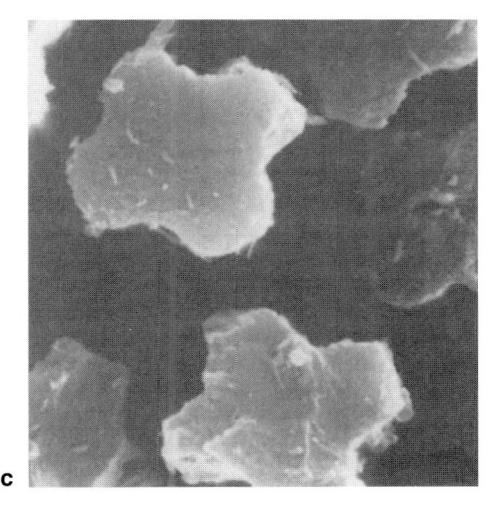

Tab. 2.6 Properties of PAN-based C fibers [23].

	Standard fibers	Air traffic applications		
		low modulus	medium modulus	high modulus
Young's modulus (GPa)	288	220–241	290–297	345–448
Tensile strength (MPa)	380	3450–4830	3450–6200	3450–5520
Elongation (%)	1.6	1.5–2.2	1.3–2.0	0.7–1.0
Elec. resistance ($\mu\Omega$ cm)	1650	1650	1450	900
CTE (10^{-6} K^{-1})	–0.4	–0.4	–0.55	–0.75
C-content (%)	95	95	95	>99
Fiber diameter (μm)	6–8	6–8	5–6	5–8
Manufacturer	Zoltec Fortafil SGL	BPAmoco. Hexcel. Mitsubishi Rayon. Toho. Toray. Tenax. Soficar. Formosa		

Tab. 2.7 Properties of pitch-based C fibers [23].

	Low modulus	High modulus	Ultrahigh modulus
Young's modulus (GPa)	170–241	380–620	690–965
Tensile strength (MPa)	1380–3100	1900–2750	2410
Elongation (%)	0.9	0.5	0.4–0.27
Elec. resistance ($\mu\Omega$ cm)	1300	900	220–130
CTE (10^{-6} K^{-1})		–0.9	–1.6
C-content (%)	>97	>99	>99
Fiber diameter (μm)	11	11	10
Manufacturer	BPAmoco. Mitsubishi Kasei		BPAmoco

2.3.2.2 Oxide Ceramic Fibers

The majority of oxide ceramic fibers consist mostly of aluminum oxide with additions of silicon oxide and boron oxide (see Table 2.8). These additions are determined by the production process selected. At higher SiO_2- or B_2O_3-contents the fibers can be produced by spinning or melting, as they have lower melting points. At higher Al_2O_3 content melting is impossible and not economic. Such fibers are produced by Precursor technology. Feed materials here are produced, by solution, dispersion (for example FP-fibers) or sol–gel processes (for example Saffil, Altex), spun precursor fibers, which are transformed into closed oxide fibers by thermal treatment [30]. The SiO_2-content has a large influence on the structure and therefore on the properties of the fibers. An overview of the properties of continuous oxide ceramic fibers is given in Table 2.8.

At very low SiO_2-content the fibers consist exclusive of α-Al_2O_3. This structure enables a high Young's modulus, which can be found in Nextel 610 fibers [22]. A

Tab. 2.8 Manufacturer and properties of commercially available oxide fibers [24–29].

Salesname	Manufacturer	Composition (weight%)	Diameter of fiber (µm)	Density (g cm^{-3})	Young's modulus (GPa)	Tensile strength (MPa)	CTE (10^{-6} K^{-1})
Altex	Sumitomo	85 Al_2O_3 15 SiO_2	15	3.3	210	2000	7.9
Alcen	Nitivy	70 Al_2O_3 30 SiO_2	7–10	3.1	170	2000	
Nextel 312	3M	62 Al_2O_3 24 SiO_2 14 B_2O_3	10–12	2.7	150	1700	3.0
Nextel 440	3M	70 Al_2O_3 28 SiO_2 2 B_2O_3	10–12	3.05	190	2000	5.3
Nextel 550	3M	73 Al_2O_3 27 SiO_2	10–12	3.03	193	2000	5.3
Nextel 610	3M	>99 Al_2O_3	12	3.9	373	3100	7.9
Nextel 650	3M	89 Al_2O_3 10 ZrO_2 1 Y_2O_3	11	4.1	358	2500	8.0
Nextel 720	3M	Al_2O_3	12	3.4	260	2100	6.0
Almax	Mitsui Mining	99.5 Al_2O_3	10	3.6	330	1800	8.8
Saphikon	Saphikon	100 Al_2O_3	125	3.98	460	3500	9.0
Sumica	Saphikon	85 Al_2O_3 15 SiO_2	9	3.2	250		
Saffil	Saffil	96 Al_2O_3 4 SiO_2	3.0	3.3–3.5	300–330	2000	

disadvantage of this structure is the high fracture sensitivity, which is expressed in the low tensile strength. The reason for this is the coarse grain structure produced by sintering. A coating, for example of SiO_2, decreases the surface roughness and leads to an increase in strength [31].

Safimax fibers are semi-continuous with a diameter of 3–3.5 µm and consist of approximately 4% SiO_2 with mostly δ-Al_2O_3 with a small amount of α-Al_2O_3. Annealing at temperatures above 1000 °C increases the α-Al_2O_3 proportion significantly, the stability of the δ-phase is limited. This leads to degradation of the properties [29, 31]. The Al_2O_3-fibers with approximately 15–20 % SiO_2 are a special group, they consist of spun, polymerized organic aluminum bonding by hydrolysis and annealing [24, 26, 29]. A typical representative is the Altex-fiber with a diameter of approximately 15 µm. The structure of these fibers is microcrystalline. Fibers with higher SiO_2-contents and additions of B_2O_3, like the Nextel fibers, have an increasingly glass-fiber character, but have a very high Young's modulus.

The advantages of oxidic ceramic fibers are:

- cost-effective manufacturing processes
- universal application possibilities
- good processability
- good resistance in air and inert atmosphere
- high stability at higher temperatures
- sufficiently low density
- low thermal expansion
- low thermal and electrical conductivity
- isotropy properties in comparison to C fibers.

A special feature of oxide fibers is the high thermal stability, which is especially important in the processing of metal matrix composites by fusion–infiltration techniques. The stability is determined by the SiO_2-content. Figure 2.13 shows the changes in strength of aluminum oxide fibers with respect to the duration of the heat treatment.

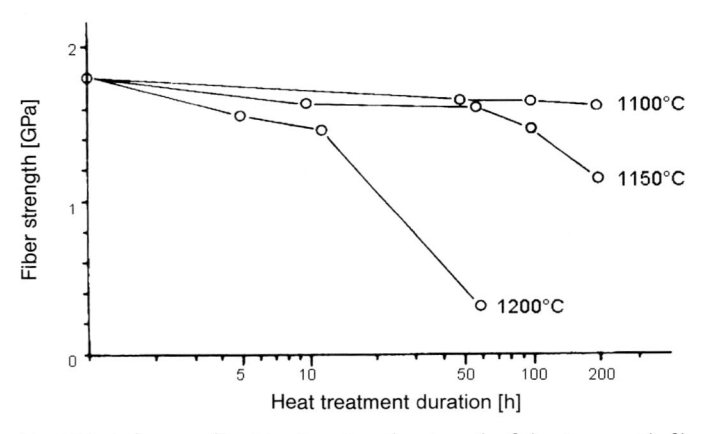

Fig. 2.13 Influence of heat treatment on the strength of aluminum oxide fibers [29].

2.3.2.3 SiC Multifilament Fibers

There are only a few developments concerning SiC multifilament fibers, for application in the area of reinforced ceramics. Some types of fibers have been developed for the application in metal matrix composites. The available fibers are manufactured exclusively using spun, polymeric precursors. Feed materials are polycarbosilane (Nicalon fibers) or polytitanocarbo-silane (Tyranno fibers). The precursor fibers are transformed into ceramic fibers at high temperatures (ca. 1300 °C) using a protective gas atmosphere [16, 30]. As a result of the applied manufacturing technique the SiC fibers show a SiO_2-layer on the surface, this is advantageous for wetting in many metal matrix systems.

The advantages of SiC multifilaments for metal reinforcement are:

- relatively low manufacturing costs
- high thermal stability in a protective gas atmosphere
- sufficient thermal stability in air
- low density
- very good corrosion and chemical resistance
- very good wettability in many metal systems
- low thermal and electrical conductivity
- low thermal expansion.

Table 2.9 shows the significant properties of common SiC multifilament fibers.

SiC fibers show a very good resistance to higher temperatures in protective gas atmospheres. They are stable up to 1000 °C in air, only above this temperature is a significant decrease in strength for long periods of strain noticeable. Figure 2.14

Tab. 2.9 Properties of Nicalon NL200 and three Tyranno fibers [32, 33].

Fiber name	Manufacturer	Content. element: wt %	Fiber diameter (μm)	Density (g cm^{-3})	Young's modulus (GPa)	Tensile strength (MPa)
Nicalon NL200	Nippon Carbon	Si:56.5 C:31.2 O:12.3	15	2.55	196	2.74
Tyranno Lox E	Ube Ind.	Si:54.8 C:37.5 Ti:1.9 O:5.8	8.5	2.52	186–195	3.14–3.4
Tyranno Lox M		Si:54.0 C:31.6 Ti:2.0 O:12.4				
Tyranno New S		Si:50.4 C:29.7 Ti:2.0 O:17.9				

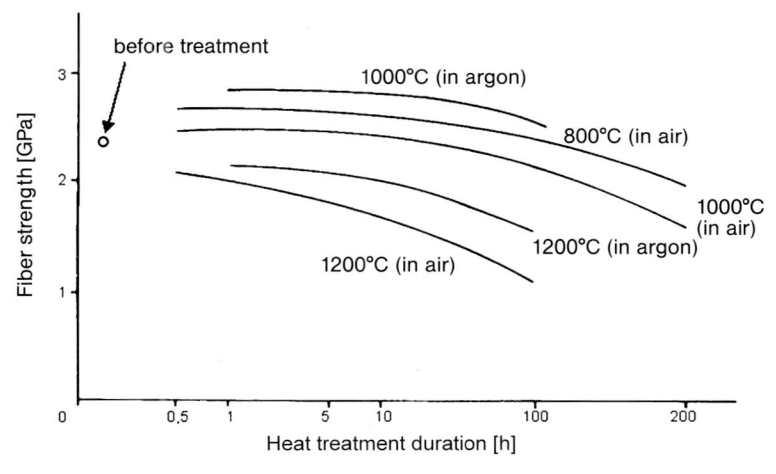

Fig. 2.14 Influence of heat treatment on the strength of SiC fibers (Nicalon) [32].

shows how the change in strength is dependent on annealing atmosphere and time for a Nicalon fiber. In principle, Tyranno fibers show similar behavior.

2.3.2.4 Delivery Shapes of Multifilament Fibers

Multifilament fibers can be delivered in various shapes. A delivery shape as a thread with up to 3000 single filaments is common. There is also the possibility to obtain fabrics in various shapes and geometries. An overview is shown in Fig. 2.15 and 2.16. There are also customised products for special needs. These delivery

Fig. 2.15 Delivery shape of Nextel Ceramic Textiles [26].

▼ **Fig. 2.16** Schematic illustration of delivery shapes of multifilaments (Plain, Satin and Reno are different fabric shapes) [32].

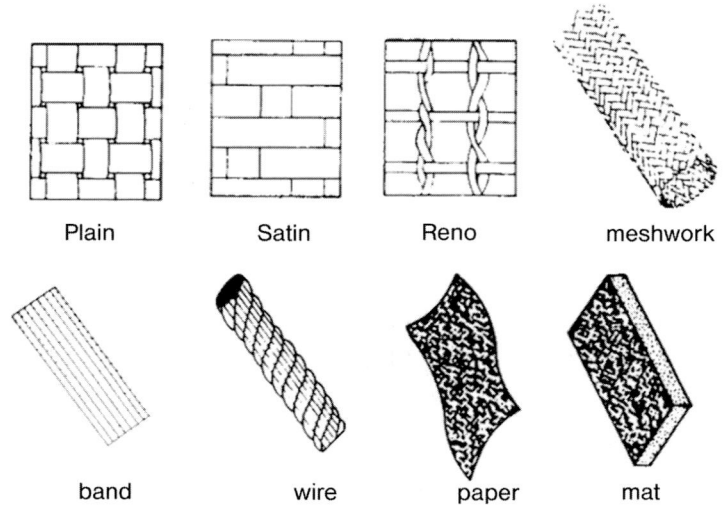

Plain   Satin   Reno   meshwork

band   wire   paper   mat

shapes are not always suitable for processing to metal matrix composites. That for so named fiber shape bodies or pre-shapes are sometimes necessary, see Chapter 3 and [34].

2.4
Short Fibers and Whiskers

Short fibers and whiskers represent a special group of reinforcement compounds, since they have the potential to provide a metal reinforcement between that of continuous fibers and particles. The reinforcement effect is indeed lower than that of long fibers, but the advantage is the change in the isotropic property profile of the produced composite material. A further advantage is the low manufacturing and processing costs.

The manufacture of short fibers takes place according to the described techniques for continuous fibers, the difference being that there is no need to manufacture continuous precursor fibers. It involves atomization (spraying) techniques to a large extent. In a few cases long fibers are chopped and milled but this is uneconomical, since lower grade short fibers are produced from higher grade long fibers. The compositions of short fibers conform to those of many oxidic long fibers. SiC short fibers do not exist at the moment. Table 2.10 shows an overview of the composition and properties of short fibers.

Whiskers are fine single crystals with a low defect density. They have diameters around 1 μm and show a very high aspect ratio. Special manufacturing techniques are applied to produce whiskers. They are grown from oversaturated gases, from solutions by chemical pyrolysis, by electrolysis from solutions or solids [40]. In many cases the creation of whiskers can be realized by the use of catalysts. There are two ways in which whiskers may grow: growth of the base plate or growth at the tip, the dominant way being determined by the manufacturing process and the composition of the whiskers. Among the many available techniques the VLS (vapor–liquid–solid) technique offers the optimal control of the manufacturing process. Figure 2.17 shows schematically the steps in the growth of SiC whiskers. A crystalline catalyst, for example steel, is positioned on a substrate in a reactor. The catalyst melts at higher temperature (ca. 1400 °C) and forms a sphere due to the surface tension. A gas consisting of hydrogen–methane and vaporized SiO_2 flows through the reactor. The fusion drop becomes oversaturated with carbon and silicon, which leads to crystallization of SiC on the substrate. The SiC whiskers grow, whereas the catalyst drop remains on the tip. After stopping the growing process the separation of whiskers and catalyst have to be separated by suitable technology, e.g. the flotation technique. In a different technique the SiC whiskers are deposited by a hydrogen reaction from methyl trichloride silane onto a carbon substrate at a temperature of 1500 °C. A combination of chlorosilane, carbon monoxide and methane as the Si and C source is also possible.

Among the possible compositions of whiskers, SiC and Si_3N_4 have proved to be particularly suitable for the reinforcement of metals. The reasons for this are the

Tab. 2.10 Manufacturer, composition and properties of well-established short fibers and whiskers for the reinforcement of metals [26, 29, 35–39].

Fiber name	Manufacturer	Composition	Fiber diameter (µm)	Density (g cm^{-3})	Young's modulus (GPa)	Tensile strength (MPa)
Saffil RF	Saffil	δ-Al$_2$O$_3$	1–5	3.3	300	2000
Saffil HA	Saffil	δ-Al$_2$O$_3$	1–5	3.4	>300	1500
Nextel 312	3M	62 Al$_2$O$_3$ 24 SiO$_2$ 14 B$_2$O$_3$00	10–12	2.7	150	17
Nextel 440	3M	70 Al$_2$O$_3$ 28 SiO$_2$ 2 B$_2$O$_3$0	10–12	3.05	190	200
SCW #1	Tateho Silicon	β-SiC$_{(w)}$	0.5–1.5	3.18	481	2600
Fiberfrax HP	Unifrax	50 Al$_2$O$_3$ 50 SiO$_2$	1.5–2.5	2.73	105	1000
Fiberfrax Mul.	Unifrax	75 Al$_2$O$_3$ 25 SiO$_2$	1.5–2.5	3.0	150	850
Supertech	Thermal Cer.	~65 SiO$_2$ ~31 CaO ~4 MgO	8–12	2.6	n. a.	n. a.
Kaowool	Thermal Cer.	~45 Al$_2$O$_3$ ~55 SiO$_2$	2.5	2.6	80–120	1200
Tismo	Otsuka Chem.	K$_2$Ox6TiO$_{2\ (w)}$	0.2–0.5	3.2	280	7000
Alborex	Shikoku Chem.	9 Al$_2$O$_3 \times$ B$_2$O$_3$	0.5–1.0	3.0	400	8000

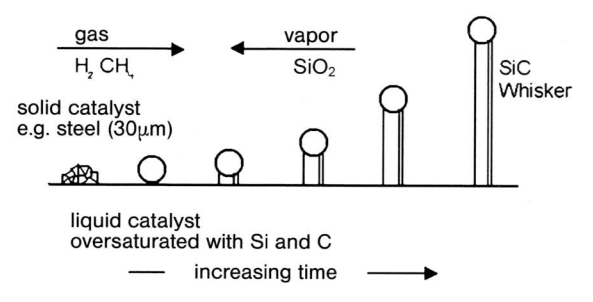

Fig. 2.17 Growth of SiC whiskers by theVLS technique [41].

good wettability of many metal systems, the very good property profile, the high stability and the low manufacturing costs. A few new developments, like potassium titanate and aluminium borate are also of interest. Table 2.10 gives information on technically interesting whiskers of various compositions. Figure 2.18 shows a SiC whisker.

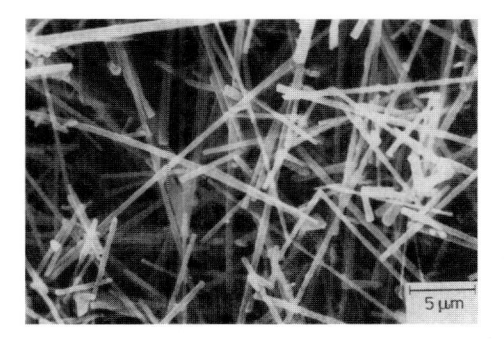

Fig. 2.18 SEM-image of an SiC whisker [42].

Whiskers have developed a poor reputation recently. The reason is that the geometry of the whisker, which is indeed responsible for its good properties, has a high hazard potential. Being very thin and light, whiskers are respirable and easy to breathe in and due to their high stability they are not degradable. They have, similar to asbestos fibers, a very high health risk and are considered to be carcinogenic [42]. There is a similar suspicion about fine mineral fibers and processing is only possible using suitable protection devices.

Short fibers and whiskers are usually deliverable in an agglomerated shape which cannot be brought into the metal matrix. Therefore it is necessary to process to fiber-form bodies, which can then be infiltrated. Such a process is described in Chapter 3. Also for powder metallurgical processing there are special manufacturing techniques needed to insert particles, long or short fibers into a matrix, see Chapter 10.

References

1 E. A. Feest, *Met. Mater.* **1988**, *4*, 52–57.
2 J. A. Black, *Adv. Mater. Process.* **1988**, *133* (March), 51–54.
3 Hollemann, Wiberg, *Lehrbuch der anorganischen Chemie*, de Gruyter, **1985**.
4 Company prospectus Hermann C. Starck, Goslar.
5 Company prospectus Wacker Ceramics, Kempten.
6 Company prospectus Electro Abrasives, Buffalo.
7 E. Dörre, H. Hübner, *Alumina – Processing, Properties, and Applications*, Springer-Verlag, Berlin, Heidelberg **1984**, pp.194–197.
8 FEPA 43-D-1984 (Federation Europèene des Fabricats de Produits Abrasifs).
9 *ASM Handbook, Vol. 7, Powder Metallurgy*, E. Klar (Ed.), **1993**, p.177.
10 F. Moll, Dissertation, Untersuchung zu den Eigenschaften SiC-partikelverstärkter Magnesiummatrix-Verbundwerkstoffe unter dem Einfluß erhöhter Temperatur und Spannung, TU Clausthal, (D) **2000**.
11 *ASM Handbook, Vol. 21, Composites*, **2001**, p. 54.
12 A.A. Griffith, *Philos. Trans. R. Soc. London, Ser.A* **1920**, *221*, 163.
13 A. Kelly, in *Concise Encyclopaedia of Composite Materials*, A Kelly (Ed.), Pergamon Press, Oxford **1989**, p.xxii.
14 M. H. Stacey, *Mater. Sci. Technol.* **1988**, *4*, 391–401.
15 J. A. McElman, in *Engineered Materials Handbook, Vol. 1, Composites*, ASM International, Metals Park **1987**, pp. 858–873.
16 A.R. Bunsel, L.-O. Carlsson, in *Concise Encyclopaedia of Composite Materials,*

A. Kelly (Ed.) Pergamon Press, Oxford **1989**, pp. 239–243.
17 T. Schoenberg, in *Engineered Materials Handbook, Vol. 1, Composites*, ASM International, Metals Park **1987**, pp. 58–59.
18 A. R Bunsell, in *Concise Encyclopaedia of Composite Materials*, A. Kelly (Ed.), Pergamon Press, Oxford **1989**, pp. 33–35.
19 R. J. Diefendorf, in *Engineered Materials Handbook, Vol. 1, Composites*, ASM International, Metals Park **1987**, pp. 49–53.
20 L.H. Peebles, *Carbon Fibers*, CRC Press, **1995**, p. 22.
21 L. S. Singer, in *Concise Encyclopaedia of Composite Materials*, A. Kelly (Ed.), Pergamon Press, Oxford **1989**, pp. 47–55.
22 S.M. Lee, *Handbook of Composite Reinforcements*, VCH, Weinheim **1992**, pp. 48–49.
23 *ASM Handbook, Vol. 21 Composites*, **2001**, p. 38.
24 Company prospectus Sumitomo, London, **2001**.
25 Company prospectus Nitivy, **2000**.
26 Company prospectus 3M, **2001**.
27 Company prospectus Mitsui Mining, Tokio, **2000**.
28 Homepage: MMC-Asses, TU-Wien, Materials Data (mmc-asses.tuwien.ac.at).
29 Company prospectus Saffil, 2001, (www.saffil.com).
30 D. D. Johnson, H. G. Sowman, *Engineered Materials Handbook, Vol. 1, Composites*, ASM International, Metals Park **1987**, pp. 60–65.
31 J. D. Birchall, in Concise Encyclopaedia of Composite Materials, A. Kelly (Ed.), Pergamon Press, Oxford 1989, pp. 213–216.
32 Company prospectus Nicalon Fiber, Nippon Carbon Co., Ltd. Tokyo, Japan, **1989**.
33 Company prospectus Ube, Tyranno Fiber, **2002**.
34 H. Hegeler, R. Buschmann, I. Elstner, Herstellung von Faserverstärkten Leichtmetallen unter Benutzung von faserkeramischen Formkörpern (Preforms), in *Metallische Verbundwerkstoffe*, K.U. Kainer (Ed.), DGM Informationsgesellschaft, **1994**, p.101.
35 Company prospectus Tateho Chemical Ind. Hyogo, Japan, **2002**.
36 Company prospectus Unifrax Corp., Niagara Falls, NY, **2002**.
37 Company prospectus Thermal Ceramics, Augusta, USA, **2000**.
38 Company prospectus Otsuka, Japan, **1999**.
39 Company prospectus Shikoku Chem. **2001**.
40 J. V. Milewski, in *Concise Encyclopaedia of Composite Materials*, A. Kelly (Ed.), Pergamon Press, Oxford **1989**, pp. 281–284.
41 K.K. Chawla, *Composite Materials, Science and Engineering*, Springer-Verlag, Berlin **1998**, p. 63.
42 F. Pott, *Proc. Verbundwerk 1990*, S. Schnabel (Ed.), Demat Exposition Management, Frankfurt, pp. 16.1–16.10.

3
Preforms for the Reinforcement of Light Metals – Manufacture, Applications and Potential

R. Buschmann

3.1
Introduction

Rising fuel prices and tighter emission standards are known to be a constant pressure on the automobile industry, leading to attempts to optimize the specific efficiency and the fuel consumption of their vehicles. According to an analysis of the Traffic Institute of the University of Michigan (Transportation Institute) weight reduction and an increase in engine efficiency are the largest contributions to reach these goals [1]. Light metal alloy composite materials, generally called MMCs (metal matrix composites), can help to reduce moving and unsprung masses within the vehicle due to their high specific properties.

In the group of MMC materials the infiltration of a porous preform, using ceramic reinforcement components in the form of short fibers and/or particles, with molten light metal alloy surely represents one of the most promising technologies with regard to the range of the attainable properties of the final composite material. Some applications, e.g. fiber-reinforced aluminum diesel pistons for trucks have been in production for over 10 years with up to several hundred thousand pieces produced annually, and it has been proven that this technology is controllable for series quantities also. Nevertheless the widespread use of MMC materials has not yet taken place. The reason for this is the manufacturing cost, since among other things special pressure casting processes such as squeeze casting are necessary and there is only limited acceptance of higher prices in the cost-sensitive automobile industry. However, hope for increased use of MMC materials comes from two different areas. The analysis in [1] forecasts that the value of weight reduction in automobiles will triple from approx. 1$ per pound at present to 3$ per pound by 2009. At the same time modern high pressure die cast machines with real time controls and the latest generation of squeeze cast machines show clear improvements regarding initial costs, cycle times and floor space requirements, thereby contributing significantly to the economic application of this technology.

Metal Matrix Composites. Custom-made Materials for Automotive and Aerospace Engineering.
Edited by Karl U. Kainer.
Copyright © 2006 WILEY-VCH Verlag GmbH & Co. KGaA, Weinheim
ISBN: 3-527-31360-5

3.2
Manufacturing Principle of Preforms

3.2.1
Short Fiber Preforms

To obtain an effective reinforcement with short fibers the preform must fulfill the following requirements:

- extreme cleanliness
- homogeneous fiber distribution
- homogeneous fiber orientation
- preferably large fiber length/fiber diameter ratio
- homogeneous binder distribution
- low binder content
- sufficient preform strength

The standard manufacturing procedure for most shaped fiber products, e.g. those used in the kiln industry for high temperature insulation, is based on the drainage of a fiber slurry by means of vacuum in a suitable mold. In order to meet the higher requirements for a fiber preform, additional process steps are necessary (Fig. 3.1).

Depending on the type of fibers used these process steps can have a different degree of complexity. Discontinuous manufactured fibers consist of two main groups: glassy, melt spun fibers, and polycrystalline fibers made by a spin sinter process. The fibers produced from a melt are predominantly alumina silicate fibers. Here the raw materials are melted with electrodes in a funnel-shaped kiln at temperatures above 2000 °C. A controlled melt stream is run out through an orifice

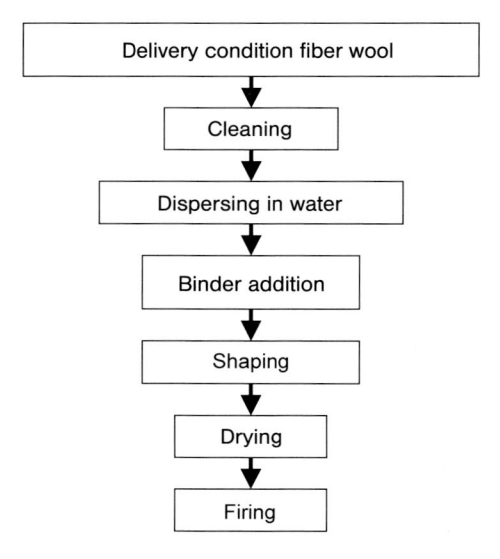

```
Delivery condition fiber wool
          ↓
       Cleaning
          ↓
   Dispersing in water
          ↓
     Binder addition
          ↓
        Shaping
          ↓
         Drying
          ↓
         Firing
```

Fig. 3.1 Process steps of the preform manufacture.

3.2 Manufacturing Principle of Preforms

at the bottom of the kiln, and then either hits a set of fast rollers rotating in opposite directions, or is blown by a vertically arranged stream of compressed air. In either case the melt stream is divided, by the centrifugal forces or the dragging effect of the compressed air flow, into fine droplets, which then, due to their high velocity, extract to fibers and solidify glassily under rapid cooling. Figure 3.2 shows the upper section of a fiber melting furnace with three molybdenum electrodes, which melt the raw material mixture. Figure 3.3 shows the lower part of the furnace with the melt stream running out and being blown into fibers by a blowing nozzle.

Fibers with Al_2O_3 content above 60 wt.% cannot be economically manufactured by this procedure, since the viscosity of the melt is too low. Here the already mentioned spin sinter procedure is often applied, in which metal-salt solutions with additives with high molecular spin implements are spun to fine and very regular fibers by drawing or blowing [2]. In a subsequent, multilevel temperature treatment polycrystalline fibers develop, e.g. alumina fibers and mullite fibers. The two manufacturing processes lead to fibers of quite different quality and properties (Table 3.1). The difference is particularly remarkable in the so-called "shot content". Shot is the portion of nonfibrous components. With the alumina silicate fibers produced from a melt, a process related drop-shaped swelling forms at the tip of each fiber. These drops break off to a large extent during the subsequent processing and then occur as glass sand in the fiber (Fig. 3.4). The shots in polycrystalline fibers predominantly result from loose bakings at the spinning device, which also get into the fiber stream (Fig. 3.5). The C fibers specified in Table 3.1 do not have shot

Fig. 3.2 Fiber melting furnace, top view.

Fig. 3.3 Fiber melting furnace, bottom view.

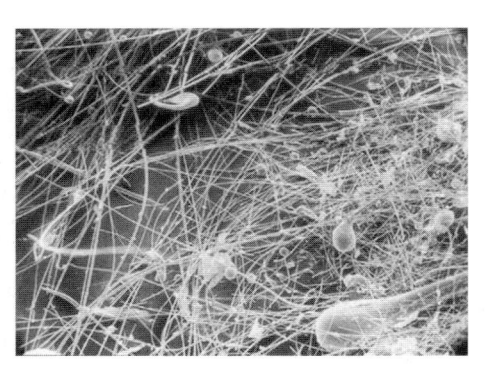

Fig. 3.4 Shot particles in aluminum silicate fibers.

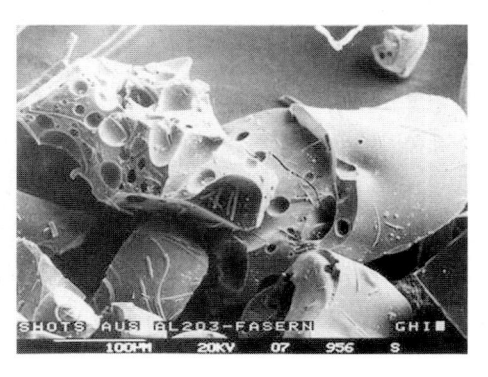

Fig. 3.5 Shot particles in aluminum oxide fibers.

Tab. 3.1 Overview of common fiber types for the preform technology.

Type	Al_2O_3	$Al_2O_3-SiO_2$	C (PAN)
Example	Saffil, Maftec	Kaowool, Cerafiber	Sigrafil
Chemistry	> 95 % Al_2O_3 < 5 % SiO_2	48 % Al_2O_3 52 % SiO_2	95 % C
Mineralogies	mostly δ-Al_2O_3	glassy	
E-Modulus, GPa	270–330	105	2240
Strength, MPa	2000	1400	2500
Fibre-Ø, µm 3	2–3	8	
Density, g cm^{-3}	3.3	2.6	1.8
Shot content, %	ca. 1	40–60	–

portions, because they are chopped or milled from continuous fibers. The shot portions in the fiber represent a substantial problem for most MMC applications. The size of the particles extends in some cases down to the millimeter range. Particularly in applications where the MMC components are stressed by thermal fatigue, shot particles can lead to catastrophic effects and to the failure of the component. Shot particles are inhomogeneities in the matrix and, at a critical size, act as crack formation centers. Figure 3.6 shows the microstructure of a fiber-reinforced aluminum alloy with a shot particle of approx. 250 µm. It can be seen clearly that cracks proceed from the particle and keep running into the surrounding matrix. Tests have proved a critical shot size of ≥100 µm, above which harmful effects arise. Therefore a process step has to be inserted ahead of the actual preform manufacturing, in which the critical shot portion is removed as completely as possible from the fibers and limited to a size less than 100 µm.

The fibers occur in the delivered state as strongly felted wool, either as "bulk" fibers with lengths of 20–60 mm, or as "milled" fibers with adjusted fiber lengths of up to 500 µm depending on the degree of milling (see Fig. 3.7). Fiber wool must be first disagglomerated and shots isolated from the fibers, before the separation can

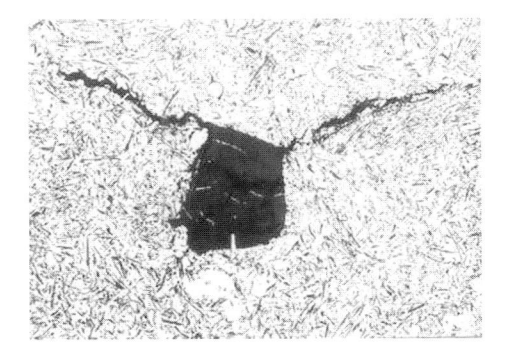

Fig. 3.6 Microstructure of an Al-MMC with shot particle.

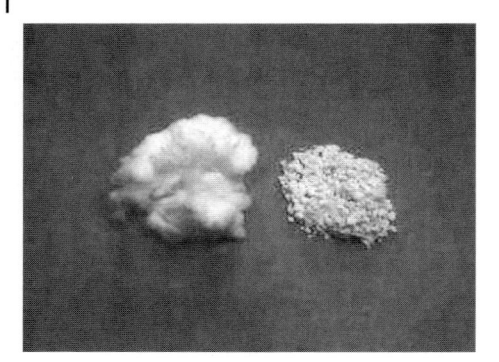

Fig. 3.7 Fiber wool: left, bulk fiber; right, milled fiber

begin. Here the fiber break rate must be considered, so that in the later preform, a sufficiently high fiber length remains. Thermal Ceramics have a proven procedure for unlocking and cleaning fibers which has been in use for years and which is able to limit shot contents to less than 10 ppm, with no shots > 75 µm. For commercial confidentiality reasons this procedure cannot be described here in more detail. The cleaned fibers are then dispersed in water, usually at low solid concentrations within in the range 10 to 40 g fibers per liter. Colloidal SiO_2 is added to the fiber suspension, in the form of silica sol, as an inorganic bonding agent. By the type of stabilization of the silica sol, e.g. with NaOH, the SiO_2 particles exhibit a negative surface charge. In water thin hydrate layers deposit on the fiber resulting also in a negative surface charge. Thus the solids mutually repel each other and accumulation of the SiO_2 particles on the fibers is prevented (Fig. 3.8(a)). The addition of a second binder component, a cationic corn- or potato-based starch solution causes agglomerization of the solids in the suspension. The long-chain molecules of the starch cause flocculation of the solids from the aqueous suspension (Fig. 3.8(b) and (c)). Under carefully controlled conditions this process produces good preform strengths with binder contents below 5%, which is sufficient for the further handling and pressure infiltration with molten metal.

The shaping of such a prepared fiber suspension takes place in a vacuum pressing. A defined quantity of suspension is filled into a mold with a porous bottom part (e.g. sieve insert) and vacuum drained up to the formation of a loose fiber cake (Fig. 3.9(a)). The fiber cake is then compressed with a press stamp to a preform with the desired height and density (Fig. 3.9(b)). The short fibers orientate themselves planar-isotropically in the plane perpendicular to the vacuum and press direction. After releasing from the mold the still moist preforms are dried at 110°C when the contained starch pastes up and provides the preform with the appropriate green strength. When fired at temperatures between 900 and 1200°C the organic starch burns completely away and the SiO_2 bonding becomes active.

For cost reasons the preform should always be designed to be able to be manufactured to the "near net shape procedure", without subsequent mechanical machining. The design possibilities of preforms with planar isotropic fiber orientation are limited due to the spring-back behavior of the fiber. An example of Al_2O_3

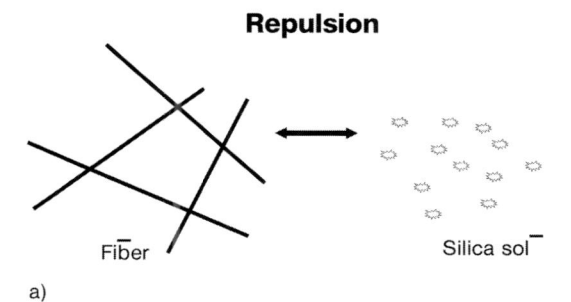

Fig. 3.8 Bond mechanism:
(a) repulsion,
(b) agglomeration,
(c) flocculation.

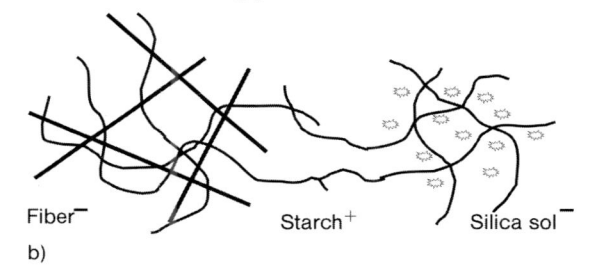

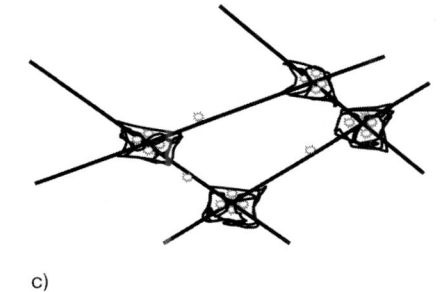

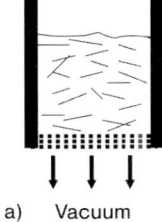

Fig. 3.9 Shaping:
(a) drainage of the fiber slurry,
(b) pressure- and vacuum supported drainage.

fibers with a mean fiber length of 100 μm, of which a preform with 15% fiber volume fraction should be manufactured, shows that an over pressing of approximately 30% is necessary to adjust the desired preform density (Fig. 3.10(a)–(c)). The spring-back of the preforms during pressure release limits the possibilities for net-shaping complex geometries such as chamfers and steps or the like. Figures 3.11 and 3.12 show the equipment for the series production of preforms in near net shape, which is in use at Thermal Ceramics.

The extent of the spring back effect depends on the type of fiber used (elasticity, strength, and fiber diameter), the fiber length of the source fiber and the preform density. During pressing of the preform shearing forces act in a direction predominantly perpendicular to the fibers lying in the press direction. If the shearing forces exceed the fiber strength single fibers will break. The greater the source fiber length, the higher the pressure needed to achieve the same preform density, resulting in a higher fiber break rate. In an internal test row the influence of the source fiber length on the attainable fiber length in a Saffil fiber preform with 20% fiber volume fraction was examined. For this test especially milled Saffil fibers 150, 300 and 500 μm in length were used and processed to preforms in the procedure de-

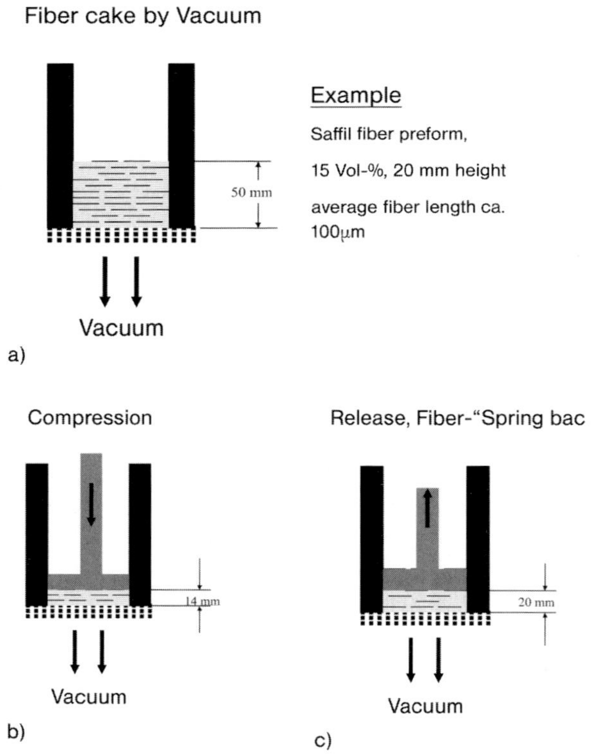

Fig. 3.10 Near-net shape process: (a) formation of a fiber cake, (b) compression, (c) release.

Fig. 3.11 Preform series production equipment, press side.

Fig. 3.12 Preform series production equipment, processing side.

scribed above. An evaluation of the resulting fiber length in the final preforms did not show significant differences despite greatly different source fiber lengths. Thus there results a direct dependence of the attainable fiber length in the preform on the preform density.

3.2.2
Hybrid Preforms

Hybrid preforms are defined as preforms which consist of a homogeneous arrangement of fibers and particles. The production of hybrid preforms takes place in a similar manner to fiber preforms, with the particular particles as an additional solid component. The selection and availability of suitable short fibers for the light metal alloy reinforcement is rather scarce. Compared to that, commercially available particles of the oxide and non-oxide ceramic range represent a substantially larger group of materials with different properties, which can clearly extend the application field for MMCs. Many ceramic powders are now mass-produced goods because they are used in large quantities in other industries such as technical ceramics or grinding media. They can be obtained with defined grain spectra and usually at relatively economical cost. A lot of work has been done on MMCs which are based purely on reinforcement with particles, e.g. castable composite materials ("cast composites"), where particles such as SiC or Al_2O_3 are stirred into the molten alloy by special procedures or composite materials are manufactured by melt infiltration of particle preforms. While the particle contents in cast composites are limited to a maximum of approx 20 vol%, MMCs manufactured with particle preforms usually have particle contents starting from 50 vol% upwards. Hybrid preforms can close the gap between both systems and make it possible to combine the advantages of both fiber and particle reinforcement. Since, with this technology, any type of fiber can be combined with any type of particle in a broad composition and volume fraction range with only few restrictions regarding the particle size, a tool is given to "tailor" the properties of light alloys within a defined range. Table 3.2 shows the most important advantages and disadvantages of the three systems: "cast composites", "particle preform reinforcement" and "hybrid preform reinforcement".

Cast composite materials can be processed with conventional casting processes. However, due to limitation in the maximum particle content a limited property potential exists. Furthermore, a great disadvantage is that no selective reinforcement is possible, only the complete component can be reinforced. The usually high hardness of particle reinforced alloys then results in extremely high and thus expensive

Tab. 3.2 Comparison of different MMC systems.

	Advantages	Disadvantages
Cast composites	• Conventional casting processes	• Particle content limited to approx. 20 vol% • No selective reinforcement • High machining effort • Expensive
Particle preforms	• Low cost preforms • Selective reinforcement	• Particle content 25–50 vol% • Complex infiltration due to high particle density and fine pore diameters • At selective reinforcement extreme property step
Hybrid-preforms	• Reasonable costs • Selective reinforcement • Wide vol% range	• Special casting processes necessary

machining efforts, which greatly limits the economic application of these materials. Particle preforms are relatively low cost materials, depending on type and grain size, and allow selective reinforcement of components. However, the infiltration with molten metal requires more complex pressure casting processes such as squeeze casting. The high particle contents limit the achievable infiltration depth and thus the size and/or thickness of the reinforcement area. In the case of selective reinforcement there is, furthermore, a drastic change in properties from the particle reinforced section to the surrounding unreinforced matrix, which leads to extreme internal stresses in the component. Hybrid preforms can be produced at reasonable cost depending on the type of fibers and particles used, but also need pressure casting processes for the infiltration. Because of the variety of possibilities regarding the adjustable preform density (approximately 10 to 50 vol%), the ratios of particles and fibers, particle sizes and type combinations, a characteristic property profile can be produced that considers the above mentioned problems of the two other systems. Figure 3.13 shows as an example the thermal fatigue behavior as a function of temperature cycles for an aluminum alloy reinforced with hybrid preforms of different porosities of 30, 50 and 70%. The investigations were undertaken by the KS Kolbenschmidt GmbH in Neckarsulm as a common research project.

The matrix used was the piston alloy KS 1275 (AlSi12CuMgNi), the hybrid preforms were combinations of Al_2O_3 fibers and Al_2TiO_5 particles. The basis alloy KS 1275 shows an average total crack length of 60 mm after 2500 temperature cycles.

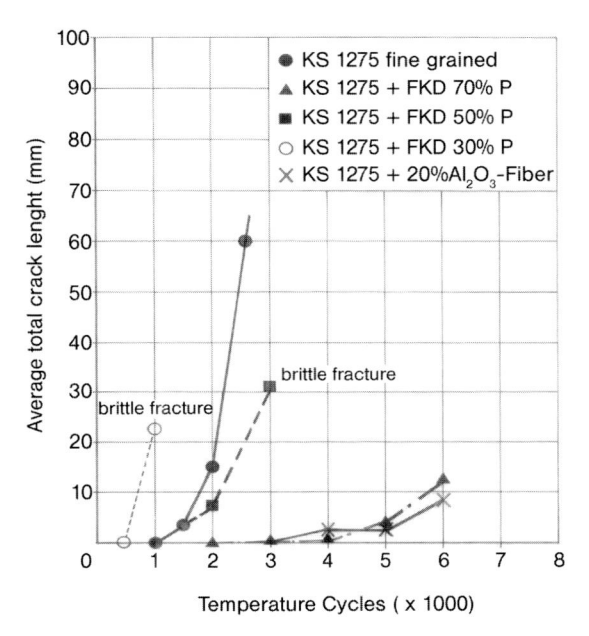

Fig. 3.13 Thermal fatigue of Al-MMCs as a function of preform porosity.

The samples with a preform porosity of 30%, thus a reinforcement portion of 70 vol%, fail by fracture after 1000 cycles. The samples with 50% preform porosity fail after 3000 cycles, also by fracture, while the variant with 70% porosity showed after 6000 cycles an average total crack length of only 12 mm and no fracture. This is in the same range as a 20 vol% Al_2O_3 fiber reinforced alloy, as shown as a reference in Fig. 3.13.

3.3
Current Applications

3.3.1
Aluminum Diesel Piston with Fiber Reinforced Combustion Bowl

The demand for improved performance of turbo-charged diesel engines, for better utilization of fuel and smaller pollutant emissions puts conventional aluminum piston alloys to the limits of their material strength, particularly in the combustion bowl rim area (Fig. 3.14). Conventional cast aluminum alloys are limited in their application to temperatures below 350 °C, whilst the use of cast iron to cover higher temperatures means a drastic weight increase. Selective reinforcement of the critical area in the combustion bowl with ceramic fibers, here the addition of 20 vol% Saffil, doubles the hot strength and fatigue strength at elevated temperatures, as shown in Fig. 3.15 and 3.16 [3]. Such pistons have been successfully used in commercial vehicles for a long time in series quantities of several hundred thousand pieces per year.

Fig. 3.14 Aluminum piston with fiber reinforced combustion bowl.

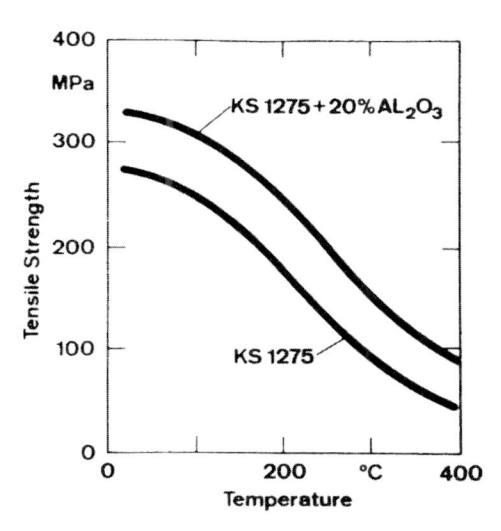

Fig. 3.15 Tensile strength of Al-Saffil MMCs as a function of the temperature.

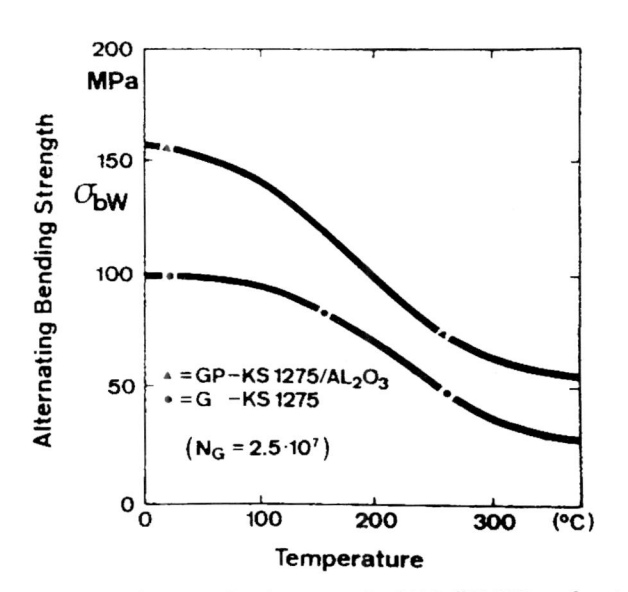

Fig. 3.16 Alternating bending strength of Al-Saffil MMCs as function of the temperature.

3.3.2
Aluminum Cylinder Heads with Fiber Reinforced Valve Bridges

Since the establishment of turbo-diesel engines with direct fuel injection in the passenger car range, their market share has been constantly rising. Significant increases in the specific performance are to be observed from engine generation to engine generation, realized by ever higher ignition pressures and the application of high pressure injection systems. At the same the aim is to build more compact engines to reduce their weight. This becomes problematic in the bridge area between inlet and exhaust valves. Between the inlet and discharge valve side a drastic temperature difference prevails which, paired with rising mechanical stresses, leads to fatigue problems in the bridge area. For conventional aluminum alloys a critical bridge width results, which cannot be reduced for material strength reasons. Selective reinforcement of the bridge area in the cylinder head with ceramic fibers (Fig. 3.17) significantly increases the fatigue strength of aluminum (compare Fig. 3.16), thus allowing for a minimization of the bridge widths. This application is at present in the final phase of an extensive testing program and, if successful, production will be undertaken shortly.

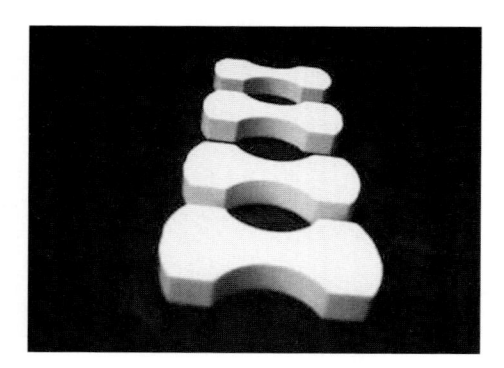

Fig. 3.17 Cylinder head preforms to reinforce the valve bridges.

3.3.3
Cylinder Liner Reinforcement in Aluminum Crankcases, Lokasil

Aluminum crankcases need reinforcement in the cylinder lining surface. The standard solution is cast-in-place gray cast iron liners. Apart from weight disadvantages due to large distances between the cylinder bores, problems due to cylinder distortion and gap formation between the liner and the surrounding aluminum are to be found with this concept, caused by the different coefficients of thermal expansion of aluminum and gray cast iron. A linerless running surface concept is given by the use of the hyper-eutectic alloy KS Alusil®, in which silicon particles form in the alloy primarily by precipitation. In a special hone process the aluminum matrix is set back to make the silicon particles protrude in the cylinder bores, thus providing a running surface structure for the pistons. Although this concept provides outstanding tribological characteristics, it has the disadvantage that the entire casting consists of a specialty alloy, whose properties are only needed in the cylinder surface. This concept is predominantly found in the large stroke volume engines of the upper segment automobiles. By melt infiltration (Fig. 3.18) of hybrid preforms consisting of ceramic fibers and silicon particles the same advantages and tribological characteristics can be obtained as with Alusil®, however, with use of a conventional remelting alloy (Fig. 3.19). This opens up the introduction of this technology into engines of the middle and lower market segments.

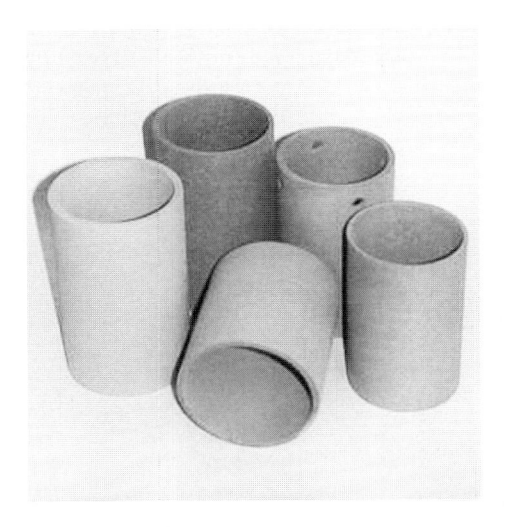

Fig. 3.18 Preforms to reinforce cylinder running surfaces.

Fig. 3.19 Crankcase part with running surface reinforcement (Lokasil).

3.3.4
Al-MMC Bearing Blocks for Crankshafts

Bearing blocks for the crankshaft are either made completely out of gray cast iron and inserted into the engine or pre-cast from aluminum and provided with a gray cast iron bearing shell. Gray cast iron bearing blocks represent a substantial excess weight in the engine, while the combination of an aluminum block with a grey cast iron bearing shell is problematic due to the very different coefficients of thermal expansion α of both materials ($\alpha_{\text{grey cast}} = (11-12) \times 10^{-6}$ K^{-1}, $\alpha_{\text{aluminum}} > 20 \times 10^{-6}$ K^{-1}). By infiltration of a hybrid preform (Fig. 3.20), consisting of ceramic fibers and special silicate ceramic particles it is possible to manufacture aluminum MMC blocks whose expansion coefficient corresponds exactly with gray cast iron. Thus the problem of different thermal expansions is removed and the employment of heavy gray cast iron bridges can be avoided.

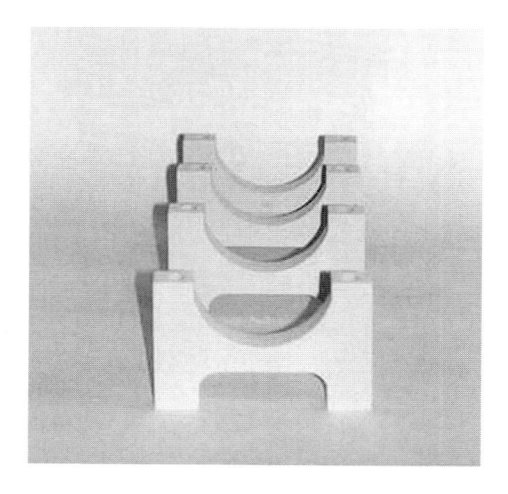

Fig. 3.20 Preforms to reinforce bearing blocks.

3.3.5
Al-MMC Brake Disks

The weight of brake components, particularly on the steering axle, has a significant influence on the handling, driving and steering behavior of a vehicle. It is therefore of interest to investigate lighterweight alternatives to gray cast iron brake rotors. Conventional aluminum is not suitable for this application due to an insufficient coefficient of friction and temperature stability. Attempts with SiC particle strengthened cast composites, e.g. Duralcan showed that the main problem is its low temperature resistance. It is reported that such brake disks can fail at temperatures from 380 to 420 °C. However, this can be significantly improved by reinforcement with hybrid preforms. On the brake disk dynamometer test stand of a brake component manufacturer aluminium MMC disks, reinforced with hybrid preforms based on ceramic fibers and SiC particles of different volume fractions, were tested (Fig. 3.21 and 3.22). The best results were obtained with a variant which was reinforced with 10 vol% ceramic fiber + 30 vol% SiC particles. A service temperature of up to 500 °C was obtained and coefficients of friction were measured, which were close to those of gray cast iron.

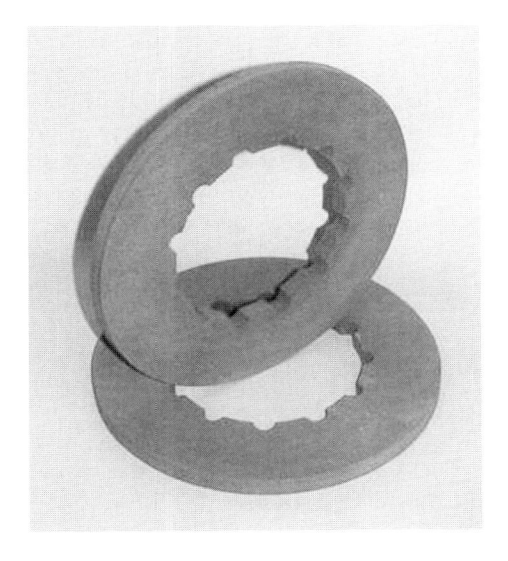

Fig. 3.21 Brake disk performs.

Fig. 3.22 Aluminum-MMC brake disk

3.4
Summary and Outlook

MMC-materials belong to a group of materials with an enormous potential to reduce the vehicle weight in general and the weight of the oscillating and unsprung masses in particular. The majority of the projects, investigations and material data originate from the range of aluminum MMCs. The aim of these developments is the substitution of heavy materials, such as gray cast iron, by aluminum, or the design of existing aluminum components more filigree, so that they can then be designed more compact. Magnesium alloys are also attaining increasing importance because of their low specific gravity. Promising results were obtained in work with carbon short fiber preforms and carbon fiber hybrid preforms [5].

Above all, the group of the hybrid preforms offers a tool to influence light metal alloy characteristics such as Young's modulus, tensile strength, fatigue behavior, creep stability, hardness, wear resistance, thermal expansion, thermal conductivity etc. and to tailor properties within certain limits. This can be further extended by the development of multiphase hybrid preforms and the adjustment of controlled property gradients in the direction of local material engineering in both the MMC component and the preform itself. Although there are so far only little accessible data, it is well-known that MMC materials are significantly superior to most single phase material systems in their oscillation and absorption behavior. A further application potential exists in the range of noise reduction: "NVH" (Noise–Vibration–Harshness).

Progress in the equipment technology for pressure infiltration and in the manufacturing of preforms, along with a certain confidence level in this group of materials and the constantly rising accepted extra costs per kilogram of weight reduction will contribute to these materials finding significantly wider application within the next 5 years. This is supported by the strong interest from the industry and the numerous development projects with potential end users.

References

1 D. Holt, Editorial, *Automotive Engineering International*, April **2000**.
2 R. Ganz, Untersuchungen zur Anwendungsgrenztemperatur von keramischen Hochtemperaturfasern des Systems Al_2O_3-SiO_2, Dissertation, Faculty of Mining and Metallurgy, Rheinisch-Westfälische Technischen Hochschule Aachen.
3 S. Mielke, N. Seitz, D. Eschenweck, R. Buschmann, I. Elstner, G. Willmann, Aluminum Pistons with Fiber Ceramics, *Proceedings of the Second International Symposium on Ceramic Materials and Components for Engines*, Lübeck, FRG, **1986**.
4 E. Köhler, H. Hoffmann, J. Niehues, G. Sick, Kurzbauende, leichte, closed-deck Aluminum-Kurbelgehäuse für Großserien, *Composite Technology*, special print of KS Aluminum-Technology AG.
5 K. U. Kainer, Development of Magnesium Matrix Composites for Power Train Application, *Proceedings of the 12th International Conference on Composite Materials*, CD-ROM V2, **1999**, paper 1263.

4
Aluminum-matrix Composite Materials in Combustion Engines

E. Köhler and J. Niehues

4.1
Introduction

The use of composite materials offers advantages when the characteristic profile of a standard material for an application is no longer sufficient. In their use in combustion engines the following objectives are of importance:

- Increase in mechanical strength (in particular at higher temperatures)
- Increase in thermal shock stability
- Increase in stiffness (Young's modulus)
- Improvement in wear resistance and tribological characteristics
- Reduction of thermal expansion.

Different concepts for the local reinforcement of construction units in aluminum combustion engines have been motor-powered tested in the past years (Table 4.1). Locally ceramic-fiber reinforced light metal pistons have become generally accepted, and cylinder surfaces have been produced by different technologies. In this chapter a cylinder surface technology, by which the cylinder surfaces in an aluminum cylinder crankcase are enriched by preform infiltration with silicon, is presented. This method develops a local metal matrix composite material. This technology went into mass production for the first time in 1996 with the cylinder crankcase of the Porsche Boxster. Since then more than 400 000 LOKASIL® cylinder crankcases for the Porsche Boxster and 911 Carrera (Fig. 4.1) have been manufactured. In this chapter this technology and its advances are discussed, and compared with old, proven concepts and new developments with aluminum cylinder crankcases.

Metal Matrix Composites. Custom-made Materials for Automotive and Aerospace Engineering.
Edited by Karl U. Kainer.
Copyright © 2006 WILEY-VCH Verlag GmbH & Co. KGaA, Weinheim
ISBN: 3-527-31360-5

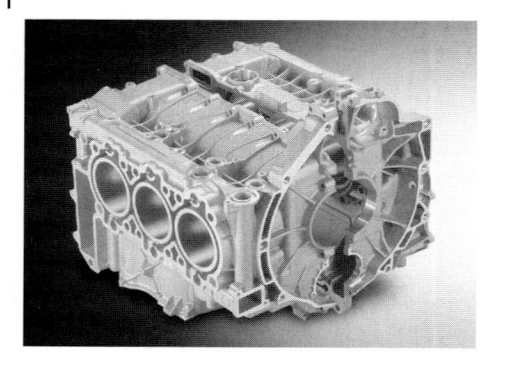

Fig. 4.1 Cylinder crankcase of Porsche 911 Carrera with LOKASIL®-technology.

Tab. 4.1 Engine components made of aluminum matrix composites (Al-MMC).

Engine component	Primary aim	Reinforcement	Present state
Cylinder head: Combustion chamber calotte	improved thermal shock resistance	short fiber	no mass production
Valve train:			
valve spring carrier	mass reduction	particle, short fiber	no mass production
rocker arm	mass reduction	particle	no mass production
Piston:			
ring carrier	mass reduction	short fiber + particle	mass production
bowl edge	improved thermal shock resistance	short fiber	mass production
Piston pin	mass reduction	short fiber, long fiber	no mass production
Piston rod	mass reduction	short fiber, long fiber	no mass production
Cylinder crankcase: cylinder surfaces	tribological properties	short fiber mixture particle short fiber + particle	mass production mass production mass production
Bedplate: crankshaft bearing area	reduction of thermal expansion	short fiber + particle particle	no mass production

4.2
Cylinder Crankcase Design Concepts and Cylinder Surface Technology

With the substitution of gray cast iron by aluminum for lightweight construction, the question for the best concept comes up, see Fig. 4.2 [1]. Thus the components of the alloy, construction, casting processes and surface technology concerning the most important criteria and the specific boundary conditions are to be optimized, giving consideration to various incompatibilities.

Whereas in cylinder crankcases of gray cast iron (a perlitic structure) the material is also suitable for the cylinder surface, this applies only in exceptional cases for aluminum. Beside this monolithic design there are still heterogeneous and quasi-monolithic solutions for aluminum cylinder crankcases (Fig. 4.3).

4.2 Cylinder Crankcase Design Concepts and Cylinder Surface Technology

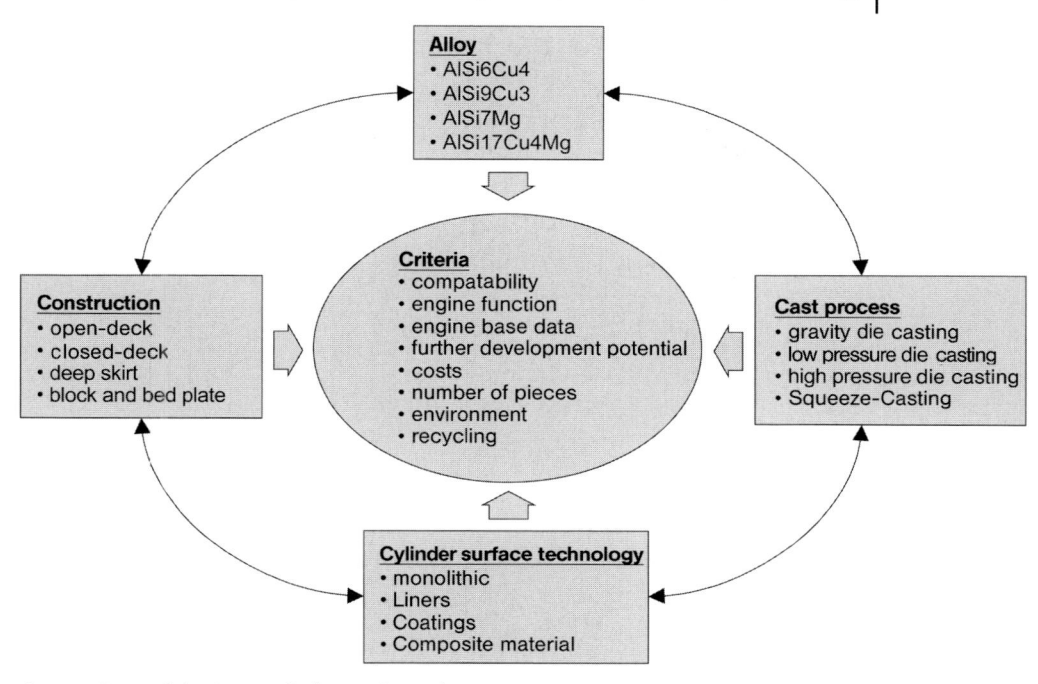

Alloy
- AlSi6Cu4
- AlSi9Cu3
- AlSi7Mg
- AlSi17Cu4Mg

Criteria
- compatability
- engine function
- engine base data
- further development potential
- costs
- number of pieces
- environment
- recycling

Construction
- open-deck
- closed-deck
- deep skirt
- block and bed plate

Cast process
- gravity die casting
- low pressure die casting
- high pressure die casting
- Squeeze-Casting

Cylinder surface technology
- monolithic
- Liners
- Coatings
- Composite material

Fig. 4.2 Parts of aluminum cylinder crankcase design concepts [1].

Aluminum-Cylinder crankcase

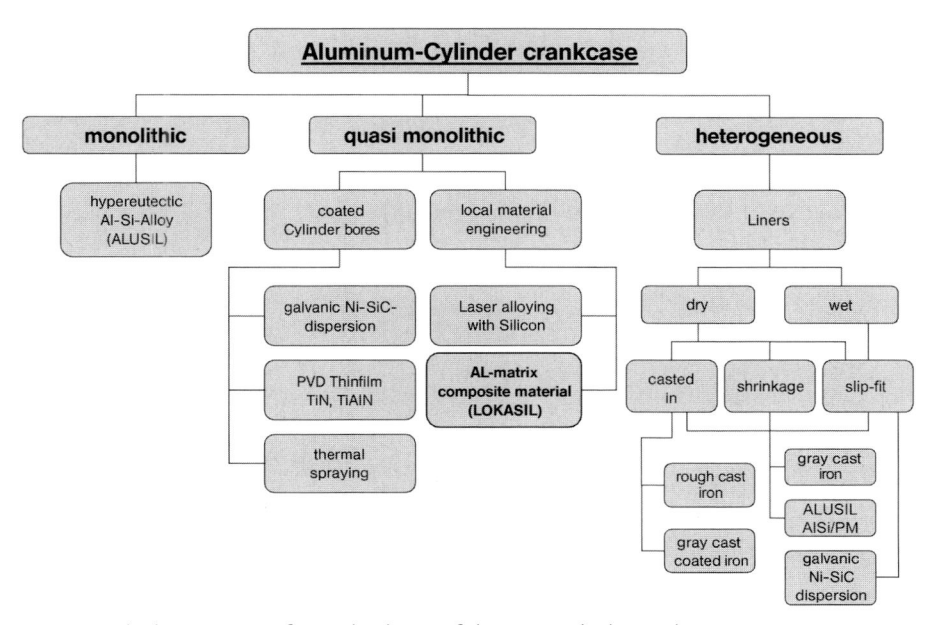

- **monolithic**
  - hypereutectic Al-Si-Alloy (ALUSIL)
- **quasi monolithic**
  - coated Cylinder bores
    - galvanic Ni-SiC-dispersion
    - PVD Thinfilm TiN, TiAlN
    - thermal spraying
  - local material engineering
    - Laser alloying with Silicon
    - AL-matrix composite material (LOKASIL)
- **heterogeneous**
  - Liners
    - dry
      - casted in
        - rough cast iron
        - gray cast coated iron
      - shrinkage
    - wet
      - slip-fit
        - gray cast iron
        - ALUSIL AlSi/PM
        - galvanic Ni-SiC dispersion

Fig. 4.3 Cylinder running surface technologies of aluminum cylinder crankcases [1].

4.2.1
ALUSIL®

One possible way to produce a monolithic cylinder crankcase is using a hypereutectic AlSi alloy. KS Aluminum Technology AG has made cylinders and cylinder crankcases of ALUSIL® (AlSi17Cu4Mg) since the beginning of the seventies. In this concept Si grains are primarily precipitated from a hypereutectic melt (Fig. 4.4). These hard particles form, after suitable machining of the cylinder surface, the support frame for the contact surface between pistons and piston rings [2]. This proven system with all the advantages of a monolithic cylinder crankcase like small mass, short overall length (minimum web width between the cylinder bores 4 mm), outstanding heat conductivity (thermal release for cylinder web and piston ring groove), small cylinder distortion (thermal and long-term distortion) and smaller piston assembly clearance due to similar heat expansion coefficients (smaller piston noise) is still in use; in particular for the cylinder crankcases in the V-building method, see the examples in Fig. 4.4.

However, the ALUSIL® concept is constrained by the state of the art in the low pressure casting process, which brings, besides many qualitative advantages, higher costs, so that this concept has not yet been able to be introduced into the mass motor market (series engines). For small to medium scale production, ALUSIL® gives price production benefits due to smaller tooling expenses (in relation to die casting) and already integrated cylinder surfaces are economically priced.

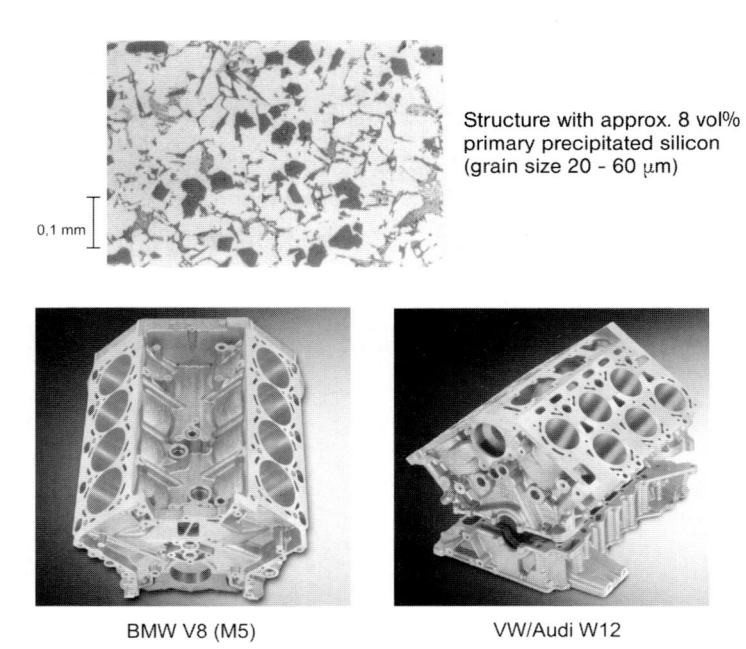

Structure with approx. 8 vol% primary precipitated silicon (grain size 20 - 60 μm)

0,1 mm

BMW V8 (M5)   VW/Audi W12

Fig. 4.4 ALUSIL® concept and two current application examples.

4.2.2
Heterogeneous Concepts

The most economical possibility to produce aluminum cylinder crankcases consists of casting-in (using classical die cast processes) liners of gray cast iron, which will, after machining, form the cylinder surface. When casting in the liners the smallest gap between the liner and the surrounding cast is obtained in the die cast process, which leads to comparatively good equivalent thermal conductivities. Liners of suitable aluminum materials can at present be cast in only in the high pressure die cast process, since they require short solidification times because of the danger of melting through. Planar merging of the aluminum liner at its exterior surface with the surrounding cast occurs only very slightly. By the use of cast liners with rough surfaces (from gray cast iron) a mechanical clamping of the liner to the surrounding cast can be achieved. Merging with the surrounding cast is clearly improved by suitable coatings on the outside of the liners but this adds to the cost. However, cast liners of ALUSIL® or hypereutectic liners, which are powder metallurgically produced, reduce some of the disadvantages of these heterogeneous solutions like higher weight, greater cylinder distortion and poorer thermal conduction, but are much more expensive than gray cast iron liners. Figure 4.5 shows some liners, which are cast-in at the KS Aluminum Technology AG at present serially.

Wet, slip fit liners, are applied in small production runs, e.g. sports car and racing engines, an exceptional case where cost is not the highest priority [3]. Here high-strength aluminum liners, which have a galvanically deposited Ni–SiC dispersion layer as a running surface, are used. An example is shown in Fig. 4.6.

rough cast iron Liner   graycast iron Liner   AlSi Liner

Fig. 4.5 Different liners to cast in.

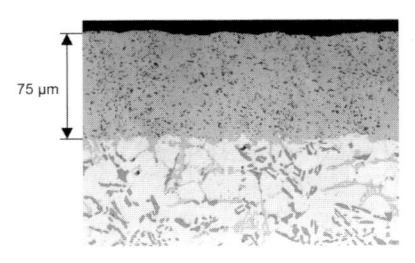

running surface
- nickel matrix
- 7 - 10 Vol.% SiC,
- SiC grain size 1 - 3 μm
- mixed hardness 610 HV

base material AlSi9Cu3
- most pore-poor

75 μm

Fig. 4.6 Microstructure image of a Ni–SiC dispersion layer.

4.2.3
Quasi-monolithic Concept

All quasi-monolithic concepts have the goal to achieve the advantages of a monolithic cylinder crankcase with smaller costs for mass production. State of the art is the galvanically deposited Ni–SiC dispersion layer (Fig. 4.6). A precondition for good adhesion of this layer is extremely small porosity of the surface to be coated. However, this tribological high-quality contact surface technology is no longer very common. A rather critical attitude to nickel, the problem of disposal of the resulting nickel sludge, and the corrosion problem when using sulfur-containing fuel are probably the main reasons for the decline in its use.

After the decline of the galvanic Ni–SiC dispersion layer different coating processes are in the ascendancy. One possibility is to apply a thin layer from TiN or TiAlN by a PVD process onto a honed cylinder surface, whereby the honed structure remains [4]. This plasma process in vacuum requires a complex preparation of the surface to be coated. High costs and lack of process security probably still prevent a break-through by this technology.

Thermal sprayed coatings are at a more developed stage. The engine of the VW Lupo FSI is the first to have gone into series production with a plasma-coated cylinder surface [5]. With atmospheric plasma spraying, not only iron based layers, but also layers with further metallic or ceramic additions are possible [6]. Figure 4.7 shows a structure image of a plasma-sprayed cylinder surface. When using low-priced powder the calculated coating costs for mass production numbers are similar to those for gray cast iron liners [6]. Initial problems such as layer adhesion and high porosity of the layer seem to be eliminated. In any case, coatings demand a high surface quality (keyword: pores), which is, according to experience, best fulfilled by low pressure permanent mold cast parts. Other casting processes (e.g. die cast, sand casting or also lost foam) do not yet have the necessary surface quality for coating in mass production with appropriate process security. This technique is relevant to cylinder crankcases of multicylinder engines.

Better layer adhesion and smaller porosity are expected with high-velocity oxygen fuel spraying than with plasma spraying [7]. Here metallic, ceramic and Cermet layers are also possible. The coating of cylinder surfaces by this procedure is still at the development stage.

To avoid adhesion problems with coatings, laser alloying has been developed [8]. Here the cylinder surface is metallurgically alloyed using a turning optics with par-

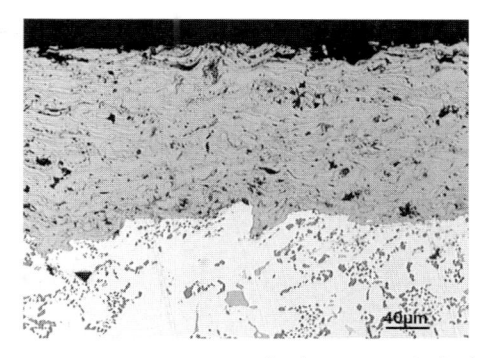

running surface
- Iron base
- mixed hardness 500 HV
- 7 - 10% porosity

Base material AlSi9Cu3
- pore-poor low pressure die casting

Fig. 4.7 Structure image of a plasma coated cylinder contact surface.

allel powder supply. Since the alloyed zone cools down very quickly, the formation of a laminated structure with fine deposited hard phases (e.g. silicon) is also achieved. However, these fine particles require a chemical process step at the surface finish after honing to set back the aluminum matrix. An objective of the further development is a clear reduction in the coating time, in order to make economic production possible.

4.2.4
The LOKASIL® Concept [9]

A hypereutectic alloy such as ALUSIL® is difficult to cast in processes like high pressure die casting or squeeze casting. It is more expensive and makes higher demands on machining than the standard alloy 226 (AlSi9Cu3) and is actually only needed in the cylinder bore. However, alternatively the silicon enrichment needed at the cylinder bores could take place only locally. Here highly porous, hollow cylindrical parts of silicon, so-called preforms, are infiltrated during casting under pressure with economical secondary alloy (A226). Thus a tribological equivalent composite material to the hypereutectic alloy ALUSIL® has been developed in connection with a very efficient casting process. The monolithic character of the cylinder crankcase is essentially maintained.

LOKASIL® can be understood as a material family. For the application in series primarily two variants are available. In the fiberless variant, LOKASIL® is an outstanding primary tribological material. This variant is used in cylinder crankcases of the Porsche Boxster and 911 Carrera series. With fiber particle composite material the higher strength requirements can be fulfilled. This benefits the realisation of a minimum web width between the cylinder bores at appropriate higher thermal stress.

Figure 4.8 gives an impression of the respective preform structure and the associated composite material structure. A small portion of alumina fibers in the basic variant forms a support frame for the embedded Si particles. With both variants the Si grain size was adapted to the grain size of the Si primary grains of the ALUSIL®. This is also very positive for the surface finish. Meanwhile both variants were developed with respect to cost reduction in the preform manufacture and higher strength of the composite material, so that today for each application a custom-made LOKASIL® solution is available. The advantages of the LOKASIL® concept are once more summarized in Table 4.2.

Typ	preform	preform structure	structure of composite
LOKASIL I 5 % Al$_2$O$_3$- Fibres + 15 % Si (30 - 70 µm) variant with increased demand of strength		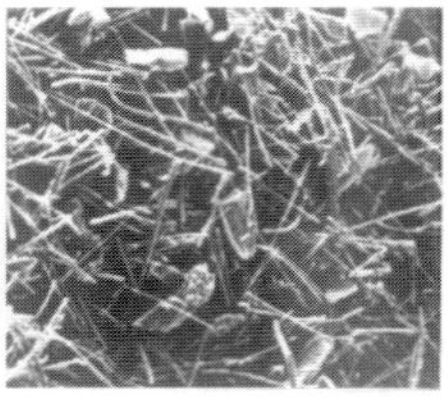 0.1 mm	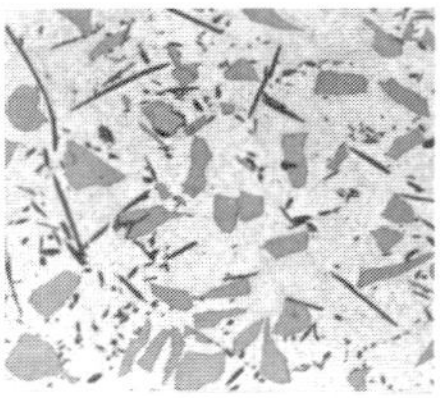 0.1 mm
LOKASIL II 25 % Si (30 - 70 µm) base variant for primary tribological application	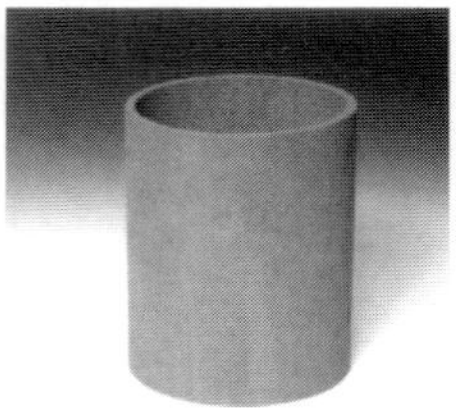	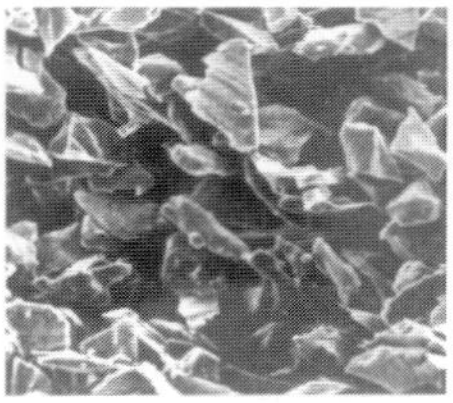 0.1 mm	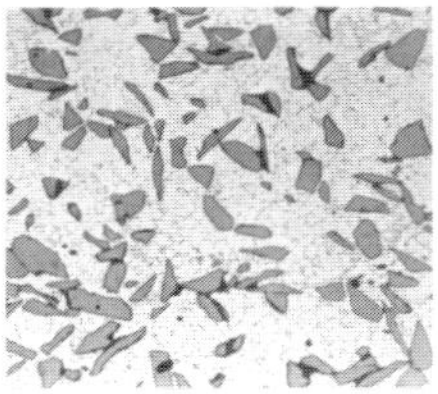 0.1 mm

Fig. 4.8 LOKASIL® cylinder contact surface variants.

Tab. 4.2 Advantages of the LOKASIL® concept [10].

Criteria	Advantages of LOKASIL	Comment
weight	without heavy castings	grey cast cast liners
compact construction	minimum web width	with fiber reinforcement
heat conductivity	gap free bonding embedded in Al casting	thermal release cylinder web and piston ring groove
cylinder deformation	low thermal and remaining deformation	no liner tolerance problems gap freedom, no bimetal effects
tribological properties	high seizure resistance, low wear, low) friction	well proven hard phase silicon (ALUSIL)
emission	low oil consumption (pos. and HC emission before converter)	optimal finish of cylinder surfaces
noise	low piston clearance	similar thermal expansion cylinder – piston
compatibility	preform replaceable by liner	technology similarity
recycling	suitable for secondary metal cycle	also with fibers

4.3
Production of LOKASIL® Cylinder Crankcases

4.3.1
Introduction

Production of metal matrix composite materials can take place in various ways. Possibilities are:

- forming (forging, extrusion of continuous semi-finished castings of composite material)
- powder metallurgical production
- thermal spraying
- laser alloying
- casting of composite melts
- melt infiltration of preforms by gas pressure infiltration or squeeze casting

In the LOKASIL® cylinder crankcases the production is by squeeze casting, where prefabricated preforms are infiltrated with the metallic melt.

4.3.2
Preform Manufacture

Fibers and particles can be processed to porous preforms with the addition of binders in the following ways:

- processing of long fibers for reinforcement and fabrics
- pressing and burning/sintering
- casting of an aqueous suspension, profiling and subsequent water removal
- gel-cast freeze drying.

Figure 4.9 shows examples of preforms of different geometry and materials.

Since the production of fiber preforms and hybrid preforms (fibers + particles) is already described in detail in other places in this book, only the production of particle preforms is described here. Pure particle preforms are used for LOKASIL®

Fig. 4.9 Examples of different preforms.

II-cylinder crankcases in series; for the achievement of a high porosity of 70–75 % the gel-casting freeze-drying method is used (Fig. 4.10) [11]. In this procedure the pores are introduced by the place holder, water. A homogeneous slurry, consisting of silicon powder, water and binders is produced first and cast into a metallic negative form of the desired geometry. After cooling the filled form, at approximately 60 °C, to temperatures below the freezing point, the parts are freeze dried in the frozen condition. That means the water is sublimed at approximately 6 mbar pressure by a supply of energy. Thus, in the places where the ice crystals were previously, the desired uniform cavities are produced, and these are then infiltrated with aluminum during the casting process.

By this process the preforms – depending on powder and water proportion as well as process parameters – will get exactly the desired porosity, pore distribution and pore structure. Finally the construction parts are solidified during a burning process at temperatures of approximately 1000 °C and cut to the required length. The resultant parts offer, despite high porosity, sufficient strength for automatic handling in the manufacturing process and form an extremely homogeneous metal matrix composite structure after infiltration with aluminum.

If smaller porosities are required, procedures like isostatic pressing, axial pressing or extruding can be applied (Fig. 4.11) [11]. With these procedures the poros-

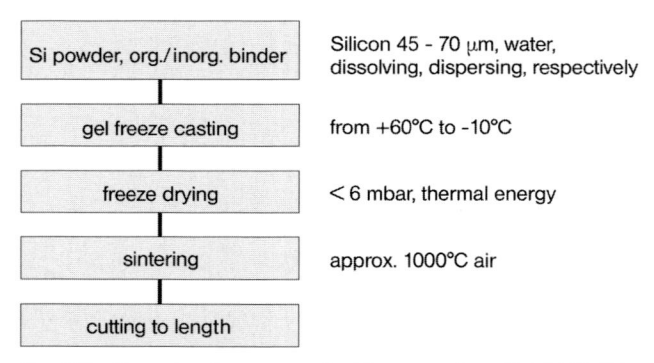

Si powder, org./inorg. binder	Silicon 45 - 70 µm, water, dissolving, dispersing, respectively
gel freeze casting	from +60°C to -10°C
freeze drying	< 6 mbar, thermal energy
sintering	approx. 1000°C air
cutting to length	

Fig. 4.10 Operational sequence of gel-freeze casting of LOKASIL II® preforms, CeramTec AG.

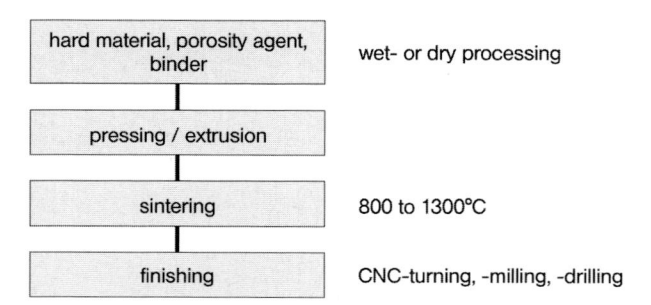

hard material, porosity agent, binder	wet- or dry processing
pressing / extrusion	
sintering	800 to 1300°C
finishing	CNC-turning, -milling, -drilling

Fig. 4.11 Operational sequence of powder technological profiling, CeramTec AG.

ities are not produced by water and the consequent expensive intermediate step of freeze casting, but by the addition of porosity mediums, which burn out during sintering. In principle nearly all kinds of hard materials can be processed with these developed procedures. Thus there is the advantage of selecting the most economical hard material reinforcements which supply the desired MMC material properties for each case.

4.3.3
Casting Process

The Porsche Boxster cylinder crankcase halves are cast by squeeze casting. This procedure combines die filling similar to the low pressure method and thus slow and a little turbulent, with the die-cast typical high pressure after die filling. The latter causes very large heat transmission coefficients, which lead to fast solidification of the casting. This permits comparable cycle times with the conventional die cast. Figure 4.12 shows a comparison of usual casting processes with regard to casting pressure and filling speed. Figure 4.13(a) shows the principle of applied squeeze casting machines. The tool, divided into a fixed and mobile mold half, is situated between the cast and the closing unit. The casting chamber is swivelled in and docked onto the fixed die half. The casting piston is in its final position after filling the form. In the casting chamber the so-called press-residual remains, from which the high pressure load results. Figure 4.13(b) shows the time curve for the most important casting parameters. Thereby the end of the mold filling is of central importance for the process control concerning preform infiltration. Here the casting piston speed must be reduced so as to keep the increase in pressure during

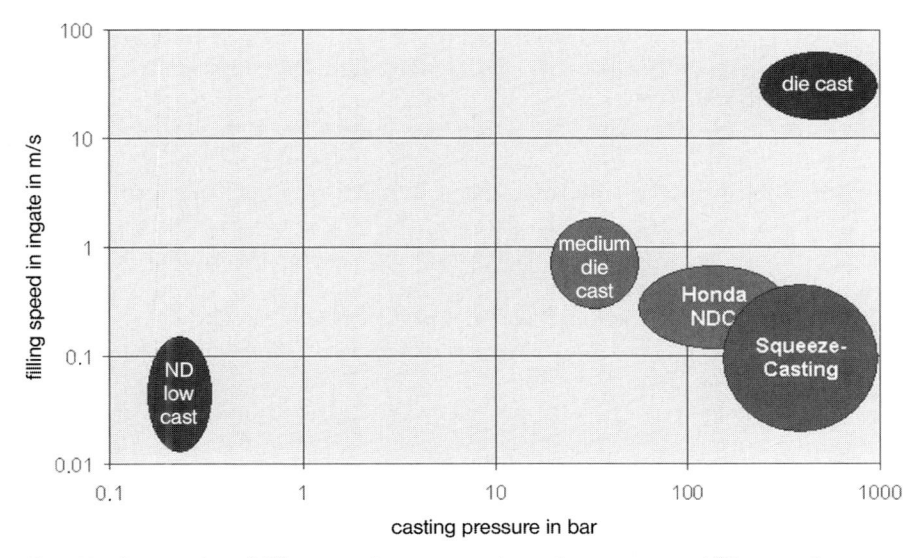

Fig. 4.12 Presentation of different casting processes by casting pressure and filling speed.

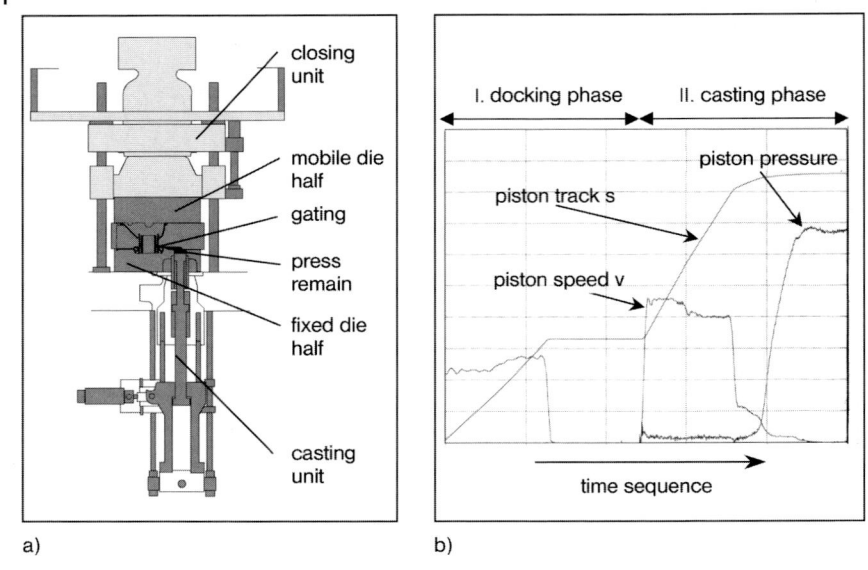

Fig. 4.13 (a) Scheme of a squeeze-casting machine and (b) time sequence of essential process parameters [10].

the infiltration within permitted values. This process is so stable that the contact surface inspection by eddy current can be reduced to a sampling inspection.

The squeeze casting machine is a component of a fully automated casting cell (Fig. 4.14), in which two robots take over all necessary handling. Whilst robot 2 cleans the die and spreads the die release agent, robot 1 takes three heated preforms (Fig. 4.15) out of the preheating furnace and sets them on tempered removable sleeves. Subsequently, these are positioned in the casting tool. After closing the die the casting process takes place, as the casting piston moves upward and fills the cavity slowly with melt. The preforms are infiltrated with melt and the casting solidifies under high pressure. Afterwards the mold opens and robot 1 takes the casting. This is inspected for completeness, marked and put down after pressing out the removable sleeves on the conveyor. The removable sleeves are then coated in an immersion bath and kept at a moderate temperature. At this time robot 2 has already begun another cycle with the cleaning of the mold.

For the first application of LOKASIL® in the cylinder crankcases halves in the Porsche Boxster engines these were cast by squeeze casting, but the extension of the procedure to horizontal working die casting machines was being studied in parallel from the very early stages of the work. The world-wide provision of such machines forced this in respect of the broader marketing of the LOKASIL® concept. Although some functional LOKASIL® cylinder crankcases had already been produced on conventional die casting machines with die-cast typical in-gate technology, process stabilization could only be achieved with real-time-controlled die casting machines. With this regulation and a modified in-gate and tool lay out design a similar process to squeeze casting is possible, as the past examples have shown, which guarantees, in principle, a safe preform infiltration with good casting quality. Series applications surely require a component specific process development.

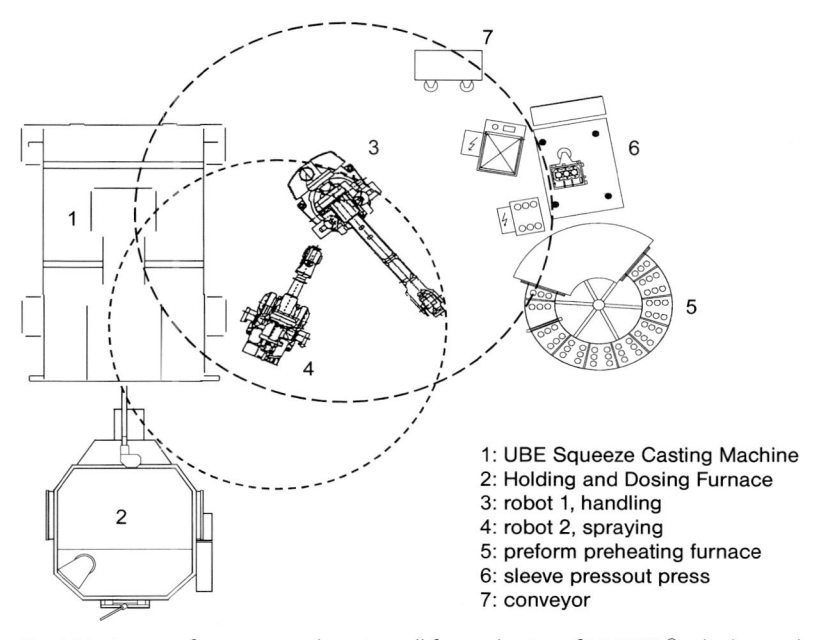

1: UBE Squeeze Casting Machine
2: Holding and Dosing Furnace
3: robot 1, handling
4: robot 2, spraying
5: preform preheating furnace
6: sleeve pressout press
7: conveyor

Fig. 4.14 Layout of an automated casting cell for production of LOKASIL®-cylinder crankcases.

Fig. 4.15 Heated preforms taken out of the preheating furnace [12].

4.4
Summary and Outlook

Monolithic and quasi-monolithic concepts are to be preferred to heterogeneous solutions due to their technical advantages. In multicylinder engines the monolithic ALUSIL® concept is also economical as the most favorable solution for middle series production numbers.

If the costs are of paramount importance in mass production, highly productive casting processes like high pressure die casting have to be applied. Casting-in of gray cast iron liners is the most economical solution, presupposing the technical disadvantages can be accepted. The proven LOKASIL® concept, which links the advantages of the monolithic cylinder crankcases with highly productive die casting processes, is surely the best solution for the advancement of preform production and composition, with only small extra costs compared to gray cast iron liners.

Plasma-coated cylinder surfaces just find straightforward introduction into series application. Their operability is beyond doubt. Thus the success of this new technology will also depend to a large extent on the two most important factors: the medium-term attainable process security and the actual costs arising per coated bore. How far the higher degree of porosity in inexpensive high pressure die cast cylinder crankcases will be acceptable cannot be answered yet.

References

1 E. Köhler, Aluminum-Motorblöcke, Aus Forschung und Entwicklung, Kolbenschmidt AG, **1995**.
2 E. Wacker, H. Dorsch, ALUSIL-Zylinder und FERROCOAT-Kolben für den Porsche 911, *MTZ*, **1974**, *35*, 2.
3 H.-J. Neußer, Kurbelgehäuse-Gießtechnik im Hochleistungsmotorenbereich, *VDI-Berichte* **2001**, 1564, Gießtechnik im Motorenbau.
4 E. Lugscheider, Ch. Wolff, Innenbeschichtung von Aluminum-Motorblöcken mittels PVD-Technik, *Galvanotechnik* **1998**, *89*(7), 8.
5 R. Krebs, B. Stiebels, L. Spiegel, E. Pott, FSI– Ottomotor mit Direkteinspritzung im Volkswagen Lupo, *Fortschritt-Berichte VDI* **2000**, 12(420), Wiener Motorensymposium.
6 G. Barbezat, J. Schmid, Plasmabeschichtungen von Zylinderkurbelgehäusen und ihre Bearbeitung durch Honen, *MTZ* **2001**, *62*(4), S. 314–320.
7 M. Buchmann, R. Gadow, Ceramic coatings for cylinder liners in advanced combustion engines; manufacturing process and characterisation, *The 25th Annual International ACERS Conference on Advanced Ceramics & Composites*, Cocoa Beach **2001**.
8 A. Fischer, Aluminum-Motorblöcke für Hochleistungsmotoren – Anforderungen und Lösungen, *VDI-Berichte* **2001**, *1564*, Gießtechnik im Motorenbau.
9 E. Köhler, F. Ludescher, J. Niehues, D. Peppinghaus, LOKASIL-Zylinderlaufflächen – Integrierte lokale Verbundwerkstofflösung für Aluminum-Zylinderkurbelgehäuse; Sonderausgabe von ATZ/MTZ: Werkstoffe im Automobilbau, **1996**.
10 P. Everwin, E. Köhler, F. Ludescher, J. Niehues, D. Peppinghaus, LOKASIL: Entwicklungs- und Serienanlauferfahrungen mit dem PORSCHE-Boxster-Zylinderkurbelgehäuse, Zukunftsperspektive, in *KOLBENSCHMIDT, Kompetenz im Motor*, **1997**.
11 E. Köhler, I. Lenke, J. Niehues, LOKASIL® – eine bewährte Technologie

für Hochleistungsmotoren – im Vergleich mit anderen Konzepten, *VDI-Berichte* **2001**, *1612*, S. 35–54

12 P. Everwin, E. Köhler, F. Ludescher, B. Münker, D. Peppinghaus, LOKASIL-Zylinderlaufflächen: Eine neue Verbundwerkstoff-Lösung geht mit dem Porsche Boxster in Serie, ATZ/MTZ-Sonderausgabe "Porsche Boxster", **1996**.

5
Production of Composites or Bonding of Material by Thermal Coating Processes

B. Wielage, A. Wank, and J. Wilden

5.1
Introduction

In modern mechanical engineering, components are generally subject to both a high strength requirement and high requirements of the surface properties, e.g. corrosion and/or wear resistance. Further, functional characteristics must often be fulfilled at the surface. The entire requirement profile may not be met by one material alone. Only by a functional separation of the tasks of the component surface and inside, which is technically realized by applying coatings onto a structural component, can the entire requirement complexity be fulfilled.

A large range of procedures can be used in principle to manufacture coatings. Besides physical (PVD) and chemical (CVD) vapor deposition, galvanic and external currentless chemical deposition, a hot-dipping process, thick-film coatings of thermal spraying and cladding welding have been utilized. Thick-film coatings are characterized by high deposition rates and short process times for manufacturing material composites. In these, partial multilayered coatings are used to ensure the stability of the composite layer. Both the substrates, which determine the component strength, and the coatings can be composite materials.

Thermal spraying and cladding procedures, which in particular find their application in protective coatings against wear and corrosion of highly stressed components, are of high economic importance and are described in detail in the following, where their applications will be discussed with examples. The economic loss as a consequence of wear and corrosion amounts to at least 8 billion Euros in the Federal Republic of Germany, so that a high value is attached to these applications. Such layers can be important to reduce friction as well as to provide thermal or electrical isolation. Finally, functional coatings with special magnetic, thermoelectric or chemical characteristics are becoming increasingly important.

5.2
Thermal Spraying

In accordance with DIN 32530 thermal spraying is characterized by the fact that the fed material is molten or solidified and spun on prepared surfaces (Fig. 5.1). The surface of the component part is thereby generally unmelted. Typical layer thicknesses are in the range from 10 µm up to a few millimeters.

The processes of thermal spraying are characterized by a powder, (filler-) wire or rod-shaped material which is conveyed into an energy source. Furthermore the European standard EN 657 "Thermal spraying" divides the spraying technology in accordance with the energy source and differentiates between jet, liquid and gas-based processes as well as electrical gas discharge. Laser spraying and molten metal spraying have at present no economic importance and thus they will be neglected in the discussion. The conventional flame spraying, detonation-, high-velocity flame and cold gas spraying belong to the gas-based processes. Arc and plasma spraying are based on electrical gas discharges.

Thermal spraying exhibits inherent advantages:

- almost any coating material is applicable,
- small thermal load on the component,
- locally limited and/or large area coating
- on-site application of many procedures is possible.

The applicability of a material for thermal spraying depends only on the requirement that a molten phase or a sufficient ductility below the decomposition temperature must exist. Practically no restrictions exist for the selection of the structural materials due to the thermal load of the substrate which is adjustable by the processing.

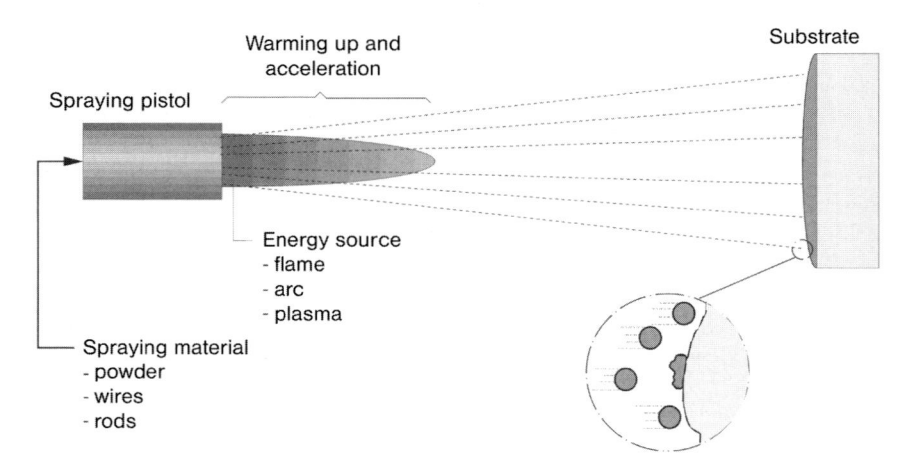

Fig. 5.1 Principle of thermal spraying.

Since thermal spraying does not result in metallurgical bonding of the layers to the base material (substrate), only a small adhesive strength results. A subsequent remelting or hot-isostatic pressing (HIP) treatment brings improvement and reduces the porosity. A compression of the layers can be achieved by simultaneous or subsequent shot peening. Adjustment to the final size and a defined surface quality usually takes place via machining.

5.2.1
Spraying Additive Materials

The spraying additive can be supplied to the spraying process in the form of powders, wires or rods. In the manufacturing processes a substantially smaller material spectrum is processable by using wires or rods than with powders. Nevertheless, approximately 3800 t wire-shaped and 840 t powdered spraying additives are processed annually. With almost 80%, zinc wire takes by far the largest part followed in importance by different steels (approximately 13%) and molybdenum (approximately 5%). For rod spraying, ceramic sinter rods are used. While spraying wires usually have a diameter from 1.6 to 3.2 mm, rods can have a diameter up to 8 mm. With the use of cored wires the spectrum of the processable materials is substantially extended. Hard materials can be used to improve the wear resistance of a metal matrix material (material of the covering). Furthermore the characteristics of the matrix can be modified by alloy formation with fillers in the coating process, without affecting the manufacturing properties of the covering material during manufacture of the wires. Folded cored wires, which permit a particularly high degree of filling and tube filler wires are also available, promoting process stability due to the closed covering.

Usually powders used for spraying have particle diameters between 5 and 150 µm. The applicable size of powders depends on the fluidity and the thermal physical characteristics. The fluidity of the powder depends on the particle form (spherical or sharp-edged) as well as the affinity to agglomerate in consequence of the increasing specific surface with decreasing particle size. The thermal physical characteristics determine which particle sizes melt off and/or whether a sufficient warming through is possible by interaction with the hot gas stream. In principle, close grain fractions positively affect the layer quality (homogeneity of the microstructure and layer thickness, porosity, deposition efficiency).

The majority of metallic spraying powder alloys are based on iron, nickel and cobalt. MCrAlY (M: = Fe, Ni, Co) alloys are of great importance for hot gas corrosion protection in combustion turbines. Furthermore, molybdenum especially is often processed as well as light metal alloys based on aluminum and titanium. Both the particle shape and the gas content depend mainly on the process control of the powder production. MCrAlY alloys are usual atomized by inert gases in a chamber with an inert atmosphere, because the hot gas corrosion resistance depends substantially on the oxide content of the layers. On the other hand self-fluxing NiCrB-Si alloys are often atomized in water, because a certain oxygen content is necessary for good melting behavior after spraying, however, coarse particles can form.

Oxide ceramics account for approximately a quarter of the total spraying powder consumption. ZrO_2-based ceramics are predominantly used for thermal insulation, while Cr_2O_3 and Al_2O_3 (TiO_2) ceramics are predominantly used for wear protection. Al_2O_3 is also used for electrical isolation. Besides melted and/or sintered and broken powders, which have sharp-edged shapes due to the milling process, spheroidized, spray-dried and sintered powders are available.

A further quarter of the powder used, can be found in Cermets, composites with a metal matrix, which prevents brittle failure, and a ceramic reinforcement component, which improves the wear resistance. The most important Cermet materials are hard materials based on WC–Co(Cr) as well as Cr_3C_2–Ni20Cr. The composite powders are usually made by spray-drying a suspension with an organic binder, whereby the bond strength of the particles is substantially improved by a subsequent sintering process. To optimize the wear and corrosion resistance for a certain hard material content, the size of the hard materials and the kind of the hard material is mainly of importance as well as the matrix composition. Due to its spherical shape the composite powder offers outstanding fluidity.

With the help of careful development of the metallurgical concept as well as the use of adapted process technology, it has recently been possible to apply hard materials for spraying applications which do not allow an initial layer production. SiC is characterized by outstanding wear protection characteristics and by low price, however, no pure SiC sprayed coatings can be produced due to its incongruent melting behavior and low ductility below the decomposition temperature. The high reactivity of all technically relevant metallic elements in the solid state opposes their application as a hard material in Cermets. Investigations on the manufacturing of SiC-based Cermets by sintering have shown that the kinetics of dissolving in an iron-based matrix alloy is strongly restrained by saturating the iron mixed crystal with silicon and carbon. At the same time the wettability of the hard material is improved, resulting in good interface bonding. For manufacturing sub-micron structured Cermet composite powders with homogeneous contribution of the SiC particles in oversaturated nickel or cobalt matrixes high-energy milling can be applied (Fig. 5.2). By processing with kerosene-operated high-velocity flame spray-

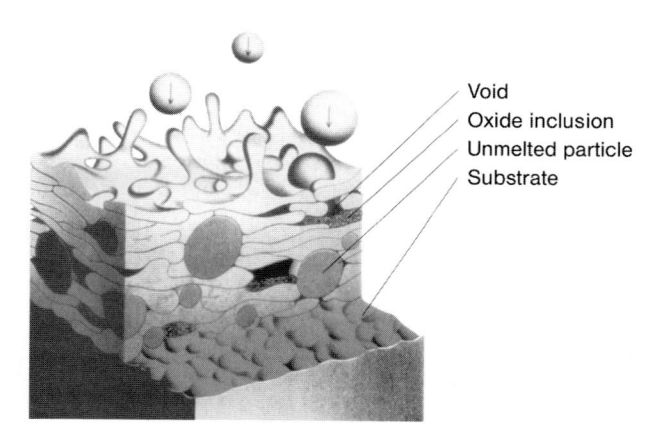

Fig. 5.2 Schematic presentation of spray coating formation.

ing technology the microstructure can be transferred into the layers. Thus hard material contents of more than 50 vol% can be realized. Hence an inexpensive alternative to the conventional WC–Co(Cr) Cermet layers is available as a result of optimization of the spraying material.

Thermoplastics (granulates) can also be processed to layers by thermal spraying. However, these coatings are only of small economic importance.

5.2.2
Substrate Materials

In principle, any material is suitable as a substrate if it can be sufficiently roughened by blasting. The most important metallic substrate materials are: unalloyed to highly alloyed steel, gray cast iron, super alloys based on Ni and Co, copper alloys and light alloys based on aluminum and titanium. Ceramics and plastics can be coated just like composites with a metal, ceramic or polymer matrix. For the coating of long fiber reinforced composites an optimization of the jet process is necessary to avoid, as far as possible, damaging the outside lying fibers. From the constructional side it is relevant to note that all surfaces should be well accessible. Undercuts cannot be coated and sharp edges and gaps should be avoided.

5.2.3
Surface Preparation

The characteristics of the substrate surface can exert substantial influence on the resulting layer adhesive strength and may be adjusted by careful pretreatment. The substrate preparation usually consists of a three step procedure: precleaning, blasting and subsequent cleaning. Precleaning primarily removes oil and fat as well as colour remains on the surface if necessary, and takes place either mechanically or chemically.

Blasting effectively activates, decontaminates and roughens the surface to be coated, in order to create suitable conditions for the adhesion mechanisms described in the following: roughing of the substrate surface leads to an increase in the defect concentration, the dislocation density and the frequency of the stacking faults. Plastic deformation in the zones near the surface leads to an increased free surface energy. In addition, the surface area is increased and offers to the impinged spraying particles the possibility of mechanical interlocking. For mechanical roughing different blast grains are used, e.g. chill casting gravel, SiC or corundum. The blast grain hardness, grain size as well as the kinetic energy of the particles and the jet angle affect the surface roughness. An optimal adhesive strength is obtained when adjusting a jet angle of 75° to the substrate surface.

A subsequent cleaning, e.g. by ultrasonic-assisted cleaning in alcohol, removes residues remaining on the surface after blasting. It also eliminates dirt and dust particles as well as fats, which cause deactivation of the surface. However, the applicability is limited by the size of the pool.

Pickling, which removes reaction layers from the surface, occasionally finds application as a chemical pretreatment. It is possible to adapt the composition of the pickles, the pickling temperature and time to the material in order to obtain the desired roughness.

5.2.4
Structure and Properties of Spray Coatings

Thermally sprayed layers differ from layers which are coated by other processes in their structure, bonding mechanism and subsequent treatment possibility. Depending on the processed materials and the applied spraying processes the layers are more or less porous.

The structure of a sprayed coating has to be seen as a stochastic procedure. Due to the particle size distribution and the variations in form of the powdered spraying materials within a grain fraction, as well as the place-dependent temperature and speed distribution in the gas stream, different conditions arise for the individual spraying particles. Each powder particle and/or each spraying particle, which is entering the gas stream, describes its own path to the work-piece surface according to its mass, density, form and speed. Due to the high number of particle trajectories a range of variations is found for the interaction between gas stream and particle, and between particles and substrate. The development of a sprayed coating is represented schematically in Fig. 5.2.

Extreme heating and cooling rates, reactions during or after the flight phase, mechanical influences at the solidification and temperature gradients through the layers develop a microstructure, which is characterized by a multitude of unstable and metastable conditions. Thermal sprayed coatings often exhibit a laminated layer structure and, depending on applied spraying process parameters and spraying additives, a more or less porous, micro-cracked, heterogeneous and anisotropic microstructure. Particles, which are incompletely melted or already solidified before impacting the substrate surface, are embedded in the layer as well as oxides or nitrides.

The spraying particles reach the substrate surface with a certain speed and temperature, which can lie either above or below the melting point. At a given, smooth substrate surface and a certain particle temperature and viscosity the amount of the kinetic energy decides the overall distribution behavior and thus the resulting form of the particles. The work performed during diffusion acts contrary to the surface energy, whose amount depends on the temperature-dependent particle viscosity. The viscosity of the spraying particles decreases with increasing temperature and causes a reduction in the work required to form the new surface, depending on the particular particle shape. Spraying particles spread out on polished surfaces radially; upon impact with a rough substrate surface or already sprayed on particles, the preferred direction is that for which the smallest work against the surface energy is necessary. The morphology of spraying particles after solidifying affects the adhesive strength of the layer on the substrate.

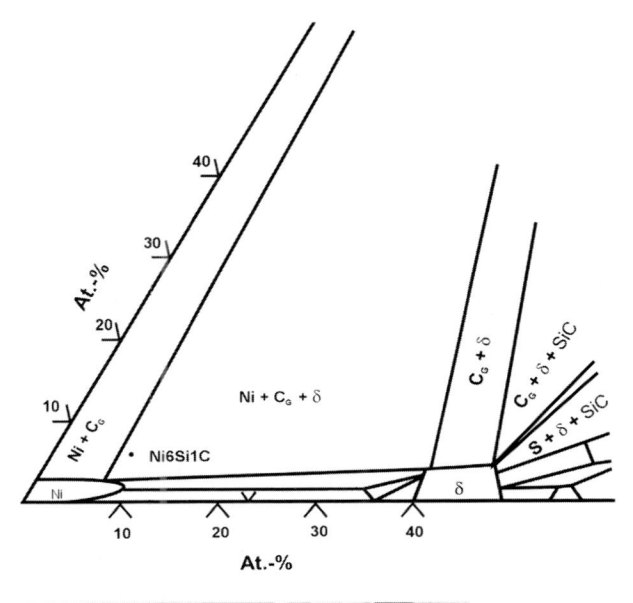

Fig. 5.3 Phase diagram of a metallurgically optimized Ni6Si1C matrix to manufacture SiC-based Cermet and SEM image of a cross section of high-energy milled Co–Si–C/70 vol.% SiC composite powder particle.

Metallic spraying particles usually form lamellas. The multilayer microstructure of the sprayed coating is caused by impurities, e.g. oxides, geometrically caused cavities and pores. Two different morphologies can occur, the so-called "pancake type" and the so-called "flower type". In particular in materials with a high material plasticity the often observed pancake type solidified particles show, besides the round form, also deviating morphology types. The flower type is characteristically found at high impact speeds at small particle viscosity. An increase in the kinetic energy leads to intensified flow and to a remerging and/or corona formation, which exhibits various appearances. Solidified oxide-ceramic spraying particles often show cracks in the vertical and horizontal direction. The horizontal running cracks are critical and limit the strength of the sprayed coating. Incompletely melted and/or already solidified spraying particles rebound from a polished surface; when a rough surface is present such particles can be pressed into the roughness of the substrate. However, only a relatively weak adhesion results.

A change in the composition and structure of the sprayed layer can occur by reaction of the spraying additive material during the flying phase of the particle. The main reactions and processes are:

- selective vaporization of a component
- reaction of metal compounds (for example decomposition of hard materials in the presence of O_2)
- formation of nonvolatile metal bonding like oxides, nitrides and hybrids in the presence of O_2, N_2, H_2 (especially in reactive metals).

Of special importance is the oxidation of the surface of metallic spraying particles on their way to the substrate. The oxides formed are usually built into the interface of the spraying lamellas of the sprayed coating. Thereby, the possibility exists of increasing the layer hardness and thus the wear resistance of the layer. Oxides can also serve as partly fixed lubricants and thus decrease friction losses.

Thermally sprayed layers can have different porosities, depending on the spraying additive and the spraying process used. The highest porosity occurs in flame and electric arc sprayings, while very compacted layers with a porosity of approximately 1% can be obtained only by high-energy processes like high-velocity flame spraying or vacuum plasma spraying as well as by cold gas spraying for some composite layers. In general, more porous layers result in the case of ceramic rather than metallic materials, as well as with the use of coarse-grained powders. Particularly for applications of corrosion protection, an appropriate subsequent treatment is often necessary. Furthermore, thermally sprayed layers show a rough surface with roughness between 5 and 60 µm. In this context the mechanical treatment is of special importance, i.e. both cutting procedures with a geometrically defined cutting edge and grinding and polishing. Due to their roughness sprayed coating surfaces provide excellent adhesion for painting and lacquering. This is used for example in flame or arc-sprayed Zn or Al layers for atmospheric corrosion protection.

Thermally sprayed layers often show relatively high residual stress, dependent on the processing. This occurs due to the thermally induced contraction during the solidification and cooling after the impact of the spraying particles on the substrate surface. Since the spraying particles already stick strongly to the substrate and/or other spraying particles, residual stress develops due to the shrinkage. Since this lower the adhesive strength of the layer, the maximum permissible layer thickness is often limited.

The developing shrinkage stress during the contraction of metallic spraying particles can be released partially by plastic deformation. In brittle materials, e.g. ceramics, a release of residual stress usually takes place via cracking. Segmenting by vertically running cracks causes e.g. in heat-insulating layers on ZrO_2-based materials an improvement in the thermal shock stability. Here special importance is given to an adapted temperature control during the spraying process.

5.2.5
Adhesion of Thermally Sprayed Coatings

The adhesion of thermally sprayed layers on metallic base materials usually results from the following physical and chemical mechanisms:

- mechanical interlocking
- local chemical-metallurgical interaction effect between sprayed coating and substrate (diffusion, reaction, formation of new phases)
- adhesion by physical adsorption
- adhesion by chemical adsorption.

The adhesion mechanism is based on forces, which allow the adhesion of a strong interface and a second phase. It is necessary to differentiate between physical adsorption which works due to electrostatic forces at a sufficiently activated (roughened) surface and particle approach to the atomic spacing, and chemical adsorption with homopolar binding forces which depend on the affinity of the metal to the adsorbed material and the activation energy of the chemical environment. Physical absorption is based on the relatively weak van der Waals bonding. In contrast to this, chemical absorption is characterized by large binding forces. However, this effect is reduced by existing surface impurities.

Mechanical interlocking as an adhesion mechanism is realized by the surface roughness of the substrate surface. The molten particles penetrate in uneven areas and undercuts due to their high thermal and kinetic energy as well as to capillary forces, then solidify and cause mechanical interlocking. This effect is increased by shrinkage stresses, which develop within the first spraying layers due to high cooling rates, when they hit the cold substrate surface.

The adhesion of the spraying particle on the base material can also take place due to mutual metallurgical influence by diffusion or chemical reaction as well as the formation of new phases. Diffusion causes a strong increase in bond energy and is an important adhesion mechanism. The reaction of the additive spraying material with the base material increases the adhesion. Adhesion of a metallic spraying particle on the substrate or on another spraying particle does not takes place over the entire particle surface, but at single microscopically visible active contact areas, which are separated by pores and oxides. At places with strong contact between the spraying particles and substrate strong diffusion or even phase transformation can eventually occur at a sufficiently high interface temperature. Processes within the spraying particles during and before the impact on the substrate, as well as between the spraying particles and the substrate after contact, influence the interface reactions. Under the conditions of a favorable element affinity and formation enthalpy, adhesion-strengthening reactions take place as well as the formation of new phases and/or solid solutions. The amount of areas with close contact zones between metallic spraying particles and a metal substrate depends on the cooling and solidification time as well as the interface temperature of the spraying particles. The adhesion of oxide-ceramic spraying particles is affected by bonding with oxides in the primer and can be increased by oxidation of the primer (substrate surface). At plasma-sprayed Al_2O_3-layers it can be proven that in the interface layer/substrate, iron and mixed oxide zones arise, whose formation is influenced by the preheating temperature.

The discussed adhesion mechanisms work individually or together with different contributions depending on the coating process, the surface preparation and the layer/base material combination. The adhesive strength is influenced by the following: the temperature of the interface, spraying particles/substrate, the particle impulse of the impact, the characteristics of the particle and substrate surface, the heat capacity of the spraying particles, the heat conductivities of the base material and spraying particles as well as the residual stresses forming in the layer. A high interface temperature increases diffusion and reactions at the interface, whereby generally the adhesion is improved. For this reason an increased substrate temperature usually positively affects the adhesion. However, oxidation reactions occur unfavorably during coating of metallic sprayed layers at high temperatures.

5.2.6
Thermal Spraying Processes

5.2.6.1 Flame Spraying
During flame spraying a fuel gas oxygen flame is used as the heat source, into which the spraying additive (powder, wire or rod shape) is inserted. Acetylene is mostly used as the fuel gas, but ethene, methane, propane, propylene, natural gas or hydrogen may also be used.

5.2.6.1.1 Powder Flame Spraying
In powder flame spraying (Fig. 5.4) only small particle speeds are obtained (<50 m s^{-1}), with a relatively long interaction time between hot fuel gases and powder particles. Powder flame sprayed layers therefore show a relatively high porosity

Fig. 5.4 Principle of powder flame spraying (Linde AG).

(5%), a high oxide content and a high gas cavity. In general, spraying powders with a diameter between 20 and 100 µm are used. For metallic spraying powders deposition rates between 3 and 6 kg h^{-1} are obtained, while for ceramics 1 to 2 kg h^{-1} is common.

The most commonly processed material group using powder flame spraying is self-fluxing NiCrBSi alloys, which are characterized by a long solidification interval. A subsequent thermal treatment of the sprayed coatings is remelting, which leads to a metallurgical bonding to steel substrates, and both molten and gas sealed coating. Remelting takes place partly using the burner, which was already applied for coating. It can also take place in furnaces or via inductive warm up, which is characterized by local warming of the edge zones of a component. The layers generally show an outstanding corrosion resistance. The hardness can be adjusted within a wide range, primarily by the content of chromium carbides and borides (between HRC 20 and HRC 65). Such coatings are applied for protective covers of shafts, roller table rolls, bearing seats, fans or rotors of barrel extruders.

5.2.6.1.2 Plastic Flame Spraying

Plastic flame spraying (Fig. 5.5) differs from the powder flame spraying in that the plastic granulates do not interact directly with the fuel gas oxygen flame. On the axis of the flame spraying pistol is a powder conveying nozzle. It is surrounded by two ring-shaped nozzle exits, whereby the internal ring gives out air or an inert gas and the outside ring the thermal source of energy, the fuel gas oxygen flame. The melting of the plastic does not take place directly via the flame, but from heated up air and radiation heat. Because of the requirement for a molten phase only thermoplastics can be converted to layers. Application areas for plastic flame spraying are railings of various kinds, grommets, drinking water containers, garden furniture or pool markings.

Fig. 5.5 Principle of plastic flame spraying (Linde AG).

5.2.6.1.3 Wire/Rod Flame Spraying

With wire flame spraying (Fig. 5.6), the wire-shaped spraying material is fed continuously to the centre of a fuel gas oxygen flame and melts. The melt is transported in the flame stream to the wire tip, where the speed is substantially increased by the addition of an atomiser gas – usually compressed air – and atomized in primary droplets. As a function of the flow conditions as well as the viscosity and sur-

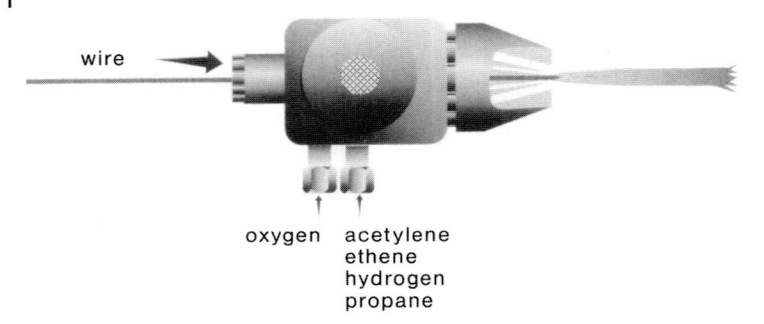

Fig. 5.6 Principle of wire / rod flame spraying (Linde AG).

face tension of the molten droplets, fine diffusion takes place in the gas stream, before the particles impact on the work piece surface.

Typical spraying materials are molybdenum, low and highly alloyed steels, as well as aluminium-, copper- and nickel-based alloys. In comparison to powder flame spraying, higher particle speeds up to 200 m s^{-1} are clearly obtained. Since the melt at the wire tip interacts with the environment for a relatively long time, an oxide content comparable to the powder flame spraying results. Since by solid solution hardening a substantial increase in the hardness can be reached, the absorption of oxygen is used during the production of molybdenum layers. Such layers find their application in the automotive industry in shift forks, synchronizers or piston rings. Compared to powder flame spraying a smaller porosity (3%) can be obtained by the avoidance of embedding unmelted particles in the layer, as well as higher particle speeds. The deposition rates of processing solid wires are usually between 6 and 8 kg h^{-1}.

For the production of ceramic layers, fibers with ceramic powder filler or sinter rods can be used. Sinter rods require cost-intensive production; however, they guarantee, contrary to laces, that spraying additive materials are completely melted. Therefore they can be used for the production of high-quality oxide-ceramic layers with defined porosity.

5.2.6.2 Detonation Spraying

Detonation spraying (Fig. 5.7) was developed in the 50s in the United States and is the forerunner of high-velocity flame spraying. Detonation spraying is a discontinuous burning process. Sequential filling of a pipe with one side open and one closed, with fuel gas (usually acetylene)–oxygen mixture at the open end and powdered spraying additive materials at the closed end takes place. After ignition of the mixture, which results in melting and accelerating the powders to high speeds towards the substrate to be coated, rinsing the pipe with nitrogen takes place. The ignition frequency can vary between 4–8 cycles s^{-1} in older systems up to 100 cycles s^{-1} with modern systems, which work with fluid-dynamic regulation of the gas injection and thus require no mechanically operated components.

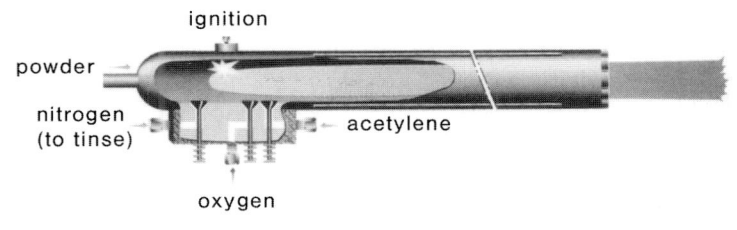

Fig. 5.7 Principle of detonation spraying (Linde AG).

Detonation spraying is characterized by relatively high process gas temperatures (<4000 °C) and extremely high particle speeds (<900 m s^{-1}). The high particle speeds lead to a compact layer structure and outstanding adhesion, and to small oxide contents due to the small interaction times of the spraying additive with the hot fuel gases.

Detonation spraying is used primarily to manufacture Cermet high-quality wear and corrosion resistant layers (e.g. WC–Co). Ceramic and metallic layers can also be produced with almost theoretical density.

5.2.6.3 High Velocity Flame Spraying

High-velocity flame spraying (Fig. 5.8) or the HVOF (high velocity oxyfuel) process represents a modification of detonation spraying by continuous burning and is now a widely used engineering process. Beside systems which use fuel gases such as acetylene, ethene, propylene, propane, natural gas or hydrogen, liquid fuel (kerosene) systems exist. In modern systems combustion chamber pressures can exceed 10 bar, so that ethene or hydrogen is preferred as the fuel gas.

High-velocity flame spraying is characterized by high spraying particle speeds and low process temperatures. Liquid fuel operated systems obtain particle speeds of up to 650 m s^{-1} at flame temperatures of less than 2700 °C. Therefore the HVOF process is especially used to process WC/Co or Cr_3C_2/NiCr materials, because due to the high impact speeds the metal matrix of the spraying powder is converted into a "pasty" condition. Thus the hard materials are not dissolved in the metal matrix and the combination of a ductile and extremely wear resistant component re-

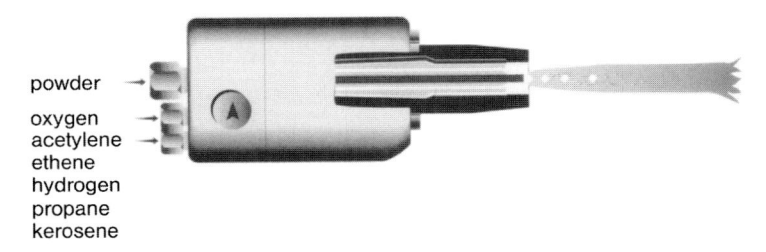

Fig. 5.8 Principle of high speed flame spraying (Linde AG).

mains. The high impact speeds cause low porosity and high adhesive strengths. The small interaction times of the particles with the relatively cold flames lead to little oxidation and gas admission of the spraying additive, so that high-quality MCrAlY layers can be produced at high deposition rates.

The temperature load of the component to be coated, which often requires the use of cooling devices, can be reduced for systems with axial powder supply by so-called flame limiters. The outside flame ranges, not necessary for the coating process, can be excluded by a screen. The diameter of the aperture plate is 4 to 6 mm smaller than the visible flame diameter. Using such devices reduces the heat entry into the substrate, for example, by approximately 55 % with an expansion nozzle diameter of 12 mm and a screen diameter of 8 mm. Advantages are: reduced cooling aggregate sizes, smaller quantities of cooling agents and, last but not least, influence on the residual stress in the manufactured layers.

High-velocity flame spraying can be applied for high-quality coatings, which can also be considered for design layout. Already the surface preparation by applying blasting to roughen the surface effects, as a consequence of induced compressive residual stress, a clear increase in fatigue strength and a small increase in creep resistance for the structural steel S355J2G3. HVOF WC Co 88-12 coatings permit a substantial improvement in both the fatigue strength and creep resistance of S355J2G3 (Fig. 5.9). Therefore, the coatings provide fulfillment of surface-specific requirements and structural strength contributions.

Due to the high noise level, which can exceed 140 dB(A), noise-insulated spraying rooms are necessary for high-velocity flame spraying. In comparison to detonation spraying, high-velocity flame spraying shows economic disadvantages due to the substantially higher gas consumption. However, the employment of detona-

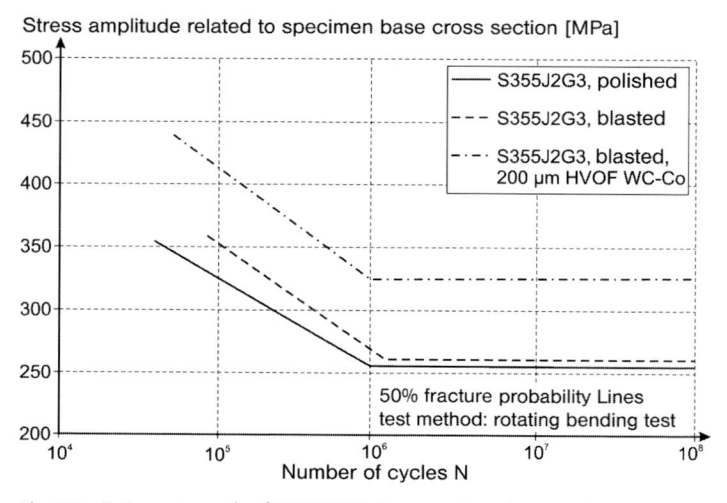

Fig. 5.9 Fatigue strength of HVOF WC–Co coated steel materials.

5.2.6.4 Cold Gas Spraying

Cold gas spraying (Fig. 5.10) is a relatively new spraying process, which permits the deposition of layers without substantial heating of the spraying powder. The particles are accelerated in a gas stream of moderate temperature (<600°C) to speeds of over 500 m s^{-1} and are then emitted at supersonic speed from a Laval nozzle. During the impact on the substrate the kinetic energy is transformed into heat energy, whereby the particles can become "forged together" to give a strong adhesive coating at almost theoretical density. Since the transformation of kinetic energy into heat energy takes place via plastic deformation, the spraying material must show sufficient ductility. Especially, metals with face-centred cubic crystal lattice such as copper, stainless austenitic steel or nickel-based alloys can be processed. Zinc is also very well processable because of its low melting point. In principle Cermet layers can also be manufactured. However, substantially lower hardnesses are obtained compared to HVOF sprayed coatings.

Since operating pressures of up to 35 bar with gas streams of 75 m^3 h^{-1} are usually used for cold gas spraying an adapted gas supply is necessary as well as a special powder feeder. Usually the process gas is heated in a spiral by resistance heating. Nitrogen is mostly used as process gas but admixtures of helium or pure helium are also used, especially when the spraying materials have high requirements regarding the heat transfer to the powder particles and the gas speeds. The high cost of helium has led to the development of plants where it is possible to recycle up to 90% of the process gas.

The substantial advantage of cold gas spraying is the avoidance of oxidation of the spraying additive. Cold-gas-sprayed copper layers reach 90% of the electrical and thermal conductivity of the cast material. Layers several centimeters thick can be produced and applied on polished glass as strong adhesive coatings. In consequence of the small particle stream, more complex structures are producible. This new technology will provide special application fields in the coming years.

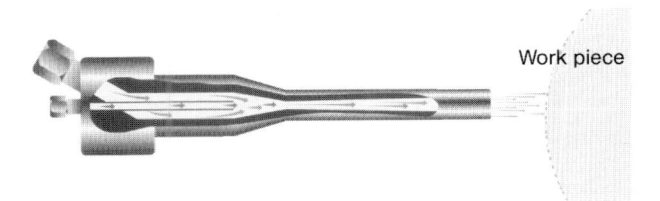

Fig. 5.10 Principle of cold gas spraying (Linde AG).

5.2.6.5 Arc Spraying

During electric arc spraying (Fig. 5.11) an arc is produced by an electric current, in which the wire-shaped spraying additive is melted. Usually two electroconductive, metallic wires, which function as electrodes, are continuously melted. Therefore a voltage in the range of 15 to 50 V is applied between the two wires which are guided at an angle to each other, whereby the distance of the wire tips continually reduces. The arc ignites when the wires are close enough and its heat melts the wire ends. An atomiser gas stream shears the melt of the wire ends and accelerates the drops onto the work piece to be coated. The size and speed of the molten droplets depend on the atomization conditions. High atomiser gas stream rates lead to fine, fast particles and thus relatively dense layers. At the same time the large specific surface of the molten droplets promotes oxidation reactions, so that generally the oxide content increases. In conventional plants deposition rates up to 20 kg h^{-1} and particle speeds up to 150 m s^{-1} are obtained. The diameter of the wires is mostly 1.6 to 3.2 mm but in high speed plants, wires up to 4.8 mm in diameter are used.

The temperature of the arc exceeds the melting temperature of the spraying additive material by far. Overheated particles result, which can cause local metallurgical reactions with the work piece surface, or lead to the formation of diffusion zones. This effect works especially for large particles with correspondingly large stored heat in the particle and leads both to good adhesive strength and good coherence within the layer. Due to the relatively small particle speeds and the large overheating of the molten droplets, the oxide content, as well as the gas loading are usually high. Usually the porosity of the layers is at least 2%.

Compressed air is usually used for atomization, but in some cases nitrogen or argon are also used. Turbulence in the atomiser gas stream substantially reduces the protective effect of inert gases. However, the application of an additional gas stream using ring shaped nozzles attached around the pistol nozzle, so-called shrouds, or by moving the spraying process into a controlled, inert atmosphere, permits a processing of high oxygen affine metals such as titanium alloys.

To manufacture high-quality layers it is necessary that the droplets impact the substrate surface under comparable conditions. Closed nozzle configurations, by which an additional gas stream in the radial direction towards the wire tips is applied, lead to a focused spray stream and are thus favorable for reaching homogeneous impact conditions of the molten droplets.

The direct use of electrical energy to melt the spraying wires makes electric arc spraying the process with the highest energetic efficiency. By employing wire-shaped spraying additives it is possible to prevent unmelted particles being thrown

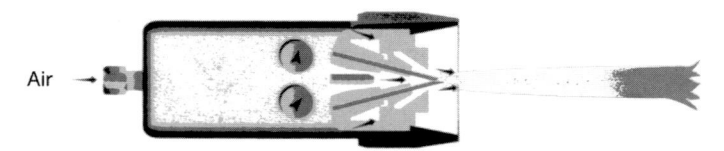

Fig. 5.11 Principle of arc spraying (Linde AG).

onto the substrate. Outstanding high coating rate efficiencies result, similar to wire flame spraying. Since the investment costs of arc spraying plants are relatively low, it represents the most economic thermal spraying technology, if the necessary layer properties can be obtained. The selection of spraying additive materials is limited to electrically conductive materials that can be produced as wires. Beside zinc-, aluminum- and copper alloys, steel, and nickel-based alloys are used. As for wire flame spraying, the application of filler wires extends the spectrum of use of the coating materials substantially.

Arc spraying in controlled, inert or reactive environments has gained no industrial importance so far and single-wire arc spray pistols are also still in the laboratory stage. In the latter the arc burns between the wire and the nozzle wall, which functions as a permanent electrode. Aligned anodically poled wires can be processed with an extremely small widening angle of the spray stream.

5.2.6.6 Plasma Spraying

5.2.6.6.1 DC Plasma Spraying

In DC plasma spraying (Fig. 5.12) an arc is ignited between a water-cooled copper anode, designed as a ring nozzle, and a similarly water-cooled, pin-shaped tungsten cathode. The gases flowing between both electrodes (Ar, He, N_2, H_2) are thereby dissociated, ionized and form a plasma jet, into which the spraying additive materials can be brought.

The outstanding advantage of plasma spraying is the high plasma temperature of approximately 6000 to 15 000 K, which makes it possible to process materials with very high melting points. Beside metals and ceramics, cermets can be used. The basic condition for processing is the existence of a molten phase. The performance of commercially available plasma spraying plants ranges from small burners with 10 kW performance, up to water-stabilised 200 kW plasma burners. Computer-aided control and regulation of the system parameters and the manipulation device, permit automatic and reproducible coatings in series production and of components with complex geometry. Increasing the spraying particle speed by us-

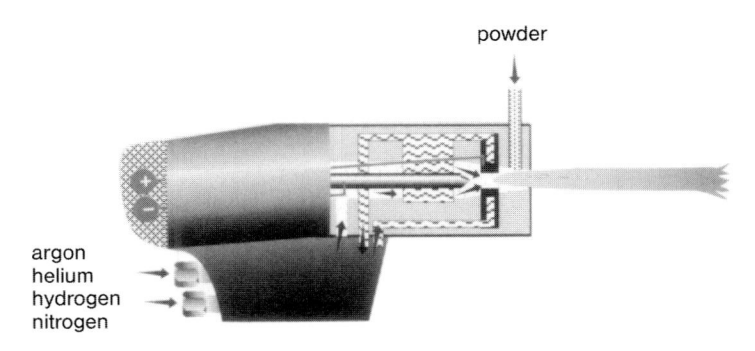

Fig. 5.12 Principle of DC plasma spraying (Linde AG).

ing high-speed electronic burners with special nozzle systems is the subject of current research. Higher gas speeds can be attained in comparison to conventional plasma spraying by, for example, application of a dynode. Investigations in this context show that the plasma gas speeds vary between 1500 and 3000 m s^{-1} at temperatures between 3500 and 6500 K.

Relocating the plasma spraying process into an inert low pressure environment leads to the vacuum plasma spraying process (VPS). The main industrial application is the coating of turbine blades of superalloys with MCrAlY alloys for hot gas corrosion protection. These layers serve as a primer for additional applied ceramic heat-insulating layers based on ZrO_2. The MCrAlY layers produced by VPS are characterized by a close, almost oxide-free homogeneous layer structure. With suitable adaptable supersonic nozzles, and internal powder injectors, improved layer qualities can be achieved, which are based on the increased particle spraying speed.

Conventional plasma burners show instabilities during the plasma production. These are caused by the movement of the arc, which causes on the one hand extreme noise and on the other hand uneven heating of the powder particles. This results in relatively unfavorable coating rate efficiency. A newly developed three-cathode burner (Triplex) does not show these disadvantages. Three arcs burn with a stationary anode root point and mutually stabilize. This results in a clearly reduced noise emission and a substantially increased efficiency.

5.2.6.6.2 HF Plasma Spraying

Production of HF plasmas (Fig. 5.13) is based on producing a high frequency magnetic field in an induction coil, which induces ring currents, provided that the plasma is already present, which run contrary to the direction of the primary current in

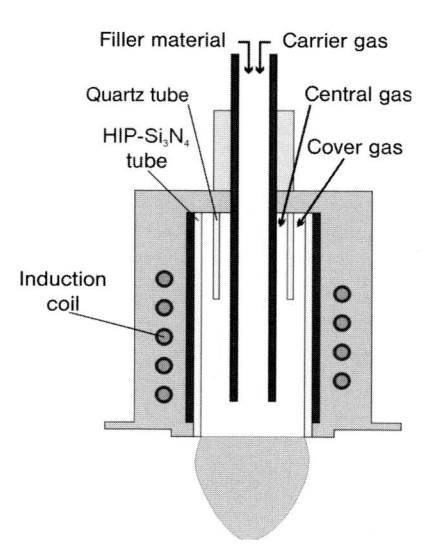

Fig. 5.13 Principle of HF plasma spraying.

the coil. The gases pass through the high frequency electrical field, become activated and dissociate almost completely into molecular gases. Electrodeless production of the plasma permits the application of a wide range of gases and makes it possible to adjust both reducing and oxidizing conditions. Beyond that, the supply of gaseous base materials for plasma-synthetic processes is possible within plasmatrons. Usually the generator performance is between 25 and 200 kW.

The large volume and relatively slow plasma jet permits high material throughput, but only relatively small particle speeds (< 50 m s^{-1}). Usually the force of gravity is used to provide a further contribution for accelerating the spraying powders. Due to the long dwell times in the plasma jet, in which temperatures up to 10000 K exist, relatively rough powders can be completely melted and reactions with the plasma gases are effectively used. However, the layers are relatively porous because of the small impact speeds. Supersonic nozzles for HF plasmatrons are at present being evaluated in laboratory conditions. However, the jet width is substantially reduced by the use of these nozzles.

The main disadvantage of HF plasma spraying burners is the limited handling conditions, since the force of gravity exerts a major influence on the particle trajectories and the injection of the electricity requires a fixed connection to the generator. Furthermore, strong electromagnetic fields develop, which prevent the application of electronically regulated handling systems. Thus the main field of application of HF plasma burners is in the area of spheroidizing and material analysis which is based on emission spectroscopy after complete evaporation in the plasma jet.

Vapor HF plasma beams are very well suited for the synthesis of layers. The axial injection of the base materials within the plasma burner prevents large material wastes and makes the application of liquid precursors possible; e.g. superconducting $YBa_2Cu_3O_{7-x}$ layers can be made by plasma synthesis (thermal plasma chemical vapor deposition, TPCVD).

Suspension plasma spraying (SPS) represents a compromise between the TPCVD process with its outstanding possibilities for material synthesis with custom-made layer structure and morphology, and the HF vacuum plasma spraying of powders with its very high deposition rates. The starting points for fast coating are solid phases in a suspension injected into the plasma, while the liquid phase of the suspension as well as the plasma gases ensure the reactivity. This technology has already been used successfully to make bioceramic hydroxyapatite layers of suspensions with nanoscale particles.

5.2.7
New Applications

Because of recent research and development some substantial new process developments and improvements in thermal spraying technology, especially for opening and extending traditional areas of application, have been developed. In electrotechnology, new areas of application result from manufacturing copper-coated components. Due to the small layer thicknesses needed the application of copper layers is, for economic reasons, limited to the well-known thin film process PVD (physical vapor deposition) and electroplating. However, advances in the area of plasma spraying allow the production of layers with thicknesses between 40 and 250 µm. Vacuum plasma sprayed copper layers show a conductivity comparable to conventional copper with sufficient layer adhesive strength.

Vacuum plasma spraying and plasma spraying with increased process pressure in relation to atmospheric pressure allow the manufacture of new materials. Due to the difference in the material properties of melting metallurgically and/or powder metallurgically manufactured materials new possibilities are opened up. Investigations of the mechanical-technological properties of sprayed materials in comparison to materials from other production techniques clarify the potential of thermal spraying technology for the manufacture of new materials. Vacuum plasma spraying is also an effective technology for applying cavitation- and erosion- resistant layers based on nickel aluminum. Vacuum plasma sprayed NiAl alloys appear suitable, due to their high melting point, high Young's modulus, low density and outstanding oxidation resistance, for applications at high temperatures. Vacuum plasma sprayed oxide and pore-free layers as well as moulded paddings of NiAl alloys show the same material structure as the base powders used as the spraying additive.

Nanostructured materials produced by thermal spraying have excellent properties regarding mechanical strength, hardness and ductility, whereby the properties of conventionally processed materials are clearly exceeded. The cold gas or high-velocity flame spraying with nanostructured powders produced by high-energy milling and/or mechanical alloying is a suitable procedure to produce dense layers as well as compact bodies of unconventional materials.

Thermal spraying technology has been introduced for the interior coating of cylinder bores in engine manufacture in the automotive industry. For this application a special burner had to be developed, which allows fast coating of cylinder bores with sizes less than 70 mm in diameter with small heat entry in the Al–Si die casting engine block as well as a reliable, long life of the burner components. It has been possible to develop coatings and liners, with superior tribological properties compared to gray cast at reasonable material costs, as well as a holistic technology which fulfils all necessary pretreatment and processing steps. The Rota Plasma System developed by Sulzer Metco rotates at a constant distance from the wall in the cylinder bore in a fixed position engine block. The grit inserted in the automated blasting process is completely removed before coating. Not only are composite powders with a steel matrix and solid lubricants used as coating materials but also low alloyed carbon steels. Since the plasma spraying process takes place at atmospheric pressure, the particles only partly oxidize. The oxides show good lubricating properties during engine operation. Open pores work as reservoirs for liquid lubricants, so that altogether smaller friction losses are obtained compared to gray cast iron bushes. To obtain these properties a careful optimisation of post-processing by honing is necessary.

Besides the ability to be integrable in production, the range of thermal spraying processes will depend strongly on the success of adjusting specific and reproducible material structure and morphology. Parameter studies for solidification clarify that for thermal spraying, the boundary conditions can be adjusted in such a way that a melt film of four to five times the thickness of particles on the substrate can develop. It is necessary to control the solidification of this approximately 10 to 50 µm thick film in such a way that an aligned structure develops. Thus the potential of materials with anisotropic properties can be fully used. Electromechanical or electrochemical surface properties can be optimized by consideration of the crystal orientation. Furthermore new possibilities in the use of thermal spraying as a production method for semi-finished material e.g. shape memory alloys arise.

5.2.8
Quality Assurance

Special importance has been given to quality assurance over the last few years, especially in Europe due to recent upgraded laws regarding product liability. Moreover, the joint European market leads to increased competition for companies and their products. Due to the high certification fees most companies and institutes, active in the sector of thermal spraying, have formed a community quality for thermal spraying, called GTS. Their quality guidelines are based on the DIN ISO 9000 to 9004 with application to the specific characteristics of thermal spraying.

In principle it applies to methods and instruments for on-line control and to control of the coating process for quality assurance and the development of nondestructive methods to investigate the layers. State of the art is controlling and adjusting the process parameters on the input of a spraying pistol by adjustment of nominal and actual values. Monitoring of these parameters is also state of the art. For example, to produce semi-finished material and/or coatings with special structure and morphology it is necessary to guarantee high process capability. This requires on-line control of the process parameters with the aim being to reproduce the energetic conditions (temperature and kinetic energy) resulting from the interaction between plasma/hot-gas jet and the particles before impact on the substrate. Therefore it is necessary to measure process parameters in the interaction zone during the entire coating process. These process parameters allow correlation of the particle speed and temperature.

Such measuring systems must be portable and adapted to a spraying pistol. Optical emission spectroscopy offers the possibility to adapt an optical system to the burner and to transfer the signals via fiber-optic cables to a spectrometer. Spectroscopic investigations show a significant emission intensity change for characteristic wavelengths of the process gases and the elements of the coating material by changing the process parameters. By local dissolved emission spectroscopy, deviations in process conditions can be sensitively monitored. Within a quality assurance system quality control plays an important role and requires the availability of different testing methods during the production process. Classical metallographic and material testing methods are destructive testing methods. To affect the coated component as little as possible by such destructive testing methods, a system to check the smallest samples possible has been developed. This system permits sampling for microstructure investigations. The samples taken have thicknesses in the range of only 0.8 to 2.5 mm, so that the damage is small and repair at the sampling area can often be omitted.

However, the constantly growing technical and economic importance of thermal spraying requires increasing application of nondestructive testing methods. The application of nondestructive material testing methods is still in the development stage. In this context, ultrasonic procedures, optical holography, eddy current technology and thermography should be mentioned. Another new method for on-line quality control in the development stage is based on sound emission analysis. The sound emitted by the spraying particles impacting the surface, processed with the help of the most modern computer technology, supplies information about the diameter, the mass, speed and viscosity of the particles. In this way constant control of the structure of the sprayed coating and corresponding parameter regulation is possible. The use of neural networks and fuzzy logic offers the potential for process control. Process controllers can be adapted to control and regulate the process parameters as a function of the desired coating characteristics.

5.2.9
Environmental Aspects

Thermal spraying is now a well known coating technology, which is characterized by a high environmental compatibility. Spraying dust and uncaptured spraying particles are extracted in special plants and do not get into the environment. In general there are no problems regarding the disposal of the waste materials. However, so far, little information is available concerning the endangerment of the plant operator during the spraying process. In this connection increasing environmental awareness has led to the development of new measuring methods, which help to overcome the information deficit. In particular the direct effects on operators, as well as new methods for the minimization of emissions, are to be examined. Such investigations become more important with the increasing requirements of environmental law. Early research work has examined different influences on the environment by atmospheric plasma spraying. The results show that apart from the dust load on changing gaseous connections, radiation emission (ultraviolet to infrared) has to be expected as well as high noise emissions.

5.3
Cladding

Cladding is characterized by the application of a strong adhesive layer on a substrate surface by the fused flux of the welding filler and also, partially, the substrate material. A metallurgical bonding between coating and substrate is obtained by partial melting of the substrate surface. The degree of the developing mixture of base and coating material is called dilution and leads to changed properties of the coating compared to those of the additive weld metal.

Cladding is characterized by outstanding adhesion of the layers, nonporous layers and an excellent heat transfer of the coating into the substrate. It produces, apart from the repair of worn surfaces, armouring (wear protection) or corrosion protection of component surfaces, as well as buffering (intermediate layers for the gradual adaptation of the substrate properties and cover layer). It is not possible to differentiate clearly between armouring and corrosion protection. Depending on the process used only small restrictions exist regarding component geometry. The coating thicknesses are single layered from 1 to 5 mm. Line tracks as well as pendulum tracks up to 50 mm width can be realized. By multilayer technology, laminar coating and form-giving cladding (cladding near to the final geometry up to "free" modeling of a component) is possible. The coatings are self-supporting and mechanically post-processable (hard machining or grinding).

5.3.1
Coating Material

Additive materials can be supplied to the cladding process as powder, wire or band, depending on the applied process. The spectrum of the processable materials for powders is very large, similar to that of thermal spraying. However, attainable surface and melt efficiencies are generally significantly smaller. The metallic powders used for cladding are mostly larger, particle diameter between 60 and 200 µm, than for thermal spraying. Depending on the area of application, substantially rougher hard material powders are used for reinforcement of the metal matrix. Besides melt atomization of metallic powders, agglomeration and sintering are also used to produce composites. Due to their spherical form, the fluidity of the relatively coarse powders is outstanding.

Wire-shaped weld additives usually have diameters between 1 and 8 mm, while the width of bands can exceed 200 mm. Solid wires and bands provide the shape with the lowest price for weld additive materials. By processing core wire, the composition of the coating can be influenced. By the use of filler wires or bands the spectrum of the coating materials can be substantially extended, similar to thermal spraying. By using sintering band electrodes still higher degrees of freedom exist regarding adjustment of the chemical composition of the weld additive. A further advantage exists in the improved homogeneity compared to filler bands. Sintering band electrodes are manufactured by rolling powders into coils of green bands followed by substantial thermal treatment in an instantaneous furnace to increase the mechanical strength by sintering. The employment of powdered source materials and the thermal treatment, usually in an inert atmosphere, lead to high manufacturing costs, so that the use of sintering band electrodes is only economic if small amounts of an expensive material are needed and there exist high requirements for the layer homogeneity.

Besides stainless steel, which contains substantial amounts of molybdenum, titanium or niobium, depending on the area of application of the coated component, to prevent pitting corrosion or intergranular corrosion, and aluminum, nickel-based alloys are also used for cladding. The aluminum content is high, especially in applications for oxidation protection. Copper alloys are also sometimes applied on the same substrates.

Coating materials for armouring can be classed as single-phase or multiphase materials. Classical austenitic manganese steels, e.g. X 110 Mn 13, are single-phase alloys. The cooling rate has a substantial influence on the structure of these alloys. Very fast cooling from approximately 1050 °C to room temperature produces a purely austenitic structure with a hardness value of only 200 HB. Impact or compression stress occurring during use leads to an increase in cold working and hardness up to approximately 550 HB in a surface layer less than 1 mm thick. The cause of this hardening is the blocking of sliding planes by fine martensite. If cold working does not occur due to a lack of operating stress, the wear resistance of the austenitic manganese steel corresponds approximately to that of normal constructional steel.

The high cooling rates of cladding of this material, which need to be guaranteed, require a special procedure. Therefore it is usually armoured with thin electrodes, using small amperages and short arc stringer bead technology. Welding in a water pool is preferable for small components. Since higher temperatures can cause brittle carbide precipitations, this material requires low application temperatures, below 300 °C. Exceeding this temperature is not permissible for plastic deforming in service or for cutting treatment, since even small amounts of heat can only be taken away extremely slowly due to the poor heat conductivity of armouring.

Martensitic chrome steel with mainly 2% carbon and up to 18% chrome is also a single-phase cladding material. The weld material, which is air-hardened from the weld heat, exhibits sufficient ductility after an annealing treatment. Depending on the carbon content, hardness values up to 500 HB can be obtained. For this reason only post-processing by grinding is possible. These coating materials can find use where strong abrasion and impact stress arise at the same time. In order to minimize the risk of hardness crack formation, slow cooling in hot sand or in the furnace must be undertaken for some alloys after welding or annealing.

Multiphase cladding materials show a relatively ductile matrix, into which hard materials such as carbides, boride and silica are embedded. Alloys based on iron, nickel or cobalt with a high content of hard material former, like chrome, tungsten, molybdenum and vanadium and the metalloids carbon, boron and silicon belong to this group (Table 5.1).

Tab. 5.1 Classification of hard metal alloys.

Matrix	Hard metal	Metal oxide	Matrix element	Hard material
Fe	Cr, W, Mo, V	C, B, Si	Mn, Ni, Co	$M_{23}C_6$, M_7C_6, M_6C, M_3C, (M–B, M–Si)
Ni	Cr, W, Mo	B, Si, C	Fe, Co, Cu, Mn	Ni_3B, CrB, Ni_3Si, (M–C)
Co	Cr, W, Mo, V, Nb, Ta	C, B, Si	Ni, Fe, Cu, Mn	$M_{23}C_6$, M_7C_3, M_6C, (M–B, M–Si)

During and after the solidification of the coated melt, carbides, preferably of the type M_7C_3, are formed. Additionally carbides of the type M_6C and $M_{23}C_6$ can develop, dependent on the carbon, tungsten and iron content. Thus, the zones influenced by dilution represent the places in which these carbides preferentially form. In nickel-based materials, which are additionally alloyed with boron and silicon, phases of the type Ni_3B, CrB and Ni_3Si form.

Dilution, usually with low-alloy base material, is decisive for the kind and quantity of the precipitating hard phases. Since the wear resistance quality depends on the coating, this knowledge is of importance. As an example the relationship can be discussed on the basis of Fe–Cr–C-based alloys. The content of precipitated carbides is determined by both the ratio of the elements Cr and C and the absolute amount of the coating. With increasing dilution these contents decrease, leading to a decreased amount of carbide. This allows adjustment of the amount of carbide from 15% to 40% by appropriate welding processing using dilution. Hardness values are between 40 and 70 HRC. On increasing the amount of layers the influence of dilution decreases.

From this, result various possibilities to vary the application properties of multi-phase coatings by alloying variants. The development of materials is not yet completed, but recent research activities aim to increase the carbide amount in the armouring in order to increase the wear resistance of such coatings.

Coatings with tungsten carbides precipitated from the melt do not obtain the excellent wear protection properties that are obtained for the same pseudo-alloys. To manufacture pseudo-alloys, NiBSi and NiCrBSi alloys have attained the greatest importance as a metal matrix in combination with tungsten melt carbides. On steel substrates hard material of 60% volume can be obtained, without crack formation, by plasma powder cladding (PTA). The hard material content will be lowered if the coated components are exposed to impact stress to reduce the residual strain and to increase the portion of the ductile metallic phase. Attention has to be paid to the selection of the welding parameters so that these carbides do not go into solution in the melt causing the matrix to become brittle. With the help of appropriate processing, hardnesses of up to 70 HRC can be realized in the coating.

Besides the carbon content, the tungsten melt carbide content and its microstructure have a substantial influence on the wear resistance of the armouring. Fine feathered structures, which can be produced by keeping the carbon content between 3.9 and 4.2 vol%, avoiding iron impurities and by a small thermal influence of the welding process, give the highest hardness and best wear resistance. Finally, the shape of the carbides can also affect the tribological properties. Because of their metal matrix, melted tungsten carbide spheroids are less strongly dissolved due to their smaller surface compared to block carbides, which are manufactured by melting and breaking. Furthermore, an internal notch effect of the carbide edges is avoided and the spheres can allow a better gliding of the counter bodies. However, an increase in the service lives could not be proven by using the substantially more expensive spherical carbides for each application.

Extensive investigations of dissolving WSC in the fused material (weld) have been carried out, especially for the development of vanadium carbide-containing wear protection alloys, since vanadium carbide shows metallurgical advantages. If it is melted by heating and dissolved in the melt, it is re-precipitated finely dispersed. Here no vanadium-rich mixed carbide forms, so that no unwanted changes in the hard phases arise. The high carbide nucleus density leads to a fine structure.

This fundamental knowledge has led to the development of a material system, which can be used for wear resistance purposes, and also for combined wear and corrosion resistance. The reason for the good wear resistance is vanadium carbide, which is embedded into a high-strength matrix. Iron-based alloys of different composition similar to steel are used as matrix materials. To ensure a martensitic hardening from the welding heat, all variants contain chrome and molybdenum. According to the required properties the contents are specified, and/or further elements alloyed. To guarantee a good weldability the alloys contain approximately 1% silicon and 1% manganese. For martensitic hardening of the matrix material 0.4% C should be available. The so-called "free" carbon content guarantees work hardness above 60 HRC. Lack of carbon, less than 0.4%, leads to the dissolution of vanadium in iron and the matrix solidifies ferritically. Carbon surplus leads to a stable austenitic residue and the material becomes brittle.

To achieve a significant improvement in the wear resistance compared to conventional tool steels, the alloys contain at least 10% vanadium. Up to an upper limit of 20% vanadium all material compositions are atomisable. Practical investigations indicate that vanadium contents from 10% to 18% are necessary to use the full potential of the property spectrum of alloyed materials.

For further improved wear resistance, 4 to 6% chromium, 1 to 2% molybdenum, approximately 1% silicon and 1% manganese is added to the basic alloys. Thus a safe martensitic hardening is ensured from the welding heat with all technically relevant cooling processes, as well as safe processing by welding. The carbide content is from 17 to 20 vol% VC for alloys with 10 to 12% vanadium and 26 to 31% vol% VC for alloys with 15 to 18% vanadium.

An increase in the amount of chromium to more than 13% at similar contents of the alloying elements as specified above, leads to chrome-martensitic variants, which are corrosion resistant to organic acids and atmospheric loads. Thus small nickel contents positively affect the plastic properties, so that the typical crack susceptibility of these materials is reduced. For precipitating chromium carbides the carbon content has to be increased.

Nickel contents from 6 to 8% cause an austenitic structure, so that the materials are highly corrosion resistant and possess an excellent wear resistance quality due to the high amount of vanadium carbide .

Powder mixtures containing up to 80 vol% vanadium carbide are realizable. However, in contrast with tungsten carbide the variation options regarding the carbide grain size are limited, since vanadium carbide can only be produced by carburising fine-grained vanadium powders and subsequent sintering. At present there are available sinter carbides in different ground sizes and agglomerated fine grain carbides which can be used, depending on the application.

5.3.2
Substrate Materials

Because the cladding procedures require melting of the component surface, only metallic substrate materials are used. Structural steels are most commonly used together with low and high alloyed steels and nickel- and copper-based materials. In exceptional cases partially protective layers are applied on components of aluminum alloys.

5.3.3
Cladding Processes

5.3.3.1 Autogenous Cladding

Autogenous cladding is the oldest cladding procedure. Usually a hand-guided autogenous burner is used as a heat source for the local melting of the component surface. Acetylene is used almost exclusively as fuel gas due to the high flame temperatures and thus the fast melting of the component surface. The weld additive can be injected in the flame of the burner in the form of powders or by external supply into the weld pool. Furthermore there exists the possibility of using wires, not only solid wires but also filler wires or cladded wires.

The equipment technology for autogenous cladding is extremely cheap and very well suited for on-site application. It is applied, for example, for armouring plough scrapes in agriculture. Because of the relatively strong dilution with the base material and the small surface and melting efficiency, this process is only rarely used.

5.3.3.2 Open Arc Cladding (OA)

Open arc cladding melts both the base material and the weld additive by a burning arc between them. The process can be both manual with hand held bar electrodes or automated with a continuous supply of cored wire electrodes. In applying bar electrodes, the covering of the core is of extreme importance, while during the use of filler wires the filling has a substantial influence on the process stability as well as the composition of the coating. The technology is characterized by the technical requirements of a small supply, since only a sufficiently dimensioned power source and a wire feed unit are necessary.

Using bar electrodes beside the core wire the covering melts by exposure to the heat of the arc and forms a slag, which covers the molten wire end as well as forming drops partly or totally on the surface. Metallurgical reactions at the interface between slag and wire have influence on the surface tension and thus the drop formation with reference to replacement frequency and mass. Electrodes with base covering promote the formation of large drops and short-circuits take place more rarely than with electrodes with titanium oxide covering, which show a fine dropped transition of thick covering. Besides the composition and strength of the electrode covering, the length of the arc, the welding current and the power source characteristic have influence on the material transition. Increasing amperage causes finer drops and decreases the possibility of short-circuits. Welding electric rectifiers with high dynamics cause higher drop frequencies than those with low dynamics.

Using filler wires, the arc attaches at the ring-shaped, outside metal body and moves along the edge, which can lead to an unstable process. The drops formed are cut and change with different flight paths into the weld pool. Generally the filling forms a more or less marked cone of the transforming material. Besides the wire structure, the compositions of the filling and the filling degree also have substantial influence on the welding parameters.

Automated open arc cladding is used in particular for armouring components with large surfaces, for example in cyclones in the cement industry, for crowns in oil exploration, and for large shovels in mining and in building material production.

5.3.3.3 Underpowder Cladding (UP)

During underpowder cladding (Fig. 5.14) an arc burns between the work piece and a continuously melting electrode. Thereby the arc burns in a cavern, which is covered by the partly melted powder and kept upright by the gas pressure formed inside. The molten powder portion solidifies to a slag layer on the seam surface and mostly bursts off automatically. The unmelted portion can be drawn off, cleaned and used again. The material transition is influenced by the welding current, the power source characteristic and the powder. The dumping height, the chemical compositions and the kind of powder production are substantial influencing variables, because they determine acceptable current load, gas permeability and the layer characteristic by pick-ups. For example, a too strong silicon pick-up for applications in some corrosive media can lead to the transition from uniform surface corrosion to selective pitting corrosion.

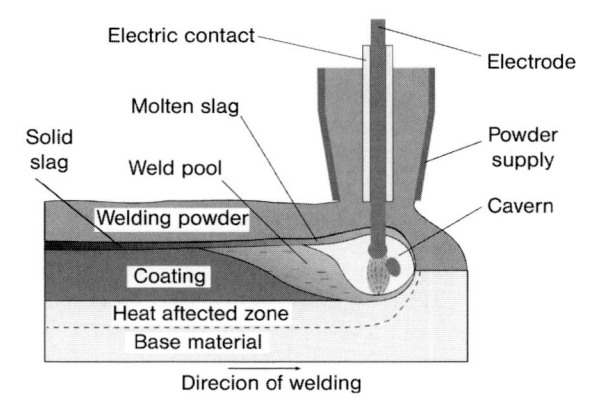

Fig. 5.14 Principle of under powder cladding (ISAF, TU Clausthal).

Simultaneously supplied wires, or bands, whose width can exceed 200 mm, are used as electrode material. By applying wide bands the arc is usually led with the help of a magnetic deflector along the lower band edge. Notably, double wire and band technology have attained industrial importance for armouring as well as for corrosion protection of structural steel units with nickel-based alloys.

Dilution of less than 10% can lead to process-specific binding defects. Thus the dilution with 13% to 40% is relatively high. To ensure corrosion resistance, three layers with a thickness between 5 and 8 mm have to be applied. This requirement causes relatively high costs for expensive welding additives. Regarding component geometry the application of the process is limited to rotationally symmetric bodies with minimum diameters of 150 mm.

Substantial process advantages are high deposition rates up to 40 kg h^{-1} and the high surface performance up to 9000 cm^2 h^{-1}. With approximately 175 J mm^{-2} the surface energy is less than for electro slag cladding.

5.3.3.4 Resistance Electro Slag Welding (RES)

Resistance electro slag welding (Fig. 5.15) is mechanized and assigned primarily for coating components with larger (wall) thickness, because the surface energy of about 190 J mm^{-2} is relatively high. Only the use of band electrodes, whose width is up to 180 mm, is of economic importance. The melting electrode is continuously supplied. Comparing the arc process with this resistance fusion welding the resistance heating of the molten slag pool serves as the heat source. The welding current can exceed 3000 A.

The welding powder must have a sufficiently high resistance for the molten slag pool to produce the necessary heat but the resistance must not be allowed to cause inadmissibly high current flows over the weld pool, to avoid short-circuits or an ignition of the arc. Usually powders with CaF$_2$ contents from 50 to 90 wt% and Al$_2$O$_3$ contents between 10 and 40 wt% are applied.

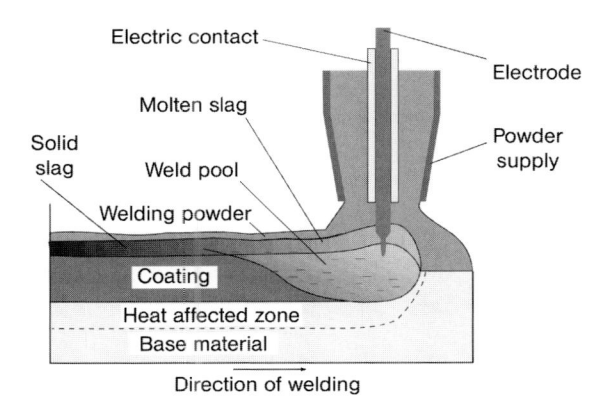

Fig. 5.15 Principle of resistance electro slag welding (ISAF, TU Clausthal).

Resistance electro slag welding has deposition rates of 15 kg h^{-1} and surface performances of approximately 5000 cm^2 h^{-1} are usual. Single layer cladding is carried out with thicknesses from 4 to 5 mm. Often in consequence of the dilution of approximately 15%, a two layer coating is necessary for cladding steel with nickel-based alloys to give sufficient corrosion protection. The process is characterized by a simple and robust principle and a uniform penetration.

5.3.3.5 Metal Inert Gas Welding

Metal inert gas welding (Fig. 5.16) works like the plasma jet-based process with a protective gas for the arc, the material transition and the weld pool. Thereby the gas does not form under exposure to heat from weld additives, but is externally supplied. Special gas-sensitive materials and large weld pools require the use of a protection chamber, inert gas shower or gas-trails to keep the inert gas in the welding area longer or to add inert gas into the solidifying and cooling pool.

With this process the arc melts partly a continuously supplied, current-carrying electrode and partly the surface of the substrate material. Both the wire and the material transition into the weld pool, which take place in the arc, are shielded from the surrounding atmosphere by a coaxially supplied process gas. Depending on the material combinations used, inert (metal inert gas, MIG) or active (metal active gas, MAG) gas mixtures are used.

Usually the burner is guided perpendicularly and oscillating to the direction of welding to obtain a constant energy input. Dependent on processing, the surface energy is between 160 and 185 J mm^{-2}. With the help of the cold wire technology, where an additional dead wire is supplied to the arc within the melting pool area, the heat entry into the component can be decreased. Dilution can be limited to 5% at the same time. Dilution of approximately 15% is mostly adjusted to guarantee a continuous binding. Similar to the cold wire technology a reduction in the heat entry into the component can be obtained by injecting powders into the weld pool.

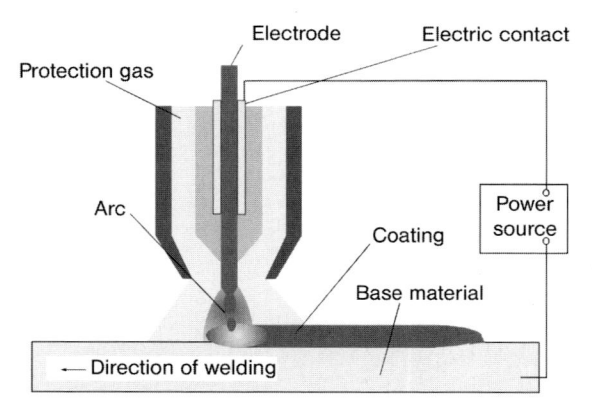

Fig. 5.16 Principle of metal protection gas cladding (ISAF, TU Clausthal).

Heat entry can be also reduced by the electric power supply. The pulsed arc technology works with a relatively low base current and in intervals with overlaid current peaks. During the base current phase a drop at the electrode tip, which has not reached the critical dissolution mass, is melted and is relieved by the Pinch effect in the high current phase. Besides the smaller heat entry due to the reduced base current, improved process stability is obtained due to defined material transition.

Metal inert gas cladding is characterized by a simple, robust equipment technology as well as an unproblematic automation. Cladding up 4 to 8 mm thick coatings means obtaining surface performances between 2000 and 3000 $cm^2\ h^{-1}$ and deposition rates of 8 to 9 $kg\ h^{-1}$. This process is applied, for example, for armouring connecting components for drill pipes in the deep drilling technique.

5.3.3.6 Plasma MIG Cladding

During plasma MIG cladding (Fig. 5.17) a wire electrode in a MIG arc is melted. The wire is supplied in the burner centre, whereby over a separate power source a concentric plasma arc burns between the work piece and a water-cooled cathode ring, which functions also as a nozzle. Applying two independent power sources and thus two independent heat sources allows adjustment of the heat control. Furthermore, the plasma arc stabilizes the MIG arc, whereby high welding currents and thus high deposition rates can be obtained with splash-free material transition. Weld pools of up to 40 mm width are possible with a small penetration and small dilution, up to approximately 10%. Due to a forward-moving cathodic cleaning via the plasma jet, the requirements for cleaning the component surface are small.

Plasma MIG cladding is characterized by high deposition rates up to 25 $kg\ h^{-1}$ as well as surface performance of 5000 $cm^2\ h^{-1}$ with relatively small surface energies of 150 $J\ mm^{-2}$. Layer thicknesses are usually 4 to 6 mm. Typical applications are

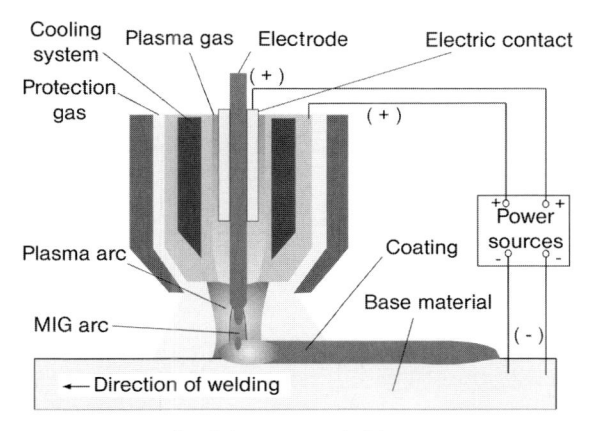

Fig. 5.17 Principle of plasma MIG cladding (ISAF, TU Clausthal).

armouring with tough or hard steel, cladding with corrosion resistant steel and cladding of copper- or nickel-based alloys of the same kind. However, due to relatively high machine effort the process has not found widespread application so far.

5.3.3.7 Plasma Powder Transferred Arc Welding (PTA)

During **PTA** welding (Fig. 5.18), a plasma jet produced by a nonmelting electrode is used as a heat source. Two arcs are used, which are controllable over separate power sources. The so-called pilot arc burns after ignition by high frequency between a rod-shaped cathode and an anodically polarized ring nozzle. The pilot arc is used for ignition of the transferring main arc between work piece and pin cathode. The thermal load is kept relatively small by a negative polarity.

Plasma powder transferred arc welding uses powdered weld additives, which are supplied to the burner by a transfer gas. Besides an internal powder supply, where the powder is injected directly in the area of the plasma nozzle into the plasma column within the burner geometry, an outside powder supply can be used, where the powder is supplied to the plasma jet outside the burner geometry. During internal supply the powder is absorbed into the plasma jet due to the injector effect on the powder/gas stream by the thermal expanding plasma gases. The powder is accelerated to high speeds. Melting of the weld additive is adjusted by the pilot arc and melting of the base material is controlled by the transferred arc.

With the outside powder supply a shorter retention period in the arc is obtained and thus there is a smaller heat transfer to the weld additive. On the one hand a smaller thermal load on the weld additive occurs and temperature sensitive materials can remain in their original structure; on the other hand the risk of binding defects and/or too high dilution exists. Generally a smaller coating efficiency is obtained with the outside supply than with the internal supply. A combination of internal and outside powder supply is partly realized. When applying Ni–B–Si

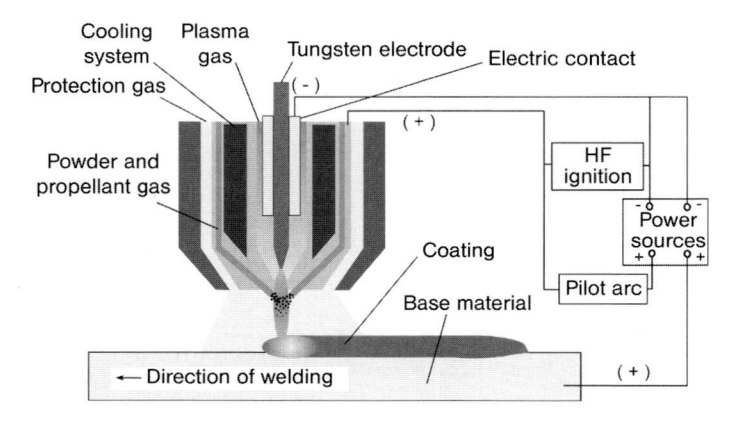

Fig. 5.18 Principle of plasma transferred arc welding (ISAF, TU Clausthal).

tungsten fusion carbide composite layers, the internal supply can be used for the metallic component and the outside supply for the carbide hard material. Thus the thermal load of the hard material is minimized and complete melting of the metallic component is guaranteed, resulting in metallurgical strongly merged hard materials with almost unchanged structure, which results in optimal wear resistance.

Inert gases, i.e argon and helium, are used for both the plasma gas of the pilot arc and the powder transportation gas exclusively. The inert gas can also contain portions of active gases to influence the pool viscosity and flow behavior, for example by the reducing effect when using hydrogen additions.

Plasma powder transferred arc welding is characterized by a high degree of automation and high surface performances up to 5000 cm² h⁻¹. Deposition rates of more than 12 kg h⁻¹ are possible. A further advantage is the high flexibility concerning the attainable layer thickness, which can be adjusted between 0.5 and 5 mm. Finally, the process is characterized by the possibility of producing defect-free coatings and interfaces with dilution of only 5%. The main application field of this process is coating of components which are subject to high corrosive and/or wear loads. Typical application areas are armouring valves in engine construction, highly stressed zones of tools in the plastic processing industry as well or protective layers on sealing surfaces of fitting parts in the (petro)chemical industry.

5.3.3.8 Plasma Hot Wire Cladding

Plasma hot wire cladding (Fig. 5.19) offers a separate adjustment for melting the base material surface by a plasma burner with transferred arc and melting the additive materials by the hot wire technology. The weld additive is usually supplied to the weld pool in the form of two crossing wires. The contact of the wires and the short-circuit over the weld pool causes a current flow of the contact tubes over the free wire ends, and thus a resistance heating to the solidus temperature of the weld additive metal. To minimize the magnetic field influence, alternating current sources are used, which are equipped with voltage limitation to avoid arc discharges (flashover). The residual energies required to melt weld additives are

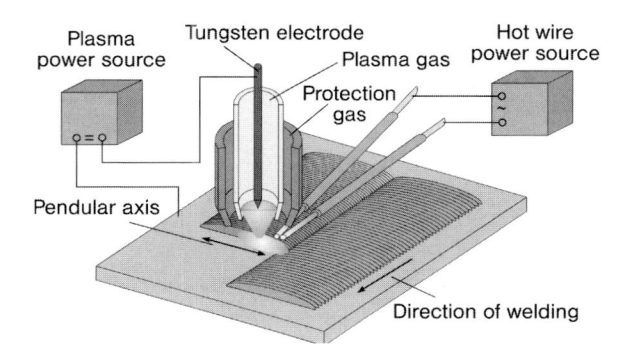

Fig. 5.19 Principle of plasma hot wire cladding (ISAF, TU Clausthal)

small, and therefore the thermal treatment of the substrate and the weld additive is mostly uncoupled. Besides wires, band electrodes can also be used as weld additives. The combination of two different wires influences the coating within a wide range.

The substantial advantage of plasma hot wire cladding exists in the combination of high deposition rates up to 30 kg h^{-1} and the performance of 7000 cm^2 h^{-1} at dilution of only 5 % for defect free coatings. Coating thicknesses are adjusted between 2 and 7 mm. The surface energy can be limited to 95 J mm^{-2} and a very small heat affected zone results. Finally, this process permits large automation.

Small dilution with relatively thin coatings permits substantial savings and makes the process economically attractive for processing expensive weld additives. For two-layered implemented coatings the same corrosion resistance properties as with rolling material or rolling cladding are obtained. Within the area of apparatus and reactor engineering, for example claddings of CrNi steel, copper- and nickel-based alloys are manufactured. Beyond that, applications exist for corrosion protection in the area of flue gas desulfuration, waste incineration, offshore industry and nuclear technology.

5.4
Summary and Outlook

Coating technologies are nowadays a decisive factor in realizing new products and making technological progress. The main focus of the applied layer systems has so far been in the range of wear and corrosion resistance or thermal insulation. Layers of special physical, electrical, chemical and electrochemical characteristics find increasing interest. Such layer developments are extremely complex and require high experimental effort. There is a need for knowledge on thermal spraying and cladding in order to predict the parameters of the coating process and the technological marginal conditions for the equipment technology and material used, regarding the desired microstructure and minimization of the experimental effort.

The goal of further scientific and technological work is the real material description of the total coating process with regard to manufacturing and material-specific characteristics and parameters. The combination of measured process variables and thermodynamic values of the processed composite layer with the simulation of the structure and morphology of the sprayed coating is crucial. It will then be possible to optimize the structure and morphology of the layers as a function of the thermodynamic material properties and to obtain process setting parameters at the same time, which are system-dependent. The experimental effort can be limited to the verification of the calculated process setting parameters, which will lead to a substantial cost and time saving in the development of new layer systems.

References

1 J. Wilden, H.-D. Steffens, U. Erning, Thermisches Spritzen, *Jahrbuch Oberflächentechnik*, Band 51, **1995**.

2 J. Wilden, M. Brune: Thermisches Spritzen, in *Moderne Beschichtungsverfahren*, Steffens (Ed.), Wilden, DGM, **1996**.

3 N. N., *Thermische Spritzverfahren für metallische und nichtmetallische Werkstoffe*, DVS-Merkblatt 2301.

4 S. Werner et al., Praktische Erfahrungen mit der Kühlung beim thermischen Spritzen. TS 93, *DVS-Berichte*, Band 152, Deutscher Verlag für Schweißtechnik DVS-Verlag GmbH, Düsseldorf, **1993**, p. 45.

5 E. Schwarz, E. Hühne, D. Grasme, R. Kröschel, Modernes, zukunftsweisendes HVOF-Spritzen mit Acetylen und anderen Betriebsgasen. TS 93, *DVS-Berichte*, Band 152, Deutscher Verlag für Schweißtechnik DVS-Verlag GmbH, Düsseldorf, **1993**, p.47.

6 J. A. Browning, Further developments on the HVIF process, *Proceedings of TS 93*, p. 52.

7 E. Ertürk, Das thermische Spritzen von reaktiven Werkstoffen in Unterdruckkammern, Dissertation, Universität Dortmund, **1985**.

8 A. Matting, H.-D. Steffens, Haftung und Schichtaufbau beim Lichtbogen- und Flammspritzen, *Metall* **1963**, *17*.

9 M. Wewel, Beitrag zum Lichtbogenspritzen im Vakuum, Dissertation, Universität Dortmund, **1992**.

10 E. Grasme, Thermisch gespritzte Aluminumüberzüge aus galvanisch wirksamen Aluminumlegierungen zum Korrosionsschutz von Stahl und Aluminumwerkstoffen im maritimen Bereich, TS 93, *DVS-Berichte*, Band 52, Deutscher Verlag für Schweißtechnik DVS-Verlag GmbH, Düsseldorf, **1993**, p. 188

11 W. Huppart, H. Dahme, D. Wieser, Verbesserung der Korrosionsschutzwirkung durch Verwendung modifizierter Spritzlegierungen. TS 93, *DVS-Berichte*, Band 152, Deutscher Verlag für Schweißtechnik DVS-Verlag GmbH, Düsseldorf, **1993**.

12 W. D. Schulz, Zum Korrosionsverhalten von Zn-, AI- und ZnAI-Spritzschichten im Kurzzeitkorrosionsversuch. TS 93, *DVS-Berichte* Band 152, Deutscher Verlag für Schweißtechnik DVS-Verlag GmbH, Düsseldorf, **1993**.

13 K. Nolde, Auftragschweißen, Moderne Beschichtungsverfahren, Steffens, Wilden (Eds.), DGM, **1996**.

14 A. Gebert, H. Heinze, S. Heydel, Verlängerung der Standzeit von Messern und Verschleißleisten durch Plasma-Pulver-Auftragschweißen von hochkarbidhaltigen Werkstoffen, *Wochenblatt für Papierfabrikation*, Bd. 126, Heft 19, **1998**, S. 946–949.

15 A. Gebert, U. Duitsch, U. Müller, T. Schubert, *Bahngesteuertes Auftragschweißen zum Verschleißschutz komplizierter ebener Konturen. 2. Fachtagung Verschleißschutz von Bauteilen durch Auftragschweißen*, 13–14 May **1998**, Halle/Saale.

16 A. Gebert, U. Duitsch, T. Schubert, U. Müller, Zum Verschleißschutz komplizierter ebener Konturen, *Praktiker*, Bd. 51, Heft 1, **1999**, S. 24–28.

17 A. Gebert, H. Heinze, U. Duitsch, Verbundwerkstoffsystem mit Vanadinkarbid, *Tagung Verbundwerkstoffe und Werkstoffverbunde*, 5–7 October **1999**, Hamburg.

18 H. Heinze, A. Gebert, B. Bouaifi, A. Ait-Mekideche, Korrosionsbeständige Auftragschweißschichten auf Eisenbasis mit hoher Verschleißbeständigkeit, *Schweißen und Schneiden*, Bd. 51, Heft 9, **1999**, S. 550–555.

19 A. Gebert, H. Heinze, *Plasma-Pulver Auftragschweißen mit zwei Brennern in Tandemanordnung. 3. Fachtagung Verschleißschutz von Bauteilen durch Auftragschweißen*, Halle / Saale, 17–18 May **2000**.

20 C. Friedrich, G. Berg, E. Broszeit, C. Berger, Datensammlung zu Hartstoffeigenschaften, *Matwissenschaft u. Werkstofftechnik* **1997**, *28*(2), 59–76.

21 A. Gebert, H. Heinze, B. Bouaifi, Neuentwickelte Fe-Cr-V-C-Legierungen für das Plasma-Pulver Auftragschweißen, Thermische Spritzkonferenz, 6–8 March **1996**, Essen, *DVS-Berichte* **1996**, *175*, 151–155.

22 A. Gebert, B. Bouaifi, E. Teupke, Neue vanadinkarbidhaltige Schweißzusätze zum Schutz gegen Verschleiß und Korrosion, Große Schweißtechnische Tagung **2001**, Kassel, DVS Verlag.

6
Machining Technology Aspects of Al-MMC

K. Weinert, M. Buschka, and M. Lange

6.1
Introduction

Materials of a low specific weight have gained importance within the last years. In this context the substitution of iron-based materials by light metals, like aluminum, magnesium and titanium is of interest. Aluminum alloys have a particular place within the structural materials due to their convenient application and manufacturing properties. However, in some areas, for example the automotive industry, conventional Al alloys cannot be fully applied because of poor heat and wear resistance. A way out is the application of Al metal matrix composites (Al-MMC), where these properties are improved by embedding hard particles and/or fibers. Thus an adjusted property profile can be produced by the combination of a light metal matrix and reinforcement phases, aimed to the construction unit strain [1, 2]. Figure 6.1 shows the application and manufacturing properties of Al-MMC as well as application examples. Application examples are particle reinforced brake compounds [3], short fiber reinforced diesel pistons and cylinder surfaces and locally reinforced cylinder crankcases [4, 5]. Further examples are whisker reinforced connecting rods [6] and particle reinforced seat rails [7].

6.2
Machining Problems, Cutting Material Selection and Surface Layer Influence

The aim during production is to achieve near net shape MMC components. However, further machining is necessary to ensure the correct function and for further reduction in weight. The main problem with machining is the wear of the cutting tool caused by direct contact with the hard particles or rather fibers in the insert. Thus abrasion and surface disruption are the dominant wear mechanisms. During abrasion the reinforcement compounds form a groove by the cutting movement into the machining and flank face of the tool. Surface disruption occurs when reinforced compounds crash into the machining surface during the cut and thus

Metal Matrix Composites. Custom-made Materials for Automotive and Aerospace Engineering.
Edited by Karl U. Kainer.
Copyright © 2006 WILEY-VCH Verlag GmbH & Co. KGaA, Weinheim
ISBN: 3-527-31360-5

Functional properties of metal matrix composites (MMC)

- Increased yield stress stiffness tensile strength, creep resistance
- Improved creep resistance at higher application temperatures
- Reduced temperature change sensitivity
- Increased damping behavior
- Increased wear resistance
- Reduced thermal expansion

Manufacturing engineering properties

- More difficult machining

Particle reinforced brake drum

Fiber reinforced engine piston

Particle reinforced cylinder surfaces

Fig. 6.1 Properties and applications of Al-matrix composites.

cause a dynamic strain. The consequences are grain eruptions in stressed areas of the tool caused by crack formation and development [8–10].

Figure 6.2 shows a comparison between the hardness of cutting materials or rather coatings and the hardness of the reinforcement of the light metal matrix composites. From this comparison the problem of machining this material group becomes clear. The cutting materials PCD and diamond coatings, or thick layers, manufactured by the CVD process, are most suitable for machining fiber- or particle-reinforced light metal matrix composites because they show a very high hardness due to the good wear resistance to abrasion. The basic condition for successful application of the diamond coating (thin films) is sufficient adhesion on the cemented carbide substrate to avoid spalling due to the tribological strain. Coated cemented carbides and cutting ceramics are only of limited applicability for light metal matrix composites with SiC-, B_4C- and TiB_2-particle reinforcement, since their hardness is equal to or lower than the hardness of the particles (2600 to 3700 HV). All cutting materials are suitable for light metal matrix composites, reinforced with δ-Al_2O_3 short fibers, with a low fiber hardness (800 HV). For machining Si particle-reinforced materials (~1250 HV) the application of coated cemented carbides is economically possible [8–12].

Figure 6.3 shows the development of wear versus cutting time for different cutting materials during turning of the spray-formed and hot-formed two-phased AlSi alloy $AlSi_{25}Cu_{2.5}Mg_1Ni_1$. The PCD tool shows the lowest flank wear due to its high hardness (Fig. 6.2). In comparison with uncoated cemented carbide (CC-MC), diamond and TiAlN coated tools show lower wear, which can be attributed to the

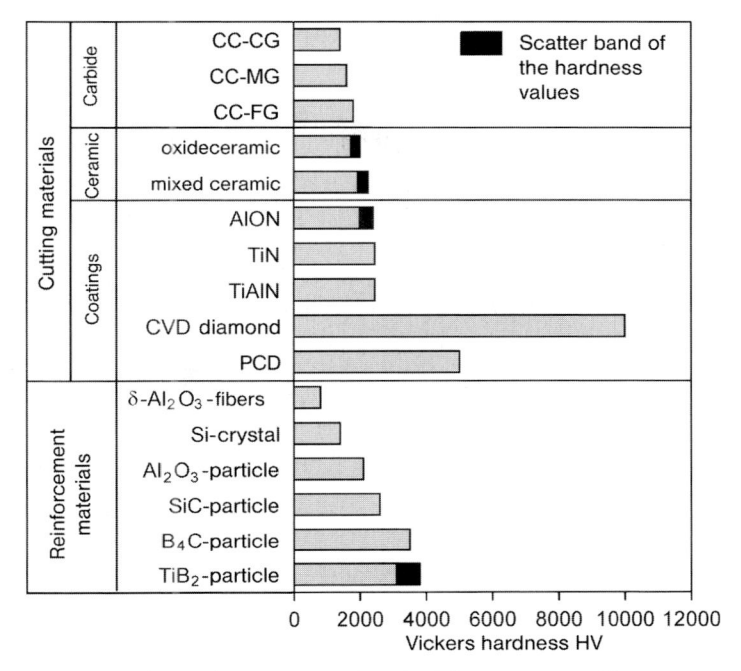

Fig. 6.2 Hardness of cutting materials and reinforcement materials.

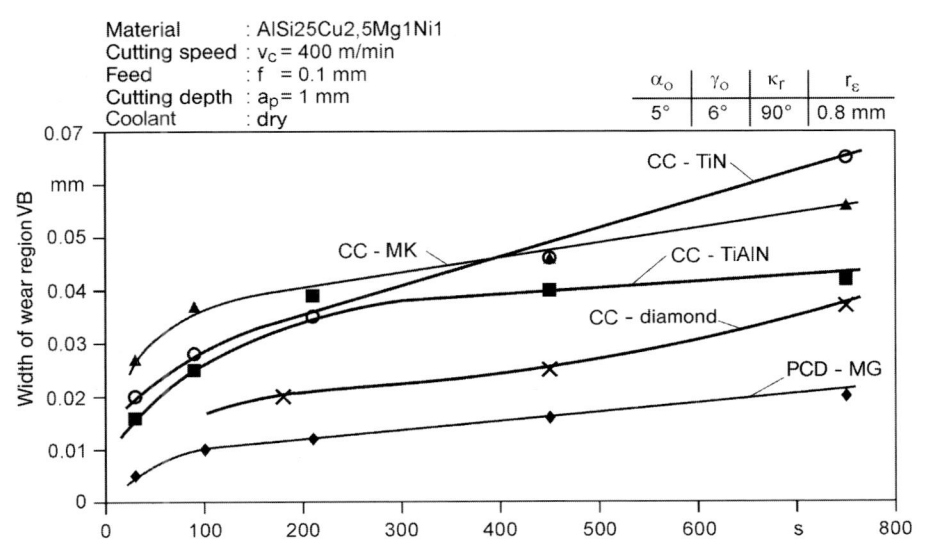

Fig. 6.3 Tool wear of different cutting materials after turning of over-eutectic AlSi alloy.

wear decreasing effect of the coating. Diamond-coated tools especially perform well during short process times due to the high hardness of the wear protecting layer. While for all tools the flank wear region width increases in the stationary phase with increasing cutting time, a progressive increase is to be noted for the diamond-coated cemented carbides. The reason for this is the flaking of the coating on the flank face and also on the machining surfaces. Thus the substrate is laid open and exposed to wear stress by the hard Si crystals of the compound material. On specially developed diamond-coated tools for machining Al materials, no wear can be detected on the tools after processing the over eutectic AlSi alloy for 750 s (12.5 min). For that tool the cutting geometry, the substrate composition, the coating adhesion and the coating thickness are optimized to the occurring strain [10, 11]. However, during machining of SiC-particle-reinforced Al alloys, coating flaking occurred during the application of inserts.

In comparison to diamond coatings the hardness of the TiAlN coating is approximately 2450 HV, whereby the resistance to abrasive acting Si particles is lower, resulting in higher tool wear. Although the hardness of the TiN coating is the same as the TiAlN coating, the flank wear after a cutting time of 750 s (12.5 min) is considerably higher. The reason for this is the high chemical affinity of TiN to Al, so that the coating dissolves by the mechanical, thermal and tribochemical strain in the contact zone, and thus the tool is not protected against wear [10, 11].

Concerning the wear of PCD tools, the PCD grain size is of special importance. To investigate the influence of the grain size of PCD tools on the tool wear during processing of different Al matrix composites, turning tests with PCD of average grain size (PCD-MG, GS = 10 µm) and coarse grain size (PCD-CG, GS = 25 µm) have been carried out on different Al matrix composites. Furthermore, a fine grained PCD type (PCD-FG, GS = 2 µm) was tested with SiC- and B_4C-particle-reinforced aluminum. The results of these tests with regard to the tool wear are summarized in Fig. 6.4. It can be seen that the grain size has hardly any influence on the tool wear during machining of δ-Al_2O_3-short-fiber-reinforced aluminum. In comparison a significant reduction of wear can be determined with increasing grain size of PCD during machining of SiC- and B_4C-particle-reinforced materials. Of special importance is the grain size ratio between the particles of the composite and the PCD grains. If the particle impacting onto the cutting edge is significantly larger than the PCD grains, the diamond bridge binding is strongly stressed by one or more PCD grains. In this way crack formation can easily occur and during further cutting, a breakout of complete grains can occur. If the composite particles are smaller than the PCD grains, the impact strain of the particle hitting the tool is predominantly absorbed by the PCD grain; the breakout of complete grains is made more difficult [9].

An important aspect during the machining of metal matrix composites besides the tool wear is the quality of the machined surface and the process caused influence on the peripheral zone. The mechanical and thermal strain of the surface layer during machining can cause fractures and cracks in the hard particles and/or fibers, plastic deformation of the metal matrix, pore formation, and changes in the residual stress profile. These surface zone changes can adversely affect the materi-

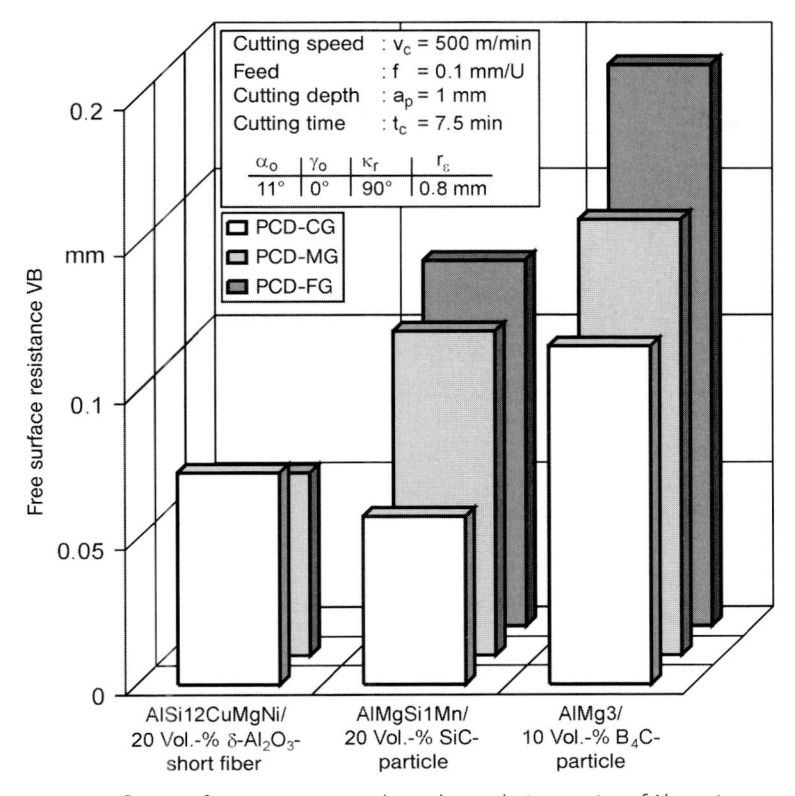

Fig. 6.4 Influence of PCD grain size on the tool wear during turning of Al matrix composites.

al properties, so that the requirements on the application of MMC compounds cannot be fulfilled. The appearance of pores and grooves in machined surfaces is characteristic for metal matrix composites. This damage occurs due to breakouts of particles and/or fibers from the surface during the cutting process. In this way further grooves emerge.

Figure 6.5 demonstrates the surface layer influence exemplified by cemented carbide (CC-MC) and a PCD machined over-eutectic AlSi alloy after a cutting time of 750 s (12.5 min). In contrast to the workpiece machined with cemented carbide, the surface layer structure shows no influence after turning with PCD, whereas even the hard Si crystals can be cut by the PCD tool. After machining with cemented carbides, plastic deformations of the surface and the peripheral zone occur and cracks are detectable in the hard phases as well as fractures in the hard phases. The Si particles are thereby aligned in the cutting direction. Similar changes in the surface layer structure can also be noticed on SiC- and Al_2O_3-reinforced metal matrix composites.

In an analogous manner to turning with cemented carbides, the composite material reacts during processing in a similar manner to coated cemented carbides.

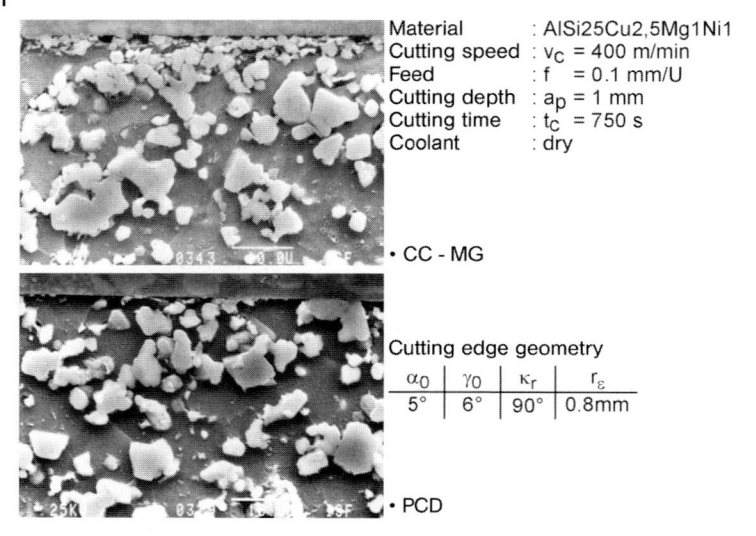

Fig. 6.5 Surface layer override after turning with cemented carbide and PCD tools.

The surface layer influence is similarly distinctive. The obtainable surface roughness increases in the order: PCD, uncoated and coated cemented carbide. After machining the average roughness values are 2 to 3 µm for PCD and 3 to 6 µm for cemented carbide tools.

6.3
Processing of Components of Metal Matrix Composites

Starting from the previous basic results for turning, drilling and milling of Al matrix composites [8–11, 13–15] tests for the machining of concrete compounds composed of Al matrix composites have been carried out. The aim was to optimize the process strategy and the parameter selection taking into account the required process quality. The compounds are used in SiC-particle-reinforced brake drums manufactured by semi-permanent-moulding, cylinder crankcases produced by squeeze casting, with local Si-particles and Al_2O_3-short-fiber-reinforced cylinder surfaces as well as *in situ* TiB_2-particle-reinforced extrusion profiles for application in passenger planes, see Fig. 6.6. During the production of cylinder crankcases highly porous, hollow cylindrical preforms are first inserted in the cylinder center sleeve of the mould, then fusion infiltration of the preform takes place by squeeze-casting under pressure with the under-eutectic secondary alloy $AlSi_9Cu_3$. This leads to a local reinforcing in the area of the cylinder surface [5].

6.3 Processing of Components of Metal Matrix Composites

Component	Brake components	Cylinder surface	Extrusion profile
Application area	Automotive industry	Automotive industry	Aircraft industry
Manufacture	Semi-Permanent-Molding	Squeeze-Casting	Extrusion
Matrix alloy	AlSi9Mg	AlSi9Cu3	AA 5083, AA 7071
Reinforcement	SiC-particle	Si-particle + Al_2O_3-short fiber	TiB_2-particle
Reinforc.-part	20 Vol.-%	15 and 5 Vol.-%	5 and/or 7 wt-%
Reinforc.-size	10 μm	30 to 70 μm (Si)	1 μm
Reinforc.-hardness	3400 HV	1250 HV	3800 HV

Fig. 6.6 Components of Al matrix composites.

6.3.1
Materials, Cutting Materials and Process Parameters

The structure of the processed MMC compounds is shown in Fig. 6.7. The brake drums consist of the alloy AlSi9Mg, in which 20 vol% SiC particles, 10 to 15 μm in size, are embedded. The cylinder crankcases consist of the alloy AlSi9Cu3, whereas the cylinder surfaces are locally reinforced with 15 vol% Si particles, 30 to 70 μm

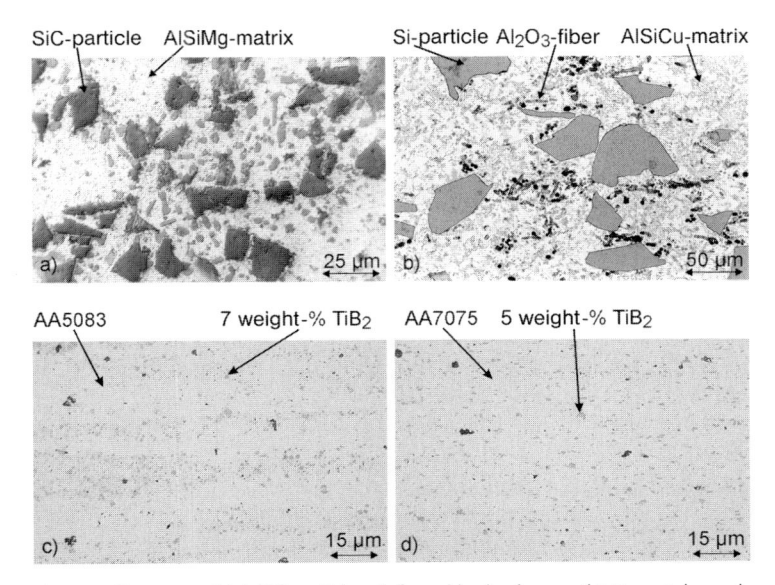

Fig. 6.7 Structure of (a) SiC-particle-reinforced brake drums, (b) Si-particle- and Al_2O_3-short-fiber-reinforced cylinder surfaces (Lokasil I), (c) and d) TiB_2-particle-reinforced extrusion profiles.

in size, and 5 vol% Al_2O_3 short fibers. The matrix of the extrusion profiles consists of the alloy AA5083 (AlMg4.4Mn0.7Cr0.15), and the alloy AA7071 (AlZn5.6Mg2.5Cu1.6Cr0.23), in which fine TiB_2 particles are embedded. The alloy AA5083 contains 7 wt% TiB_2 and the alloy AA7071 5 wt% TiB_2. The average size of the TiB_2 particle is approximately 1 µm, and they are aligned in the structure in the direction of transformation.

A list of process parameter values which are varied when machining MMC compounds is summarized in Table 6.1. For turning brake drums and drill finish cylinder surfaces, polycrystalline diamond (PCD) or a cutting material with a diamond thick-film layer produced by CVD technology (CVD-diamond) is used. The properties of the cutting materials PCD and CVD-diamond are shown in Table 6.2 [16]. The Knoop hardness of CVD-diamond is 85 to 100 Gpa, significantly higher than that of PCD which shows a Knoop hardness of 50 to 75 GPa. The cutting material produced with the CVD technology consists of 99.9% diamond, whereas PCD contains a small part of Co, approximately 8 vol%. To verify the capability of cutting materials based on diamond, uncoated and coated cemented carbide is used as cutting material to drill finish cylinder crankcases. The coatings used are commercial TiN and diamond coatings. Cemented carbide tools are not used for turning of brake drums and finish boring of cylinder surfaces. During the drilling of reinforced extrusions, uncoated full cemented carbide and PCD tools were used. For milling TiAlN and diamond-coated end mills were used. Processing was carried out with different PCD types, which differ in their grain size.

6.3.2
Evaluation of Machinability

To evaluate machining, the criteria of tool wear and surface roughness have been used. Wear and surface roughness were evaluated in a scanning electron microscope. Additionally metallographic analysis of the processed surface peripheral zones was carried out to check for possible structure changes like plastic deformation and cracks in the reinforcement phase. The measurement of the deterioration depth in the hard phases was carried out with an optical microscope with an additional microhardness measuring system. To determine the roundness of brake drums after turning and after finish boring of the cylinder surfaces, measurements were carried out with a roundness measuring device and a 3D-coordinate measuring machine. The measurement of the cylinder surfaces was done at three drilling depths: 10, 50 and 85 mm at three levels.

6.3.3
Turning of SiC-particle-reinforced Brake Drums

Figure 6.8 shows the development of the tool wear with cutting time after the application of the cutting material PCD and CVD-diamond during turning of brake surfaces at a cutting speed of 500 m min^{-1} and a feed rate of 0.3 mm. Additionally, the tool wear is documented by scanning electron microscopic images. Direct com-

Tab. 6.1 Process parameters of machining of MMC components.

Work piece	Material	Process	Cutting material	Lubrication concept	Cutting parameter v_c, m min^{-1}	f, f_z, mm	a_p, mm	a_e, mm
Brake drum	AlSi9Mg + 20 vol% SiC	straight turning and facing	PCD-CG CVD-diamond	dry MQL	[500, 1000]	[0.1, 0.7]	0.5	–
Cylinder crankcase	AlSi9Cu3 + 5 vol% Al$_2$O$_3$ (Lokasil I)	pre-boring + 15 vol% Si	CC CC-diamond PCD CVD-diamond	dry CC-TiN	[100, 750] emulsion	[0.08, 0.12]	1.2 to 1.4	–
		finish bore processing	PCD CVD-diamond	dry	[500, 2500]	[0.1, 0.2]	[0.1, 0.3]	–
Extrusion profiles	AA5083	drilling + 7 wt% TiB$_2$	CC	dry PCD-MK	100	0.1	–	–
	AA7075 + 5 wt% TiB$_2$	milling	HM-TiAlN HM-Diamond PCD-FG PCD-MG PCD-CG	dry	[100, 560]	[0.1, 0.2]	2	[0.5, 1]

Tab. 6.2 Properties of polycrystalline diamond (PCD) and CVD diamond thick film [3].

Characteristic	Unit	PCD	CVD-diamond
Cutting material thickness	mm	1	0.5
Density ρ	g cm^{-3}	4.1	3.51
Young's modulus	GPa	800	1180
Knoop hardness	GPa	50–75	85–100
Cross break strength	GPa	1.2	1.3
Fracture toughness K_{IC}	MPa m$^{1/2}$	8.89	5.5
Coefficient of thermal expansion α	10^{-6} K^{-1}	4	3.7
Heat conductivity λ	W m^{-1} K^{-1}	500	750–1500

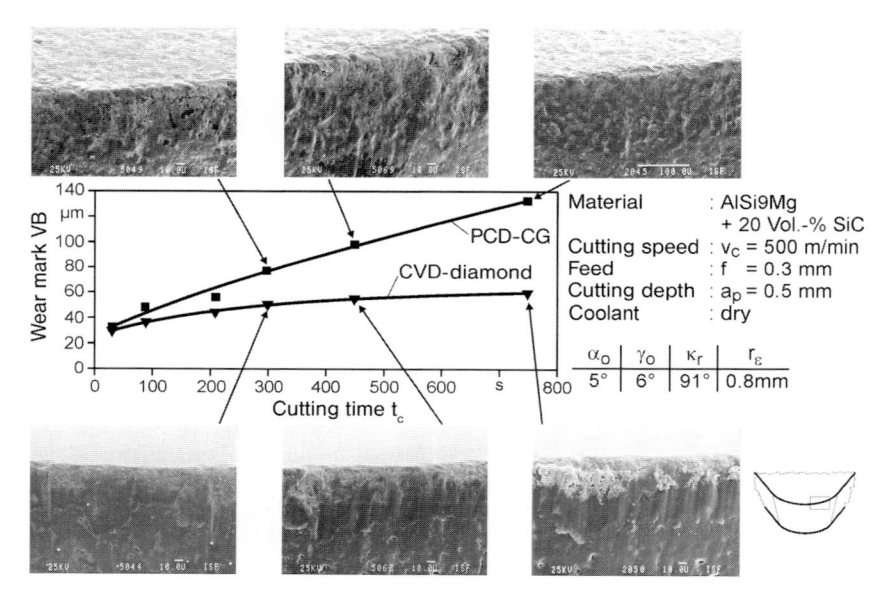

Fig. 6.8 Width of wear region during turning with polycrystalline diamond (PCD) and CVD diamond thick layer inserts.

parison between both cutting materials shows that CVD diamond has a better wear resistance than PCD.

Due to the significantly higher hardness of CVD-diamond cutting materials, this material compared to PCD is less damaged by the SiC particles during turning. The significantly higher hardness of diamond cutting tools compared to SiC particles makes an abrasive wear attack by the reinforcement phase of the Al composite at the tool impossible. Here surface damage has to be assumed. The detection of grooves on the cutting face and/or flank face is caused by self-grooving of broken out diamond grains or parts of grains, which slide over the chip and flank face during cutting movements causing grooves to develop.

The gradient structure of CVD diamonds leads to spalling of only small areas of the cutting edge due to the surface disruption of the cutting material in the working area. The grain size at the cutting face is up to 2 μm. The grain size increases to 25 μm inside the substrate [16]. At isotropic PCD large areas could spall so that the wear ratio is higher. Furthermore, the tendency to form material deposits of the Al alloy at the chip and flank face of the insert is lower for CVD diamond than for PCD. This is caused by the lower adhesion tendency between the Al alloy and the cutting material of CVD diamonds, in comparison to PCD. However, CVD diamonds do not contain Co, which could interact with the compound material.

Figure 6.9 shows examples of scanning electron microscope images of CVD diamond- and PCD inserts after a cutting time of 120 s. To show the above described wear mechanisms, the typical wear appearance of each tool can be seen. At the tool face of the PCD tool, spalling is seen directly at the cutting edge, which is the result of continuous impact of SiC particles on the chip surface during turning. Thus the surface roughness of the insert at the tool face increases. The CVD diamond tool does not show any damaged rake face, however, spalling occurs at the cutting edge. Furthermore the image shows the fine cutting material grain structure of the CVD diamond cutting material.

With increasing cutting speed, the thermal and mechanical strain of the cutting material increases. This can lead to higher tool wear and pseudo-chip formation can occur at the tool due to the increasing process temperature or large material deposits. The process quality is negatively influenced by this procedure. To reduce the tendency of material deposit formation, tests using CVD diamond cutting tools at higher cutting speeds using minimum quantity lubrication (MQL) have been carried out. Figure 6.10 shows the influence of the cutting speed on the tool wear during turning of SiC reinforced Al brake drums. At higher cutting speed the tool wear rises due to the increasing thermal and dynamic cutting material strain. At a cutting speed of 750 m min^{-1} a wear mark of over 100 μm is measured after a cutting time of 750 s (12.5 min). By further increasing the cutting speed up to

Material : AlSi9Mg + 20 Vol.-% SiC
Cutting speed : v_c = 500 m/min
Cutting depth : a_p = 0.5 mm
Cutting time : t_c = 210 s
Coolant : dry

α_o	γ_o	κ_r	r_ε
5°	6°	91°	0.8mm

- CVD diamond (f = 0.3 mm)
- Spalling on the cutting edge

- PCD (f = 0.2 mm)
- Spalling on the chip surface

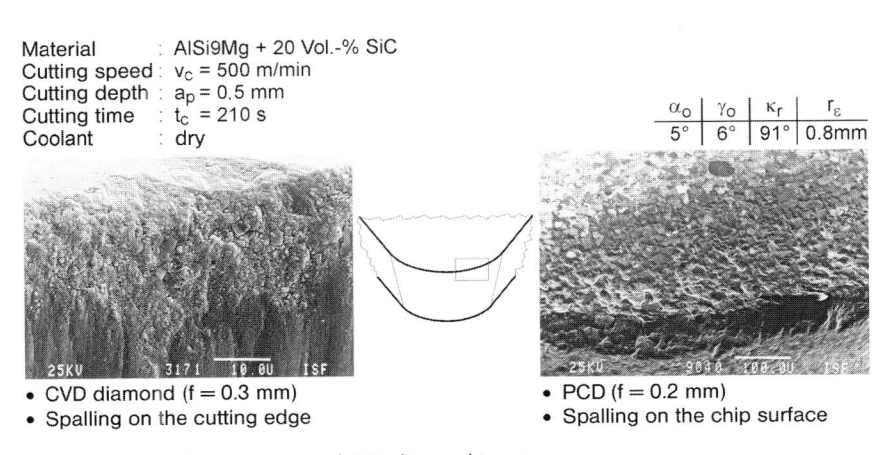

Fig. 6.9 Wear and tear on PCD and CVD diamond inserts.

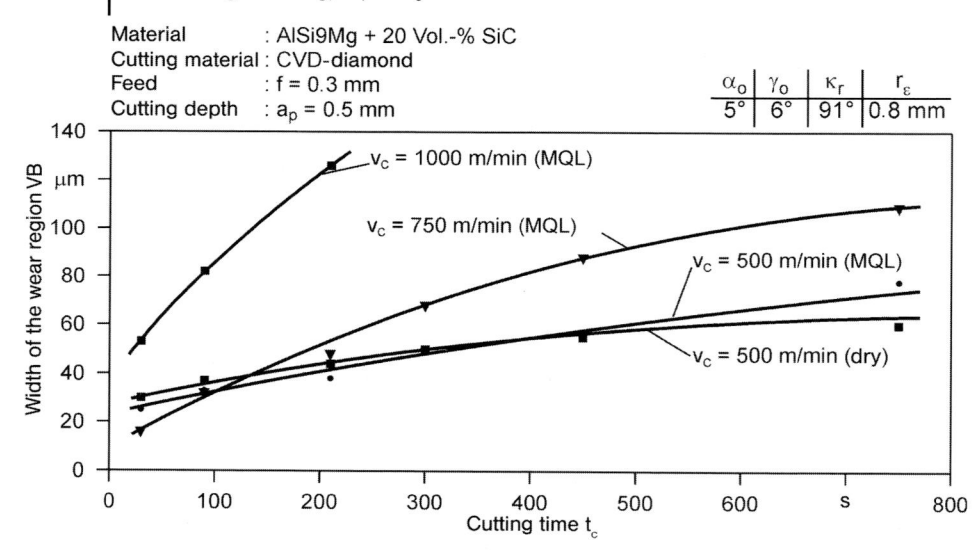

Fig. 6.10 Influence of the lubrication concept and cutting speed on the tool wear.

1000 m min^{-1} extreme flank wear can be seen after short cutting times. Thereby turning tests had to be stopped after a cutting time of 210 s (3.5 min). The measured wear region had already reached a value of 130 µm. The influence of the cooling lubricant concept on the tool wear is not clear for turning tests.

Figure 6.11 shows the influence of the feed rate (f = feed per tooth) on the flank-wear region width after turning with PCD inserts. The measured wear mark is only present for tools with an edge radius of 1.2 mm. The significance of the influence of the edge radius on the wear is not clear. With increasing feed rate the flank wear rises whereas there is no change in the wear appearance and thus wear mechanism. During the increase of feed the amount of SiC particles which are impacted onto the cutting face or slide over the chip and flank face does not increase, but the strain of the cutting material increases due to the higher chip thickness. For example the cutting force during application of a tool with an edge radius of 0.8 mm rises from 52 N (f = 0.1 mm) to 249 N (f = 0.7 mm), so that the tool wear increases in the same direction.

Furthermore, Fig. 6.11 shows the influence of feed rate and edge radius of the insert on the average surface roughness. The roughness of the brake surface is not allowed to exceed an upper limit of average surface roughness of 11 µm after finishing. The feed rate of finishing the brake surface depends on the edge radius and may not exceed 0.25 mm for an edge radius of 0.8 mm and 0.3 mm for an edge radius of 1.2 mm. During the pre-machining, e.g. removing the mold and the finished shape chamfer at the brake surface, as well as during processing the center, during inner facing and during processing the outside, significantly higher feed rates can be set. If the values for the feed rate do not exceed 0.5 mm, the brake drum manufacturing is process-sure. By applying CVD diamond and PCD inserts,

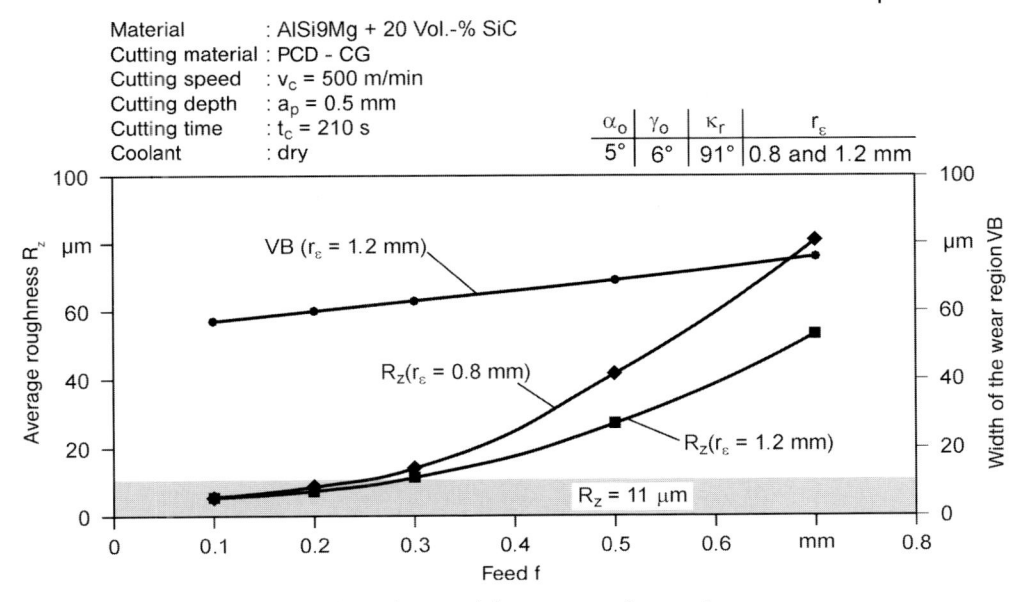

Material	: AlSi9Mg + 20 Vol.-% SiC
Cutting material	: PCD - CG
Cutting speed	: v_c = 500 m/min
Cutting depth	: a_p = 0.5 mm
Cutting time	: t_c = 210 s
Coolant	: dry

α_o	γ_o	κ_r	r_ε
5°	6°	91°	0.8 and 1.2 mm

Fig. 6.11 Influence of the feed on the tool wear and the average surface roughness.

small spalling can occur occasionally at a feed of 0.7 mm. Processing SiC-reinforced brake drums with high feeds means, short process times and thus high productivity because of the low dependence of the tool wear on the feed rate.

Figure 6.12 shows the surface and the peripheral zone after turning brake drums with PCD in the area of the surface after a cutting time of 210 s (3.5 min). The scanning electron microscope image of the surface shows a typical surface topography, which is present after processing of SiC-reinforced Al alloys. Pores can be seen in the areas where the hard SiC particles have been separated. SiC fragments can be found partly in the pores. The surface and the peripheral zone of the Al alloy is altogether slightly plastic deformed. This is made clear by optical microscope images of the peripheral zone. Near the processed surface only broken SiC particles can be seen.

Starting with the tests to select the cutting parameter and cutting materials, which are carried out with the machining of the brake surface of a SiC-particle-reinforced brake drum, a turning process took place in the area of the center, the fixing holes and the brake surface as well as the processing of the outside. Figure 6.13 shows the clamped, processed brake drum with the processing steps and the selected cutting conditions. The brake drum is clamped in a conventional three-jaw chuck, which is adjusted to the outside geometry of the drum. With this clamping a roundness of 14 µm can be reached in the area of the brake surface after finishing. As tools CVD diamond inserts with a corner radius of 1.2 mm are used. The complete turning process takes place at a cutting speed of 500 m min^{-1} in dry machining. In the area of the center of the fixing holes and when machining the outside a feed rate of 0.4 or 0.5 mm can be used. To process the brake surface the feed rate needs to be reduced during the final process step to 0.3 mm to reach an average surface roughness of 1 µm.

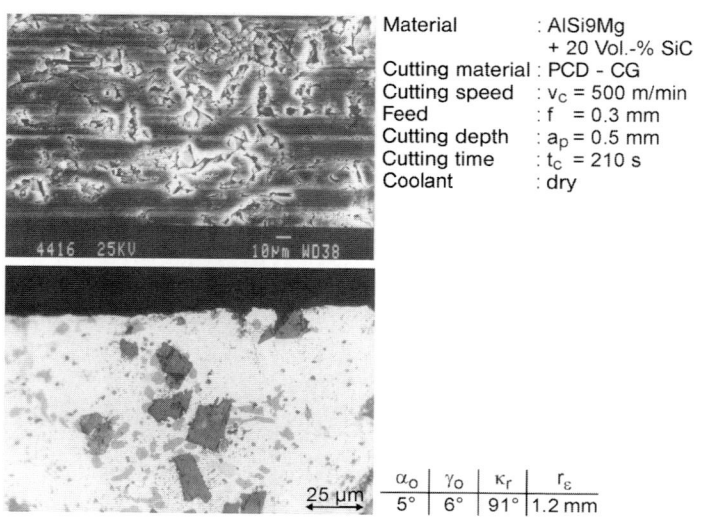

	α_O	γ_O	κ_r	r_ε
	5°	6°	91°	1.2 mm

Material : AlSi9Mg + 20 Vol.-% SiC
Cutting material : PCD - CG
Cutting speed : v_c = 500 m/min
Feed : f = 0.3 mm
Cutting depth : a_p = 0.5 mm
Cutting time : t_c = 210 s
Coolant : dry

Fig. 6.12 Surface formation and surface layer after turning of the brake surface.

Clamping: radial

Process operations:
- Inside longitudinal turning
- Inside facing
- Outside facing
- Outside longitudinal turning

Requirement for the brake surface:
- Roughness : $R_{z,max}$ = 11 µm
- Roughness : t = 8 µm
- Concentricity (to the hub): t_k = 0.1 mm
- Cylindricity : t_z = 0.05 bis 0.1 mm

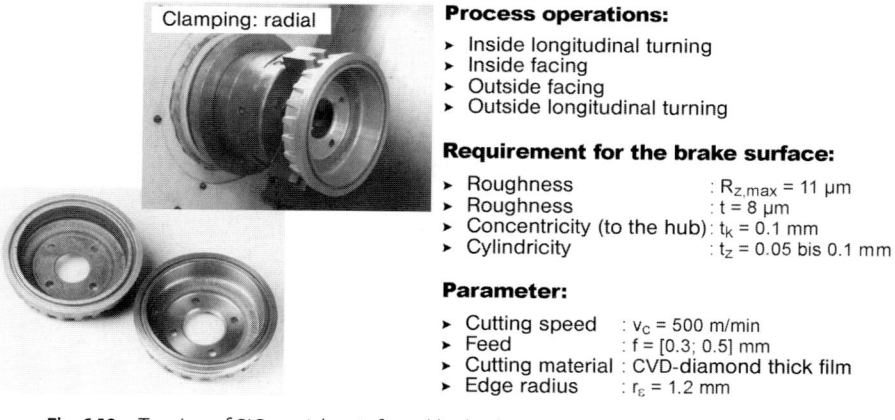

Parameter:
- Cutting speed : v_c = 500 m/min
- Feed : f = [0.3; 0.5] mm
- Cutting material : CVD-diamond thick film
- Edge radius : r_ε = 1.2 mm

Fig. 6.13 Turning of SiC-particle-reinforced brake drums: process operations, requirements, and process parameters.

6.3.4
Boring of Si-particle- and Al$_2$O$_3$-fiber-reinforced Al Cylinder Surfaces

Figure 6.14 shows the cylinder crankcase that was selected for tests (4-cylinder in open deck pattern) with local reinforced cylinder contact surfaces, as well as the boring tools used for the tests. A double-edged tool is applied for the pre-boring operation, which has been adjusted to a process diameter of 83.4 mm to remove the cast surface and the draft angle. The cutting depth is between 1.2 and 1.4 mm. The final finish boring operation is a premachining for honing, the process when fine boring has been adjusted to reach the necessary average surface roughness of 5 to 9 µm during the first hone step. The roundness is therefore limited to a value of 15 µm. The diameter of 83.8 mm is the process diameter including honing measurement.

Fig. 6.14 Cylinder crankcase with cylinder surfaces locally reinforced with Si- and Al$_2$O$_3$ particles and boring and pre-boring tools.

6.3.4.1 Pre-boring Operation

During the pre-machining under dry cutting conditions heavy material disposals and pseudo-chip formation occurred. Thus the pre-boring operation had to take place with the aid of emulsion. Figure 6.15 shows scanning electron microscopic images of uncoated and TiN- or diamond-coated tools after pre-boring finishing a

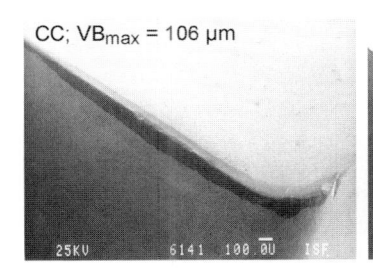

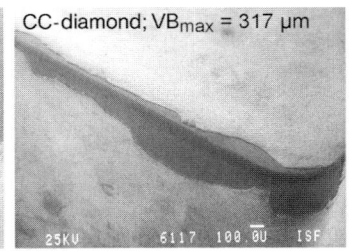

Fig. 6.15 Wear of coated and uncoated cemented carbide tools.

cylinder crankcase (4 cylinder surfaces). From these images of the wear region it becomes clear that customary wear resistance coatings, like TiN and diamond, show higher flank wear after the drill boring operation than uncoated cemented carbide. Using TiN tribochemical coating, dissolution takes place during machining [10–12], while the diamond coating spalls at an early stage during drill out operation due to the low adhesion on the cemented carbide substrate.

The present investigations to drill out cylinder surfaces with cemented carbide tools show that an economical processing does not appear to be possible. On the one hand this is caused by the strong tool wear. On the other hand the machining has taken place relative to recent basic investigations at a cutting speed of 100 m min^{-1}, whereby the cutting time (primary processing time) of the pre-machining of cylinder crankcases at the selected cutting parameter is 554 s (9.2 min). Because of the high wear resistance of PCD this cutting material can be used at significantly higher cutting speeds. Experiments to vary the cutting speed and the feed rate have shown that the optimal cutting parameters during pre-machining of cylinder surfaces with PCD are a cutting speed of 400 m min^{-1} and a feed per tooth of 0.12 mm. At a cutting speed of 500 m min^{-1} significantly higher flank wear occurs and thus there is a strong material wear down at the inserts whereby the feed per tooth had to be reduced to 0.08 mm. The cutting time of pre-machining of cylinder crankcases with more favorable cutting parameters is 92 s (1.5 min).

Further investigation concerning the performance characteristics of the cutting materials PCD- and CVD-diamond during the pre-boring operation took place at a cutting speed of 400 m min^{-1} and a feed per tooth of 0.12 mm. Figure 6.16 shows a comparison of wear development for both cutting materials during processing of

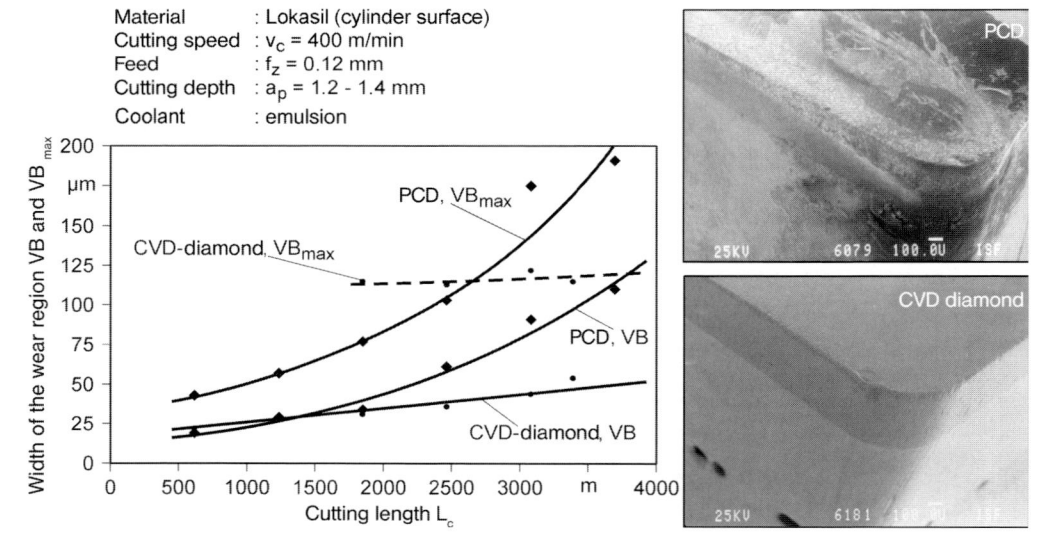

Fig. 6.16 Tool wear during pre-boring with cutting materials with polycrystalline diamond (PCD) and CVD diamond thick film.

MMC cylinder surfaces. Progressive wear development can be seen during boring with PCD inserts. Already after pre-boring of 6 cylinder crankcases (24 cylinder surfaces) a maximum width of wear region of approximately 200 µm has been reached. The wear of tools during application of inserts with a CVD diamond thick layer develops linearly, contrary to PCD inserts, and at a significantly lower level. Due to a small spalling, which appeared after 3 cylinder crankcases (12 cylinder surfaces), a wear mark of approximately 120 µm could be measured at the insert, which did not increase during further processing. The scanning electron microscope images in Fig. 6.16 make clear the strong difference between both cutting materials in the tool wear development after pre-boring. Furthermore the image of the CVD diamond tool shows the occurrence of the cutting edge chipping.

6.3.4.2 Finish Bore Processing

In a similar way to the turning process of SiC-reinforced brake drums and to the pre-boring operation of MMC cylinder surfaces, it is also shown for finish boring that the tools with a CVD diamond thick layer show lower wear than PCD tools after machining. In Fig. 6.17 the cutting speed related influence of the the tool wear is shown after finish boring of reinforced cylinder surfaces during the use of CVD diamond thick layer inserts in dry cutting. Additionally, scanning electron microscope images show wear and tear. The most useful cutting speed for the application of CVD diamond thick layer tools is 500 m min^{-1}. Processing at higher cutting speeds leads to strong, progressive flank wear and to spalling of the flank face even at short cutting lengths. This spalling already appears at a cutting speed of 600 m min^{-1}.

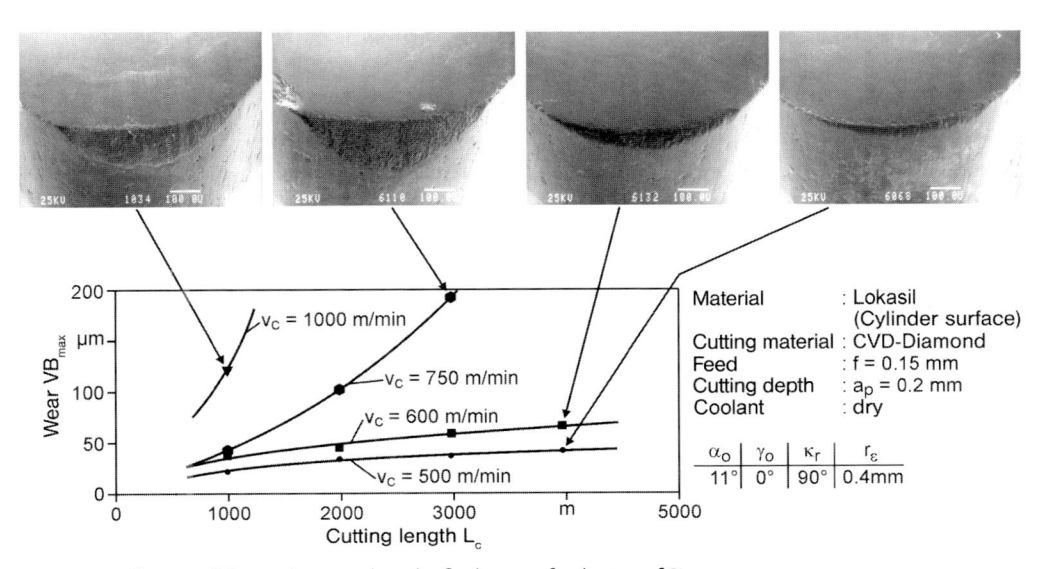

Fig. 6.17 Influence of the cutting speed on the flank wear after boring of Si- and Al$_2$O$_3$-reinforced cylinder surfaces.

6 Machining Technology Aspects of Al-MMC

Figure 6.18 shows the average and theoretical surface roughness after finish boring of the cylinder surfaces as a function of feed and cutting speed. The curve of the average surface roughness versus the feed or cutting speed is always above the theoretical surface roughness. To reach the required average surface roughness of 5 to 9 µm after finish boring of cylinder surfaces, a feed rate between 0.1 and 0.15 mm has to be selected, based on the measured surface roughness at a cutting speed of 500 m min^{-1} and a tool corner radius of 0.4 mm.

After finish boring at a cutting speed of 500 m min^{-1} spalling always appears at the processed surface, see Fig. 6.19, especially at places where the hard Si particles are separated from the surface. The damage depth, up to which the Si particles in the peripheral zone show cracks or are broken out of the surface, can be up to 30 µm. Random tests to avoid spalling on processed surfaces show that at very high

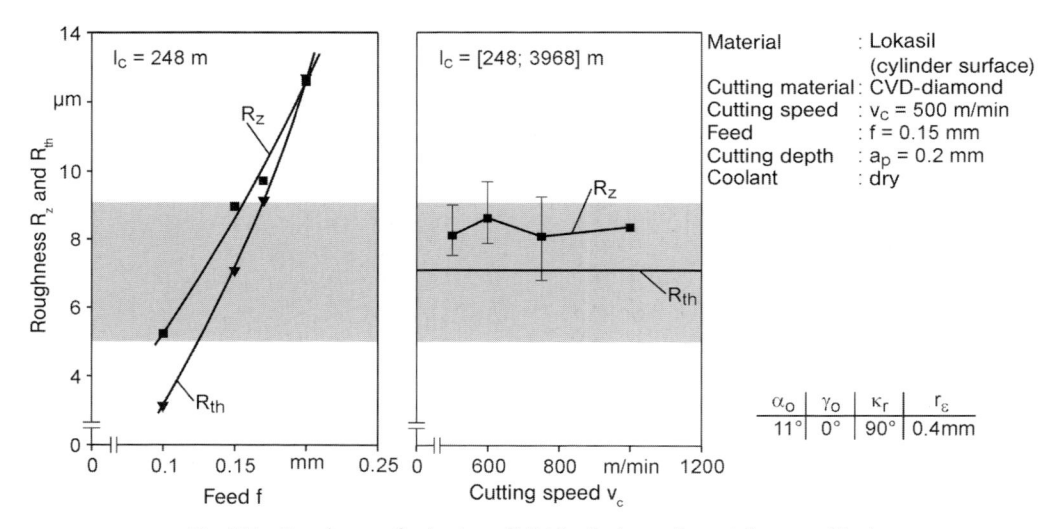

Fig. 6.18 Roughness after boring of MMC cylinder surfaces, influence of feed and cutting speed.

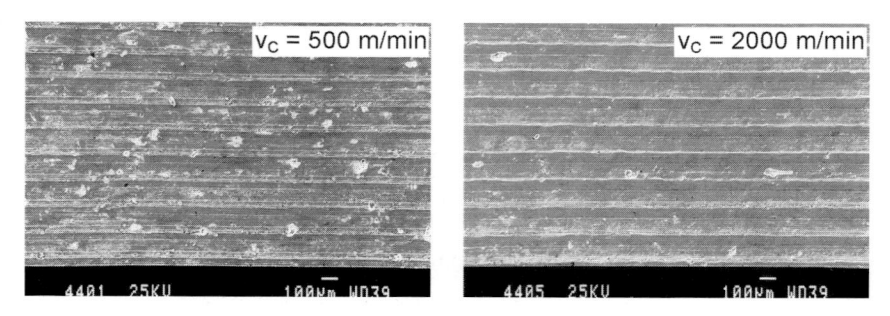

Fig. 6.19 Surface after boring at various cutting speeds.

cutting speeds of 2000 to 2500 m min^{-1} the spalling occurring during chip formation can be significantly reduced, see Fig. 6.19. However, at these high cutting speeds extremely high tool wear can be noticed, which excludes the finish boring of cylinder surfaces at these speeds. Plastic deformation of the Al matrix after the process with cutting speeds of 500 to 2500 m min^{-1} is not observed in metallographic peripheral zone samples.

Besides the surface parameter, the shape of the processed cylinder is an important requirement for finish boring operation. Thus a maximum roundness of 15 μm should not be exceeded for the following honing of all four cylinders. At the two inner cylinder surfaces a roundness in dependence on the drilling depth between 3 μm ($L_z = 10$ mm) and 5 μm ($L_z = 90$ mm) is reached at a cutting speed of 500 m min^{-1}, a feed rate of 0.15 mm and a cutting depth of 0.2 mm. The two outside cylinders show a significantly worse roundness due to the open construction of the cylinder crankcase, which exceeds the critical value for the roundness of 15 μm, particularly at a drilling depth of 50 mm. To reach the required roundness also at the outside cylinder surfaces, investigations to determine the influence of the cutting depth on the roundness have been carried out. Figure 6.20 shows the roundness in dependence on the cutting depth during finish boring. Due to the roundness value of 55 μm after pre-boring of the two outside cylinders, an enhancement of the roundness occurs with increasing cutting depth during the finish process operation. At a cutting depth of 0.3 mm the required roundness of 15 μm is reached at the outside cylinders. Pre-boring in two cuts is a further possibility to improve the roundness. With that machining concept the mechanical strain of cylinder surfaces can be reduced, so that the roundness after the pre-machining could be better.

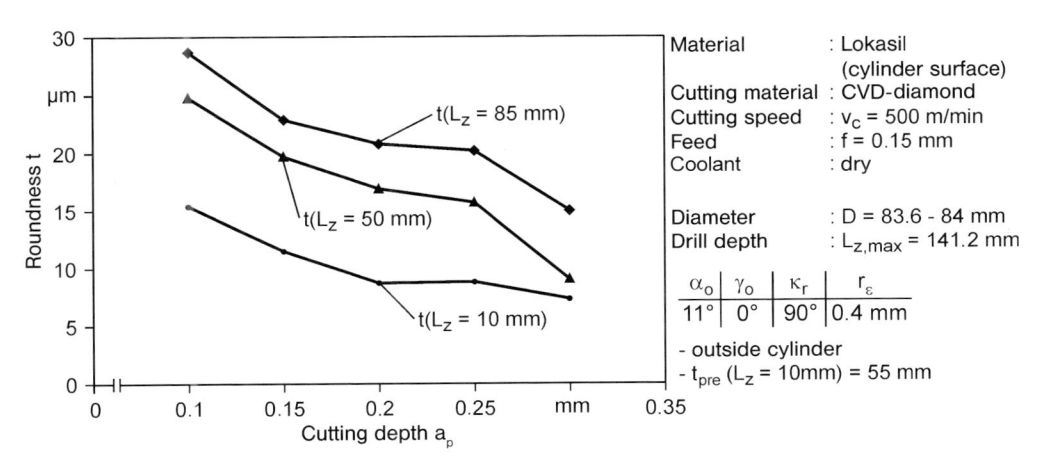

Fig. 6.20 Roundness after boring of reinforced cylinder surfaces – influence of the cutting depths.

6.3.5
Drilling and Milling of TiB₂-particle-reinforced Extruded Profiles

Machining operations have to follow the extrusion process of profiles for application of TiB_2 –particle-reinforced extrusion profiles in the aircraft industry with the aim of mount fixing possibilities and for weight reduction. Therefore drilling and milling processes can be used. Process problems occur due to the extremely hard particles embedded in the Al matrix, and also due to the thin walled extrusion profiles, which increase the problems of vibration loads due to the discontinuous cut during the milling process. This can lead to increased tool wear due to spalling at the cutting edges, especially during machining with the brittle PCD cutting material.

6.3.5.1 Drilling

Figure 6.21 shows the tool wear after drilling of TiB_2-reinforced extrusion profiles at a cutting speed of 100 m min^{-1} and a feed rate of 0.1 mm against the drilling length and in dependence on the numbers of drill holes, respectively. There the influence of the cutting materials (cemented carbide and PCD) and of the TiB_2-content of the Al matrix is shown. While the alloy AA7075 with a content of 5 % TiB_2 shows good machining behaviou during processing with an uncoated cemented carbide tool, significantly stronger tool wear occurs during drilling of the alloy AA5083 with a content of 7 % TiB_2. The wear mark at the flank face reaches a value of 200 µm after 50 drill holes, whereas after processing of the material with 5 % TiB_2 a wear mark of 50 µm is measured. Although both Al composite materials dif-

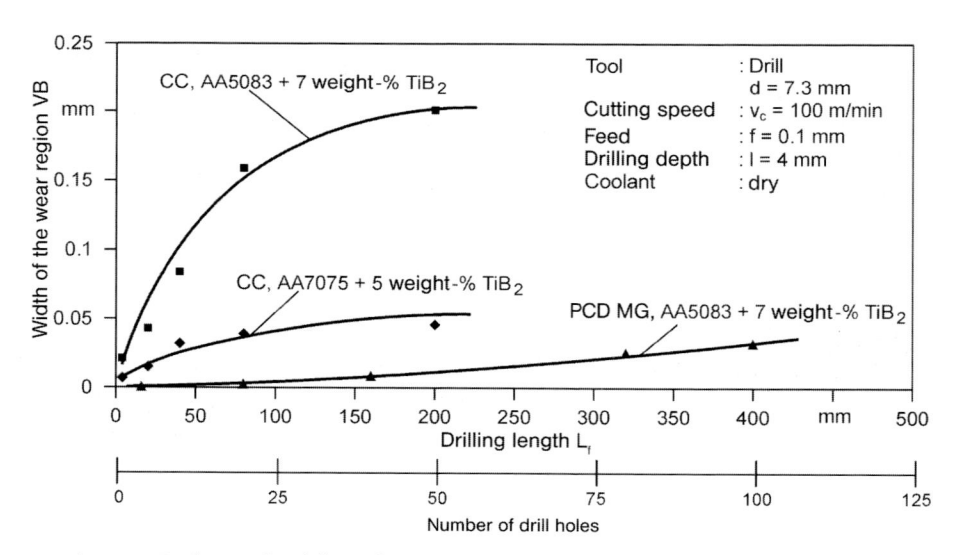

Fig. 6.21 Tool wear after drilling of TiB_2-particle-reinforced extrusion profiles with cemented carbide and PCD tools.

fer in both the TiB_2 content and also the type of Al-matrix, the strong difference in wear behavior of cemented carbide drills is due to the TiB_2 content. The influence of the type of Al matrix on the tool wear is assumed to be not that significant.

When comparing the wear mark after processing the Al alloy with a content of 7 % TiB_2 with cemented carbide and PCD drills (see Fig. 6.21) it becomes obvious that the PCD tools in comparison to cemented carbide drills show a significantly more convenient wear behavior. The width of the wear region after processing 100 drill holes is approximately 30 µm, even lower than after drilling the Al alloy with 5 % TiB_2 during application of cemented carbide tools after 50 drill holes.

The average surface roughness after drilling the Al alloy with 7 % TiB_2 with a PCD tool is between 3 and 7 µm, in which smaller surface parameters are measured at the end of the test series due to the increasing tool wear. When using cemented carbide tools an average surface roughness value between 1 and 3 µm is reached. The better surface roughness compared to PCD drills can be explained by the significant marked wear of the cemented carbide tool and by a significantly higher drilling torque. With increasing drilling length the drilling torque increases only slightly with the application of PCD drills and reaches a value of almost 1 N m after a drilling length of 400 mm (100 drill holes). With application of cemented carbide tools, the drilling torque after a drilling length of 200 mm (50 drill holes) is 2.2 N m.

The strong wear condition and the high drilling torque lead to plastic deformation of the drill hole peripheral zone when using cemented carbide tools, see Fig. 6.22. The peripheral zone after processing the first drill hole shows a similar strong alignment of the structure in the cutting direction as after processing the 50th drill hole. The drill hole peripheral zone after drilling with PCD tools shows a signifi-

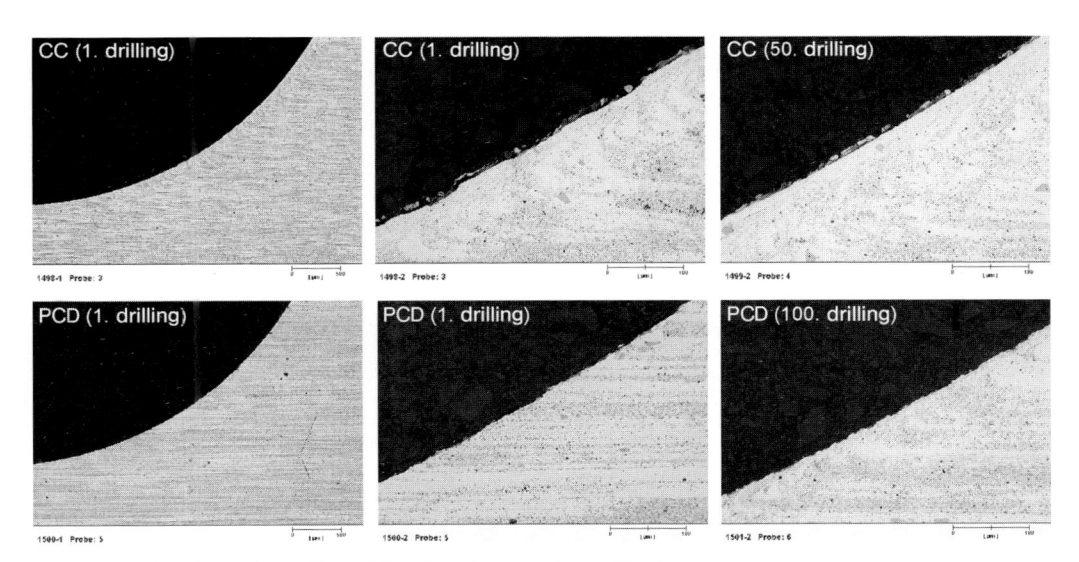

Fig. 6.22 Surface layer after drilling of the alloy AA5083 with 7 wt % TiB_2.

cantly lower influence due to the smaller drilling torque and the lower tool wear. Metallographic investigations of the peripheral zone of the first drill hole show no change in the structure. After 100 drill holes only a slight alignment of the structure in the cutting direction is noticeable.

6.3.5.2 Milling

Due to the interrupted cut during milling of thin walled extrusion profiles, the clamping device which is used to clamp the work pieces to the machining table shows a significant influence on the wear development. Fixing the 700 mm long profiles on the machining table with three jaws already causes spalling by the use of a cemented carbide shaft milling cutter after a single milling. If the work pieces are clamped over the full length by one ledge no spalling occurs due to the low oscillating strain. To reduce the oscillating strain during milling, most of the tests are carried out in up-cut milling. The cutting edge contact takes place in the area of the minimum chipping thickness, whereas the chip thickness increases continuously with further cutting procedure until the edge comes out of the insert. The cutting insert in the work piece proceeds with the biggest chip thickness during synchronous milling, so that the process is initiated by the strongest oscillation due to the strong insert impulse.

Figure 6.23 shows the development of the width of the wear region in dependence on the feeding path during peripheral milling of the alloy AA5083 with a content of 7 % TiB_2 with cemented carbide and PCD tools. Furthermore, scanning electron microscope images of the wear marks of both milling tools, taken at the end of the process, are shown. The cemented carbide tool shows a significantly higher flank wear after milling at a cutting speed of 100 m min^{-1} in contrast to the PCD milling cutter. After a feed length of 2.1 m the width of wear region is 70 µm.

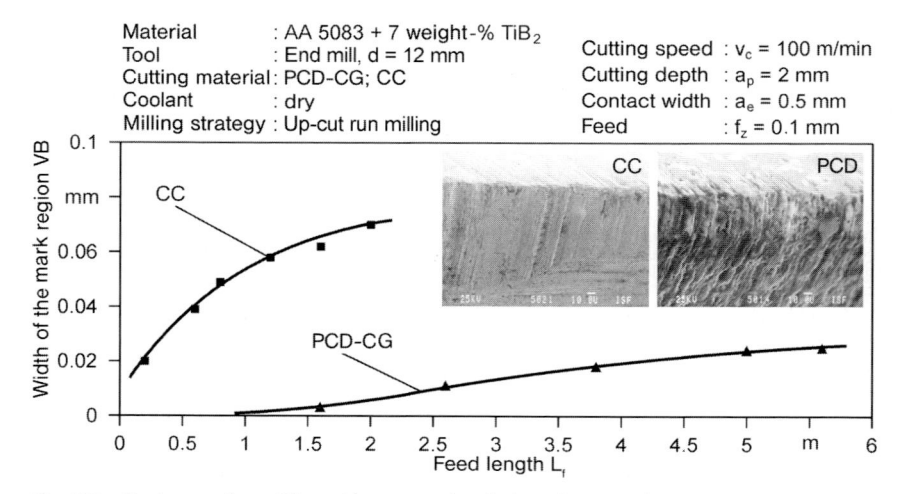

Fig. 6.23 Tool wear after milling with cemented carbide and PCD tools.

6.3 Processing of Components of Metal Matrix Composites

In contrast to that the PCD tool reaches a flank wear value of 25 µm after a feed length of 5.6 m. The convenient wear behavior of the PCD tool can also be seen in scanning electron microscopic images of both tools. Spalling at the cutting edges occurs on neither of the tools.

The influence of the cutting speed, feed and contact width on the tool wear for processing with PCD milling cutters is shown in Fig. 6.24. The three curves at cutting speeds, 100, 250 and 560 m min^{-1}, show declining, nearly parallel developing wear. Thus it can be seen that with increasing cutting speed a slight tendency towards higher tool wear occurs. During the increase in the cutting speed from 100 to 560 m min^{-1} a positive effect takes place from a process technological view. There the cutting time, which is used for the processing of one work piece, is reduced by a factor of 5.6. Because the tool wear does not increase significantly at a cutting speed of 560 m min^{-1} compared to lower cutting speeds, the TiB$_2$-particle-reinforced Al alloy is more economical for machining at higher cutting speeds. Wear development during the milling process of Al alloy is not significantly influenced by the feed and the contact width, so that appropriately high values of cutting parameters can be chosen for processing.

Figure 6.25 shows the influence of the milling strategy up-cut and synchronous milling on the application behavior of TiAlN-coated cemented carbide and PCD tools. During processing with the PCD milling cutter it can be seen that, due to the high hardness and connected low ductility, the up-cut milling turns out to be a better strategy than synchronous milling. While, for the average width of wear, a wear development can be determined which is independent of the milling strategy, only during synchronous milling does spalling occur. The reason for this spalling is to be found in the high input impulse of PCD cutting edges in the TiB$_2$-reinforced work piece. Due to this cutting edge spalling during synchronous milling, a maxi-

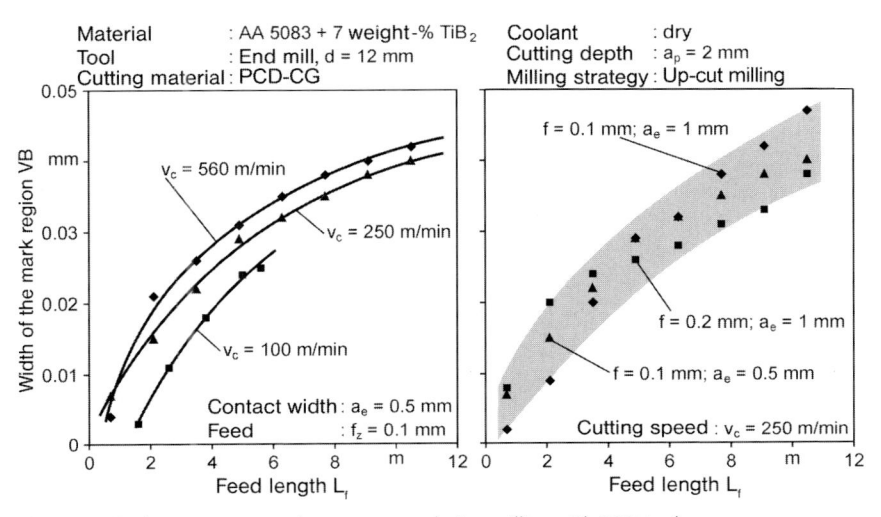

Fig. 6.24 Tool wear versus cutting parameter during milling with PCD tools.

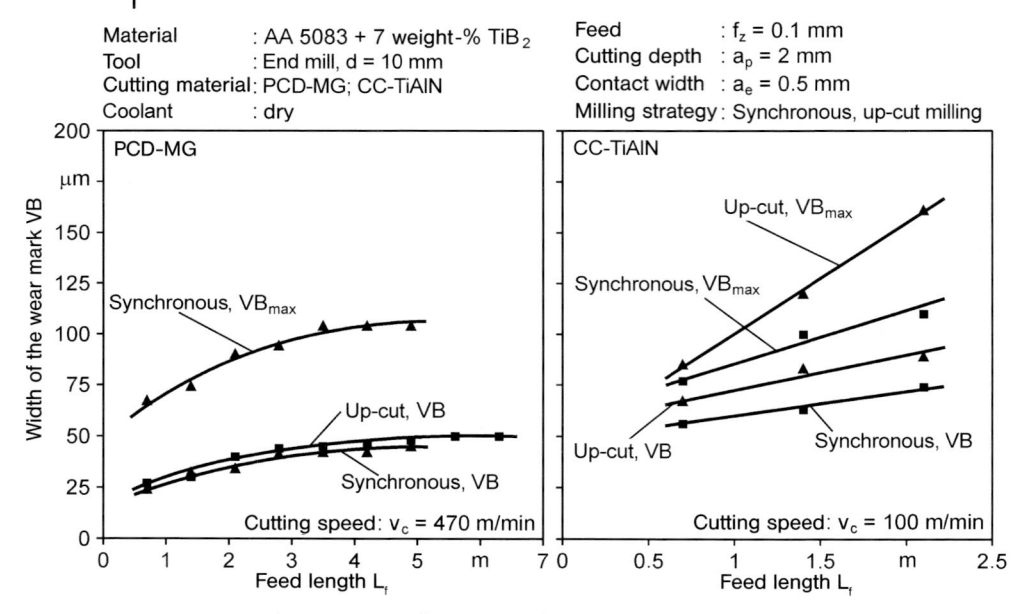

Fig. 6.25 Tool wear versus milling strategy during milling with TiAlN-coated cemented carbide and PCD tools.

mum width of wear region of over 100 µm is measured after a feed length of 4.9 m. Due to the better ductility of cemented carbide, compared to that of PCD, different behavior is shown during milling with TiAlN-coated cemented carbide tools. Lower tool wear at synchronous milling, compared to up-cut milling, can be noticed. Spalling at the tool edges occurs both during up-cut and synchronous milling. Thus, in both cases, a maximum width of wear region which is clearly higher than the average width of wear region can be measured.

The use of a diamond-coated cemented carbide shaft milling cutter does not show any improvement in the application behavior when machining TiB_2-reinforced extrusion profiles compared to TiAlN-coated tools, due to flaking of the diamond coating during the milling process. It is not possible to determine any influence of the grain size of PCD (2 and 10 µm) on the average width of wear region after up-cut milling. However, cutting edge spalling occurs with application of fine grained PCD types with a grain size of 2 µm after a feed length of 2.8 m. For this reason the maximum width of wear region reaches a value of almost 100 µm at a feed length of 6.3 m.

A change in the microstructure of the peripheral zone, for example in the form of a plastic deformation, could not be detected by metallographic investigations after processing of extrusion profiles with cemented carbide and PCD milling cutters. The average surface roughness that was measured parallel to the feed direction is always between 1 and 2 µm after up-cut or synchronous milling with PCD tools. With the application of cemented carbide tools an average surface roughness of 1 to 2 μm is also measured for synchronous milling. After up-cut milling the values range from 5 to 15 μm due to strong material wear down on the tools. The worn down material on the tool can be chipped off when inserting the cutting edge into the work piece.

6.4 Summary

The main problem during the machining of metal matrix composites is the strong tool wear, caused by direct contact between the hard reinforcement phase of the composite and the cutting edge. For this reason very hard cutting materials are particularly suitable for machining. During processing of SiC reinforced Al alloys it can be seen that coarse grained PCD types show lower wear compared to fine grained PCD. The present basic knowledge on the process of machining metal matrix composites of SiC-particle-reinforced brake drums, local Si-particle and Al_2O_3-short-fiber-reinforced cylinder crankcases and TiB_2-particle-reinforced extrusion profiles is transferred to the processing of concrete compounds. To evaluate processing tool wear and process quality (surface, peripheral zone and roundness) is used. The application of new cutting materials with a diamond thick layer, produced by CVD technology, leads to significantly lower tool wear compared to PCD during turning of SiC-reinforced brake drums and during drill finishing of Si- and Al_2O_3-reinforced cylinder surfaces. During drilling and milling of TiB_2-reinforced extrusion profiles a significant superiority of PCD tools is shown compared to the applied cemented carbide tools. For machining of MMC compounds more convenient process parameters can be used.

References

1 K. Schulte, Faserverbundwerkstoffe mit Metallmatrix: Auf Hochleistung getrimmt, *Industrie-Anzeiger* **1987**, *48*, 23–28.
2 W. Tillmann, E. Lugscheider, Möglichkeiten des stoffschlüssigen Fügens metallischer Verbundwerkstoffe, *Schweißen und Schneiden* **1994**, *46*, 543–549.
3 D. Richter, Bremsscheiben aus Keramik-Partikel-verstärktem Aluminum, *Aluminum* **1991**, *67*, 878–879.
4 S. Beer, W. Henning, M. Liedschulte, MMC-Motorenkomponenten: Herstellung und Eigenschaften, in *Spanende Fertigung*, K. Weinert (Ed.), Vulkan-Verlag, Essen, **1997**, pp. 333–346.
5 E. Köhler, F. Ludescher, J. Niehues, D. Peppinghaus, LOKASIL-Zylinderlaufflächen – Integrierte lokale Verbundwerkstofflösung für Aluminum-Zylinderkurbelgehäuse, *ATZ/MTZ-Sonderausgabe Werkstoffe im Automobilbau*, **1996**.
6 E. Tank, Erfahrungen bei der Erprobung von experimentellen Motor-Bauteilen aus faser- und partikelverstärkten Leichtmetallen, *Metall* **1991**, *44*(10), 988–994.
7 K. Weinert, M. Buschka, M. Liedschulte, D. Biermann, U. Huber, J. Niehues, Mechanische Bearbeitung von Komponenten aus Leichtmetallverbundwerkstoffen, in *Verbundwerkstoffe und Werkstoffverbunde*, K. Schulte, K.-U. Kainer (Eds.). Wiley-VCH, Weinheim **1999**, pp. 207–212.

8 D. Biermann, Untersuchungen zum Drehen von Aluminummatrix-Verbundwerkstoffen, Dissertation, Universität Dortmund **1994**.

9 K. Weinert, D. Biermann, M. Buschka, M. Liedschulte, Be- und Verarbeitungstechnologien für Verbundwerkstoffe, in *Metallische und Metall-keramische Verbundwerkstoffe*, F.-W. Bach, H.-D. Steffens (Eds.), KONTEC Gesellschaft für technische Kommunikation mbH, Hamburg, **1999**, pp. 133–169.

10 K. Weinert, M. Buschka, J. Niehues, A. Schoberth, Spanende Bearbeitung von Bauteilen aus Al-Matrix-Verbundwerkstoffen, *Materialwissenschaft und Werkstofftechnik* **2001**, *32*, 447–461.

11 K. Weinert, D. Biermann, M. Buschka, Drehen sprühkompaktierter, übereutektischer AlSi-Legierungen, *Aluminum* **1998**, *74*(5), 352–359.

12 K. Weinert, D. Biermann, M. Buschka, M. Liedschulte, Machining of Spray Deposited Al-Si Alloys, *Prod. Eng. – Ann. Germ. Acad. Soc. Prod. Eng.* **1998**, *V/2*, 19–22.

13 D. Meister, Untersuchungen zum Vollbohren partikelverstärkter Aluminumlegierungen mit Wendelbohrern. Dissertation, Universität Dortmund, **1991**.

14 S. K. Changwaro, Fräsen von Faserverbundwerkstoffen mit Aluminummatrix. Dissertation, Universität Dortmund, **1990**.

15 P. Müller, Untersuchungen zum Bohren von Faserverbundwerkstoffen mit Aluminummatrix. Dissertation, Universität Dortmund, 1988.

16 R. Hay, CVD-Diamantwerkzeuge eröffnen neue Möglichkeiten, Der Schnitt- & Stanzwerkzeugbau, **1998**, p. 4.

7
Mechanical Behavior and Fatigue Properties of Metal-matrix Composites

H. Biermann and O. Hartmann

7.1
Introduction

Swords have been manufactured, for example in Japan, for many decades by the layer-wise superposition of hardened and soft steel with subsequent massive transformation, for example by folding and hammering. In this way the opposing demands on the properties of the blade material, i.e. hardness so that the blade can be sharpened and high toughness, have been met. Nowadays, composites offer, from a technical view, similar advantages. By combination of several materials a compromise between the properties of each single component can be achieved, whereby the properties of the composites are mostly better than the properties of the homogeneous components. The selection of the combination is determined by the properties of each component. Thus, the combination, for example, of a ductile metal and a brittle ceramic, results in (compared to the pure metal) a composite with a higher Young's modulus, and also lower ductility.

In the last quarter of the last century more and more metal-matrix composites (MMC) were developed for technical applications. The better mechanical properties, compared to nonreinforced matrix materials, led to an impulse in materials research and to a considerable increase in techniques to manufacture composites. The decrease in manufacturing costs mostly helped in the application of some types of composites in industrial applications, mainly discontinuously reinforced MMCs. Nevertheless, MMCs are niche market materials, which, however, can replace other construction or functional materials due to their special properties [1].

Technical applications are mostly combined with cyclically varying mechanical stresses. These result in fatigue damage which is responsible for numerous failure cases. The mechanical parameters of a material, which are relatively easy to determine at a quasi-static load, are not easily transferable to properties of cyclic loading. Therefore, knowledge of the influence of the microstructure and the composition of a composite on the fatigue properties is of special interest.

The results presented in this chapter are a part of an extensive investigation of the mechanical behavior of metal-matrix composites [2]. The influence of the ma-

Metal Matrix Composites. Custom-made Materials for Automotive and Aerospace Engineering.
Edited by Karl U. Kainer.
Copyright © 2006 WILEY-VCH Verlag GmbH & Co. KGaA, Weinheim
ISBN: 3-527-31360-5

trix strength and the reinforcement morphology on the cyclic deformation and fatigue behavior is considered in detail. In a nonreinforced material (hardened Al alloy AA6061) as well as with particle- and short-fiber-reinforced composites with an aluminum matrix, isothermal, total-strain controlled cyclic deformation tests have been carried out with three different composites. These differ either in the shape of the reinforcements (particles or short fibers (Saffil fibers) of Al_2O_3) or the matrix strength (AA6061 alloy or technically pure aluminum Al99.85).

7.2
Basics and State of Knowledge

7.2.1
Thermal Residual Stresses

Thermally induced residual stresses are unavoidable during cooling from the processing temperature to room temperature, because the thermal expansion coefficient of aluminum is larger than the coefficient of the, mainly ceramic, reinforcements. Thus, during cooling, different thermal contractions occur in the phases. The residual stresses, which occur, are in equilibrium with each other and hence [3]:

$$f \cdot \langle \sigma_R^{th} \rangle + (1-f) \cdot \langle \sigma_M^{th} \rangle = 0 \tag{1}$$

In the composite, tensile residual stresses in the metal matrix (M: matrix) ($\langle \sigma_M^{th} \rangle > 0$) and compressive residual stresses in the ceramic reinforcement phase (R: reinforcement) ($\langle \sigma_R^{th} \rangle < 0$) inevitably occur, see for example Refs. [4, 5]. After Eq. (1), thermally induced stresses develop which are higher within the reinforcement phase than in the matrix, if the volume fraction of the reinforcement phase, f, is less than 50%. These residual stresses can be measured experimentally, for example by X-ray or neutron diffraction, or also determined theoretically. The thermally induced stresses depend on various parameters of the composite, for example the volume fraction, size, morphology or arrangement of the reinforcement phase and, of course, the strength of the matrix and thus also the heat treatment.

7.2.2
Deformation Behavior of Metal-matrix Composites

Figure 7.1 shows schematically the quasi-static stress–strain behavior of MMCs. The metallic matrix thereby shows simplified ideal elastic–plastic behavior. The ceramic reinforcement deforms in a purely elastic way until it fails. In comparison to nonreinforced material (matrix), a higher stiffness (Young's modulus), a higher strain hardening (increase in the σ–ε curve after the start of plastic deformation), a higher strength and a lower strain to fracture are found in the composite.

By a simple approach, the Young's modulus, which depends on the properties of the components and their portions, can be estimated [6, 7]. It is differentiated

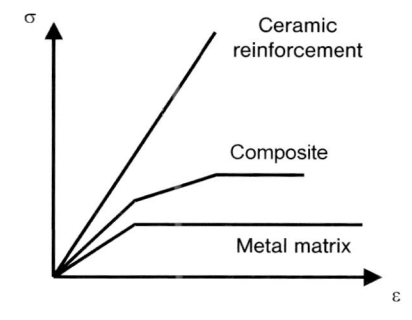

Fig. 7.1 Schematic representation of the deformation behavior of composites, resulting from the mechanical behavior of the single phases.

between the parallel and serial arrangement of reinforcements, which present the limits of the expected Young's modulus of the composite, E_{MMC}:

$$E_{MMC} = f \cdot E_R + (1-f) \cdot E_M \quad (2)$$

$$E_{MMC} = \frac{E_M \cdot E_R}{f \cdot E_M + (1-f) \cdot E_R} \quad (3)$$

Figure 7.2 shows the dependence of the Young's modulus of the composite on the volume fraction of the reinforcement (f) for both possible borderline cases. Therefore, composites with continuous arrangement of the reinforcements (long fibers) reach higher Young's moduli than those with discontinuous reinforcements (short fibers or particles).

In comparison with nonreinforced metallic materials the higher work hardening of composites results from the stronger hindering of dislocation movements by the presence of the ceramic phase. This causes a smaller grain size (Hall–Petch hard-

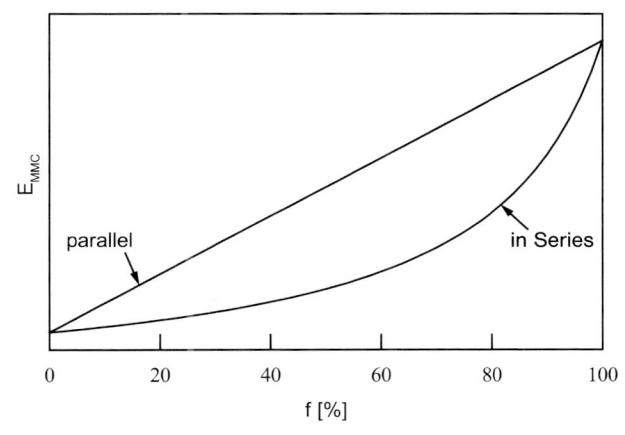

Fig. 7.2 Influence of the volume fraction of the reinforcement f on the Young's modulus after Eqs. (2) and (3). The shown lines set the limits of composite properties of parallel (Eq. (2)) and of serial (Eq. (3)) arrangement of the reinforcements.

ening) and a higher dislocation density, already created before the deformation by thermal residual stresses.

Figure 7.1 shows that the stress of the reinforcement phase in the composite at a given macroscopic load is considerably higher than the stress in the matrix (load transfer). A failure of the reinforcement already occurs at a lower external stress than in the homogeneous ceramic material. The damage acts contrary to the increase in the strength of the matrix due to the plastic deformation and leads to an earlier failure (in terms of strain) of the composite compared to the nonreinforced matrix material.

Figure 7.3 shows the stress–strain curves of nonreinforced Al matrix materials AA6061-T6 and Al99.85 and of particle- or short-fiber-reinforced composites. All composites show a higher Young's modulus than the corresponding matrix material. The composites show a higher degree of strain hardening. In the case of short-fiber reinforcement with the AA6061-T6 matrix, the curve of the MMC lies partly below the curve of the nonreinforced material due to the earlier starting damage of the fibers on reaching the yield stress.

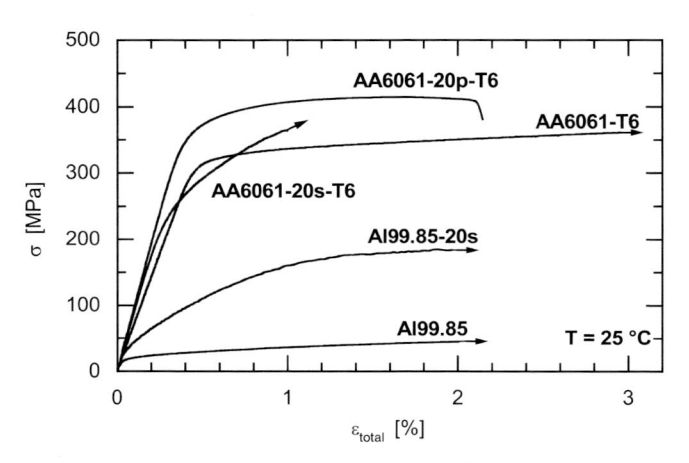

Fig. 7.3 Stress–strain curves of AA6061-T6 and Al99.85 and of their particle- (20p) and/or short-fibre (20s) reinforced composites. The arrows indicate that the deformation of the sample was stopped before failure.

7.2.3
Determination of the Damage in Composites

The damage of ceramic reinforcements (especially the fracture of particles) can be analyzed by optical microscopy on metallographic sections. Lloyd [8] has investigated the polished and electrochemically marked surface of a reinforced AA6061 alloy, containing $f = 10$ and 20 vol.% SiC particles, with a scanning electron microscope (SEM) after defined deformation steps and correlated the number of damaged particles with the strain or stress.

A commonly used method to analyse the damage in MMCs over the last few years has been the *in situ* deformation in a scanning electron microscope (SEM) [9–12], in which local changes at the polished surface of a sample can be observed. Beside qualitative results, for example the position of the first occurrence or the greatest observed damage, quantitative damage values can also be determined. However, this *in situ* method allows only conclusions on the damage at the surface of a sample, which could be quite different from that in the interior of the material [13].

Another method to analyse damage is the measurement of the elastic stiffness of a sample, which can be determined in mechanical tests by partial unloading. Thus, the decrease in stiffness is combined with the deformation-induced damage of the reinforcements [14–22], since the load-carrying fraction of the reinforcements is decreasing. Kouzeli et al. [19] have investigated the relation between the damage caused by deformation and the decrease in density. From this, they have defined a resulting damage parameter which has been related to the values determined from the unloading tests.

Singh and Lewandowski [23] have discovered a relationship between the change in the Poisson's ratio and the damage caused by mechanical deformation. The strain has been measured both in the deformation direction and transverse to it. With increasing deformation the Poisson's ratio increases, whereas a significant strong increase has been found at a threshold stress (strain).

Microtomography presents an investigation method in which the microstructure of the sample can be imaged in a nondestructive way (compare [13, 24–29]). With this technique an MMC sample up to 2 mm thick is irradiated with a monochromatic, coherent synchrotron X-ray beam and the intensity contrast caused by the different absorption of the two phases (weakening of the X-ray beam) is measured by a CCD camera. Not only the intensity difference but also the phase contrast can be used for imaging (called holo-tomography). During the rotation of the sample, several hundred two-dimensional images are produced which can be recalculated into a three-dimensional image of the composite using information on the rotation angle. With this data any (virtual) cross-section through the material can be shown. The advantage of this technique is that a nondestructive insight into the material interior is possible, and the sample can be further deformed after the measurement. Hence the development of damage can be presented in more detail. The lateral resolution of this technique is nowadays a maximum of 0.7 µm.

Acoustic emission analysis (see for example Refs. [30–36]) is a further nondestructive investigation method to quantify damage. In this the emitted sound is detected on the sample surface, which, for example, occurs during rupture of particles. The recorded signals can be assigned to the energy emitted by the damage (elastically stored strain energy) which is a measure (at least relatively) of the size of the particles using their intensity and frequency.

The damage parameter D commonly used in literature is the change X_0-X of a parameter X related to the damage, normalized on the initial parameter X_0:

$$0 \leq D = \frac{X_0 - X}{X_0} = 1 - \frac{X}{X_0} \leq 1 \qquad (4)$$

D has, thereby, values between 0 (no damage) and 1 (maximum damage). In most cases the difficulty is to determine a critical parameter of this damage measurement because it is necessary to differentiate whether the received parameter is representative of the total sample or not. In many publications, for example Refs. [20, 37–40], it is shown that the fracture probability of the particles increases with increase in the particle size, and thus the biggest particles fail first. This is caused by the increase in the defect probability of a particle with increasing particle volume and by the increasing strain energy.

However, *D*, defined after Eq. (4), is a parameter which describes the global (average) damage – provided that several evenly distributed statistical areas of the sample are investigated, and does not admit any statements about the local damage fraction or the classification into a special group (for example the orientation of the particles with respect to the stress axis). Moreover, a sample fails at that part where the highest damage has been found [18, 30, 41, 42]. Since this part is not always known, a value measuring the integral damage of the entire sample is not necessarily suitable to predetermine the failure of components or samples.

7.2.4
Basic Elements and Terms of Fatigue

Cyclic load is the temporal variation and repeating plastic or purely elastic deformation of a component. By microstructural processes, like, for example, the localized deformation or locally arising "over loading" and thus localized damage, a weakening (fatigue) of the material, and finally a failure of the component takes place. In fatigue tests such a load is transferred in a reproducible way to the standardized component (fatigue specimen). So, for example, low cycle fatigue (LCF) tests are carried out in strain control that means the temporarily varying strain of the sample follows a given course. Figure 7.4 shows the corresponding test proce-

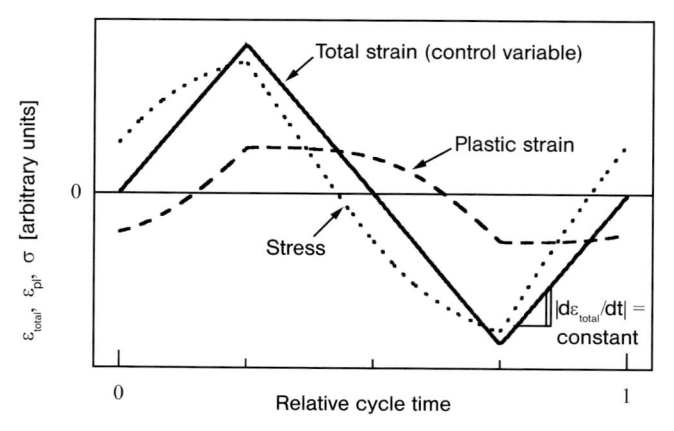

Fig. 7.4 Course of the signals of stress, total strain and plastic strain during one cycle ($N > 1$) at total strain control.

dure with a triangle-shaped signal course of the total strain ε_{total} and the resulting material reaction, the stress σ. Figure 7.5 shows the stress-strain hysteresis loop. The characteristic parameters of this are:

$\varepsilon_{total,\,min}$, $\varepsilon_{total,\,max}$ — Minimum or maximum total strain
σ_{min}, σ_{max} — Minimum or maximum stress
$\Delta\varepsilon_{total} = \varepsilon_{total,\,max} - \varepsilon_{total,\,min}$ — Total strain range
$\sigma_m = (\sigma_{max} + \sigma_{min})/2$ — Mean stress
$\Delta\sigma = \sigma_{max} - \sigma_{min}$ — Stress range
E_T — Sample stiffness after load reversal in tension
E_C — Sample stiffness after load reversal in compression

Often the term amplitude is used instead of range, which is half the total range (for example: stress amplitude $\Delta\sigma/2$). All parameters depend on the present microstructural state of the material and they can change during the course of fatigue loading. The presentation of the stress or strain amplitudes as a function of the number of cycles N is called a cyclic deformation curve (one cycle corresponds to one repetition of the cyclic deformation). Figures 7.6(a) and (b) show schematically the possible courses of a cyclic deformation curve for the case of a total-strain controlled test, which can be mostly divided into three regions. Depending on the initial condition showing higher strength (for example by massive deformation due to the manufacturing process) or lower strength (for example by a recovery or recrystallization annealing done before the test), the strength decreases or increases during the first cycles, represented by the change in the stress amplitude. These changes of strength in regions I are called initial cyclic softening and hardening of the material, respectively. Due to the repeating mechanical deformation, a cyclically stable microstructure appears, for example a constant density and arrangement of dislocations. Such a condition gives a constant course of the cyclic deformation curve, compare region II in Fig. 7.6. Due to the mechanical load, the material is

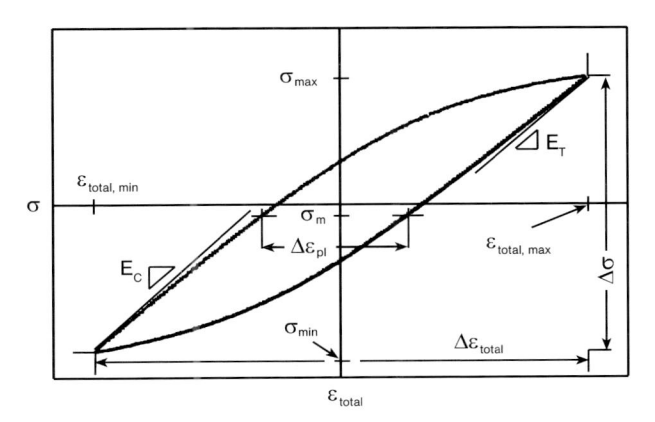

Fig. 7.5 Mechanical parameters of a hysteresis loop.

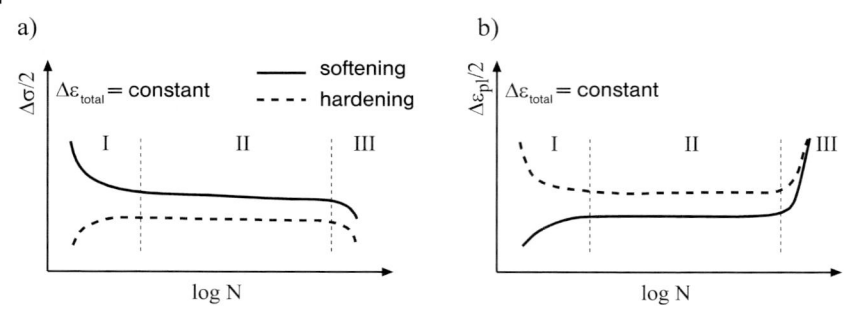

Fig. 7.6 Schematic representation of cyclic deformation curves. Characterization of the cyclic behavior of two materials with cyclic softening (solid line) and cyclic hardening (dashed line) during total strain control ($\Delta\varepsilon_{total}/2 =$ constant) with (a) the stress amplitude ($\Delta\sigma/2$) and (b) the amplitude of plastic strain ($\Delta\varepsilon_{pl}/2$).

damaged, mostly by localization of the strain, and fatigue cracks occur. Their growth leads to a rapid decrease in the stress amplitude (region III). Figures 7.6(a) and (b) show that the changes in stress and plastic strain amplitude are contrary.

A further possibility to describe the fatigue behavior is a plot of the stress amplitudes of several tests versus the corresponding (plastic) strain amplitudes of cyclic saturation. The resulting curve is known as a cyclic stress–strain (CSS) curve, see Fig. 7.7. It contains information on the strength for a given strain amplitude. Furthermore, the comparison with the stress–strain curve of the tensile test gives information about the cyclic softening/hardening behavior.

The fatigue life of a material (number of cycles to failure) depends on the stress or strain amplitude under cyclic loading. The dependence of the deformation parameter on the fatigue life is recorded in double-logarithmic so-called Wöhler diagrams (stress-controlled tests) or in strain-Wöhler diagrams (strain control). In the Wöhler diagram the stress amplitude is given versus the number of cycles to failure. At strain control a double logarithmic diagram of the total, plastic and elastic strain amplitudes in the saturation condition are plotted against double the number of cycles to failure, $2N_F$. If there is no cyclic saturation, values are taken of the

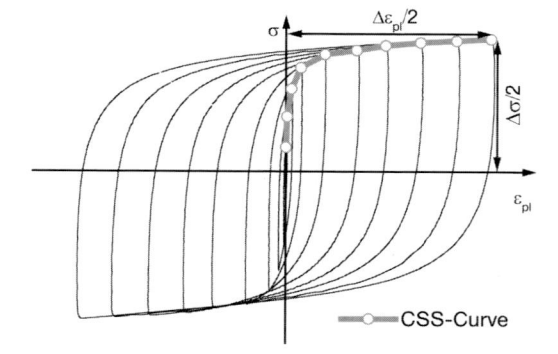

Fig. 7.7 Connection between stress–strain hysteresis loops of cyclic saturation and the cyclic stress-strain (CSS) curve.

hysteresis at maximum strength ($N = N_{\Delta\sigma\,max}$) or at half the number of cycles to failure ($N = 0.5\,N_f$). The relation between plastic strain amplitude and the number of load reversals until failure of a sample was first reported by Manson [43] and Coffin [44] and they, independent of each other, set up the relation between N_f and $\Delta\varepsilon_{pl}/2$:

$$\Delta\varepsilon_{pl}/2 = \varepsilon_f'(2N_f)^c \qquad (5)$$

Here ε_f' is the fatigue ductility coefficient and c the fatigue ductility exponent. In an analogous way Basquin [45] discovered already at the beginning of the 20th century the relation between elastic strain amplitude $\Delta\varepsilon_{el}/2$ and fatigue life:

$$\Delta\varepsilon_{el}/2 = (\sigma_f'/E)(2N_f)^b \qquad (6)$$

where σ_f' is the fatigue strength coefficient and b the fatigue strength exponent. Because of the decrease in fatigue life with increasing elastic or plastic strain amplitude, the exponents in Eqs. (5) and (6) are always negative.

Basically three types of damage are possible by cyclic loading of discontinuously reinforced materials: (i) fracture of the reinforcement, (ii) failure of the interface between the reinforcement and the matrix (delamination) and/or (iii) failure of the matrix. A combination of all types mostly takes place with one dominateing. Figure 7.8 shows schematically the possible different paths of fatigue cracks depending on the failure type for the example of particle-reinforced composites (after [46]). A fatigue crack therefore preferably runs through damaged areas. The type of the damage depends on the relation between the local load, and each (fracture) strength of the matrix or the reinforcement as well as the interface adhesion between both phases. The latter is strongly influenced by the interface reaction between elements of the reinforcement and the matrix. Depending on the system of the MMC, different interface products can result from the thermal prehistory and manufacturing technique. Their properties, especially their adhesion to the matrix and to

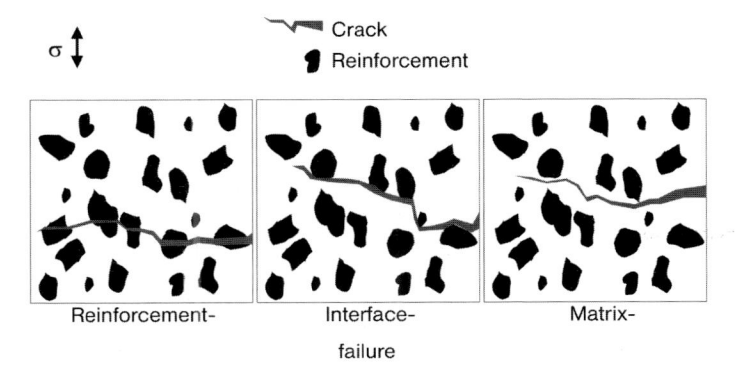

Fig. 7.8 Schematic representation of possible crack paths of different failure cases.

the reinforcement, influence the properties of the composite, especially the ability to transfer load from the matrix to the reinforcement, since the strength of the composite depends strongly on that [47].

The load of the material at the crack tip is influenced by the macroscopic load, given by the stress range $\Delta\sigma$, as well as the length of the crack, a, and is characterized by the stress intensity factor range ΔK:

$$\Delta K = K_{max} - K_{min} = \Delta\sigma \cdot Y \cdot \sqrt{\pi \cdot a} \qquad (7)$$
$$\text{with } K_{max} = \sigma_{max} \cdot Y \cdot \sqrt{\pi \cdot a}$$
$$K_{min} = \sigma_{min} \cdot Y \cdot \sqrt{\pi \cdot a}$$

Y is thereby a value dependent on the sample geometry. It follows from Eq. (7) that the load at the crack tip increases with increasing crack length.

At small stress intensity levels the reinforcements are not damaged and thus the cracks bypass them. The size of the plastic zone in front of the crack tip is influenced (reduced) by the particles and the crack growth is thereby hindered. At higher stress intensities a large plastic zone forms in front of the crack tip, in which the reinforcements break due to stress concentrations (process zone), through which the crack runs. The influence of the plastic zone in front of the crack tip depends, at a given stress amplitude, on the crack length. Li and Ellyin have shown, that cracks only a few hundred micrometers in length run along the particles in front of the crack tip [48], because from this length onwards a value of ΔK is reached, at which the particles or the interfaces to the matrix are damaged. Similar results have also been found by Wang and Zhang [49].

The resulting crack growth rate da/dN at a given load is determined by using standardized samples, where the crack growth with the cycle number N is measured at a given stress intensity factor range ΔK. The double-logarithmic graph showing da/dN against ΔK presents a crack growth curve, as is seen in Fig. 7.9. Therefore it is worth mentioning that crack growth demands a sufficiently large value of the stress intensity factor range. Below this critical threshold value ΔK_0 no crack growth is found. At larger values of ΔK the curve is characterized by a linear course, which can be described by the Paris law:

$$\frac{da}{dN} = C\Delta K^m \qquad (8)$$

At high values of ΔK the crack growth rate increases to a great extent until an over critical crack growth occurs and thus failure of the sample takes place.

Fatigue cracks develop in metal-matrix composites mostly at damaged reinforcements near to the surface area [50, 51]. Results of Chen *et al.* [52], Habel et al. [53] and Lukasek and Koss [54] show that, related to N_f, a crack initiation starts earlier at higher stress amplitudes than at low amplitudes (high fatigue lifes), in which small cracks occur relatively late. Due to damage occurring within the first cycles at low-cycle fatigue, microcracks around 10 µm length occur already after 10% of the fatigue life; however, at small amplitudes, with a resulting high number of cycles to failure, a crack initiation only occurs at the end (70–90%) of the fatigue life.

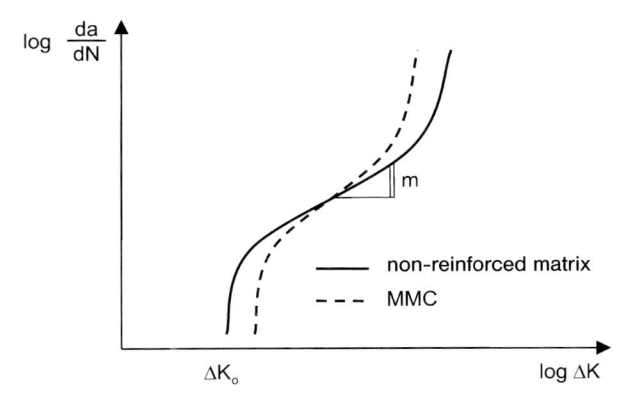

Fig. 7.9 Dependence of the crack growth rate da/dN on the stress intensity factor range ΔK for the nonreinforced material and the MMC.

In general the threshold value of the cyclic stress intensity factor ΔK_o is slightly increased by the addition of ceramic reinforcements [55, 56]. However, the critical crack growth starts earlier than in the nonreinforced matrix [57, 58], see Fig. 7.9. The crack growth rate increases thereby with an increase in volume fraction of reinforcements [51, 59–62].

7.2.5
Fatigue Behavior of Composites

In technical applications the components are designed in such a way that the relation between maximum stress and yield point is significantly smaller than 1. The material is then subjected to HCF loading (high-cycle fatigue). Various tests of fatigue behavior apply stress-controlled tests, because here the control parameter (stress amplitude) can be controlled more easily than the very low strain amplitude. In the area of short fatigue lives the maximum strains are higher, so that in this case the strain control is often applied. However, the circumstance that the Young's modulus and often the hardening exponent as well as the yield point are increased by the addition of a phase with higher Young's modulus makes it necessary to consider the type of test guidance if a MMC is compared to nonreinforced material [63]. This context is shown in Fig. 7.10, in which the hysteresis loop of a nonreinforced material is compared to the curve of a metal-matrix composite for various cases of control type. It can be seen that during total-strain control the MMC is exposed to a higher load (stress) and moreover the plastic strain range $\Delta\varepsilon_{pl}$ is higher than in the nonreinforced material. However, comparing the hysteresis loops with each other under the aspect of stress control (constant $\Delta\sigma$) shows that both the plastic and also the total strain amplitudes are significantly lower. The nonreinforced material is therefore, in tests with identical stress amplitude, under higher plastic strain amplitudes and therefore a lower service life is reached

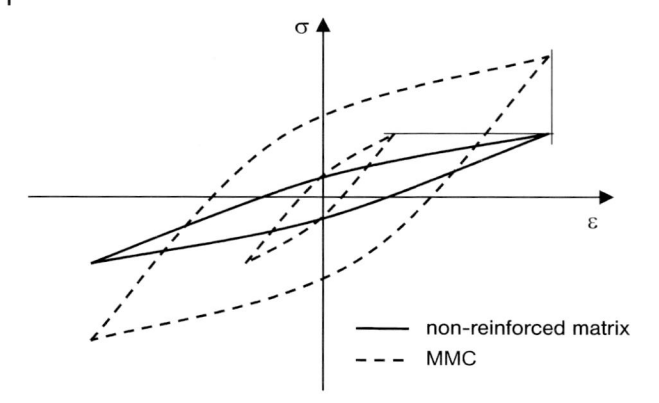

Fig. 7.10 Comparison of hysteresis loops of a nonreinforced matrix material (solid line) with hysteresis loops of reinforced material (dotted lines) at identical total strain amplitude (outside hysteresis) and identical stress amplitude (inner hysteresis).

[64–66]. Under total-strain control this is mostly the other way round [61, 64, 67].

Due to the increasing fracture probability with an increase in ceramic reinforcements at given stress, the fatigue life for MMCs can be less than that of nonreinforced materials; this is also true in stress-controlled tests, when the reinforcements are sufficiently large [40, 68]. Under stress control the effect of fatigue life increase with an increase in the reinforcement volume fraction is more pronounced at smaller stress amplitudes, because the MMC is damaged by particle fracture to a smaller degree from the beginning of the deformation onwards. It should also be considered that plastic deformation occurs within the MMC in a strongly localized manner. This could be proven by finite element calculations [69–72]. For this reason the microplastic strain in the matrix is locally higher than the hysteresis opening of the macroscopically measured plastic strain.

There have been several investigations of the influence of the average size of reinforcements on the fatigue behavior [40, 68, 73, 74] and it was found that the number of cycles to failure decreases with increasing particle size. Two reasons are mostly given: analogous to uni-directional behavior [69] the fraction of damaged reinforcements increases at a given stress with increasing amount of reinforcements, due to the increase in the probability that one critical defect initiating rupture of the sample will be present [62]. Moreover, the distance between the reinforcements, and thus the free length for dislocation glide movements, is also influenced by their average size, which affects the formation of dislocation arrangements (dislocation cell structures, slip bands) [75]. A current overview on the fatigue behavior of discontinuously reinforced composites is given by Llorca [76], some of the results presented in this work with additional data on fatigue tests on continuously-reinforced composites were given elsewhere [77].

7.3 Experimental

7.3.1 Materials

The investigated material is an age hardenable AA6061 alloy (AlMgSiCu), which is reinforced by Al_2O_3 either with f = 20 vol% particles (commercially available MMC from Duralcan company), or f = 20 vol% Saffil short fibers. These MMCs are denoted in the following by AA6061-20p (p: particle) or AA6061-20s (s: short fiber). A nonreinforced AA6061 alloy serves as a reference material. Furthermore, an additional composite of short-fiber-reinforced, technically pure (therefore very soft) aluminum Al99.85 (Al99.85-20s) was investigated. The samples with the AA6061-matrix were hardened by the T6 precipitation heat treatment (30 min solution annealing at 560°C, followed by water quenching, 16 h hardening at 165°C for AA6061-20p and 8 h for AA6061-20s). The gauge length of the round samples was polished mechanically with a diamond suspension of finally 1 µm after T6 heat treatment. Further details on the preparation can be found in Refs. [2, 78, 79].

Figures 7.11 and 7.12 show the reinforcement structure of the composites for the particle-reinforcement and the short-fiber-reinforcement (both short-fiber-reinforced MMCs have the equivalent Al_2O_3 reinforcement preforms), respectively, in sections parallel to the axes of the samples. The particle-reinforced MMC has been extruded into rods. Figure 7.11 shows the orientation of the approximately

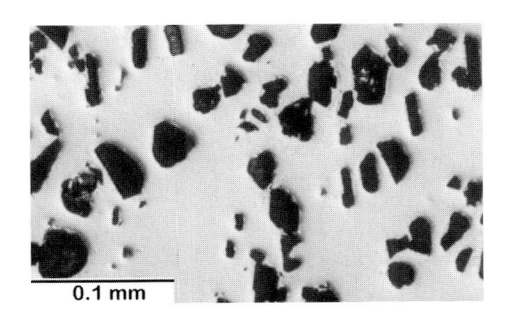

Fig. 7.11 Optical micrograph of the particle-reinforced composite AA6061-20p.

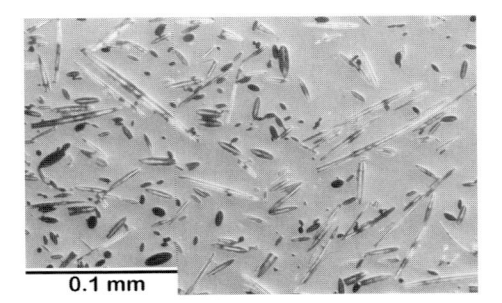

Fig. 7.12 Optical micrograph of the short-fiber-reinforced composite AA6061-20s.

13 µm long, square-shaped and almost evenly distributed particles with a preferential orientation in the extrusion direction (vertical). The short-fiber-reinforced material (plates) was produced by infiltration of a porous preform and shows a so-called random planar arrangement of about 3–5 µm thick and 200 µm long fibers; i.e. the fibers are oriented parallel to a plane, in which the sample axes also lie.

7.3.2
Mechanical Tests

The mechanical tests were carried out in a servo-hydraulic universal testing machine with total-strain control ($|d\varepsilon/dt| = 0.002$ s^{-1}), where the strain was measured with a clip-on extensometer directly at the gauge length. The symmetrical tension–compression cyclic tests were carried out with amplitudes of total strain, $\Delta\varepsilon_{total}/2$, between 0.001 and 0.01 at room temperature (see also Ref. [2]).

7.4
Results and Comparison of Different MMCs

7.4.1
Cyclic Deformation Behavior

The cyclic deformation curves of the nonreinforced AA6061-T6 alloy and of the composites with AA6061-T6 matrix are shown in Fig. 7.13 (compare also for more details Refs. [2, 77, 79, 80]). The numbers next to the curves indicate the total-strain amplitude for each test. The nonreinforced material shows cyclic hardening at only $\Delta\varepsilon_{total}/2 = 0.008$. At low strain amplitudes stable cyclic behavior is observed. Addition of a ceramic reinforcement phase causes an increase in strength.

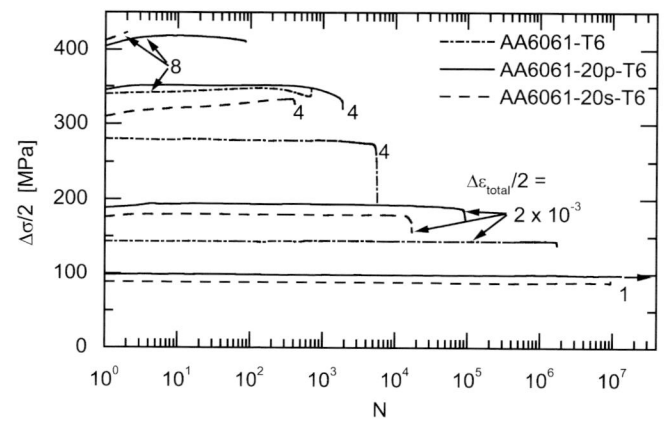

Fig. 7.13 Cyclic deformation curves of AA6061-T6 and composites with AA6061-T6 matrix at different total strain amplitudes [2].

The cyclic deformation curves of the composites at a given strain amplitude are therefore above those of AA6061-T6, whereas the particle-reinforced MMC shows higher stress amplitudes than the short-fiber-reinforced composite. Furthermore, an initial cyclic hardening occurs at total strain amplitudes of $\Delta\varepsilon_{total}/2 = 0.002$ and higher, which the nonreinforced material does not show. At small strain amplitude, the cyclic deformation behavior of the composites is cyclically stable.

The amount of the initial cyclic hardening depends on the plastic deformation of the alloy; the hardening is based on the mutual (interfering) interaction of the dislocations. Since plastic deformation occurs in localized shear bands and thus a locally higher dislocation density is present than macroscopically measurable, composites harden to a larger extent than the non-reinforced alloy (at given total-strain amplitude). The short-fiber reinforced composite hardens more than the particle-reinforced composite.

In the particle-reinforced MMC, a cyclic softening occurs at higher strain amplitude (see Figure 7.13, only at $\Delta\varepsilon_{total}/2 = 0.004$), which is not caused by the growth of a macro crack, but by softening processes within the matrix [2].

Figure 7.14 shows cyclic deformation curves of short-fiber-reinforced MMCs with different matrix strength. Due to the low strength of the matrix at comparable strain amplitude, the curves of Al99.85-20s are below the AA6061-20s-T6 curves. Furthermore, Al99.85-20s shows a significant cyclic hardening during the first cycles, which becomes more distinctive with increasing strain amplitude. However, with the MMC AA6061-20s-T6, cyclic hardening appears only from a total strain amplitude of $\Delta\varepsilon_{total}/2 = 0.004$ and higher. The regime of initial cyclic hardening of Al99.85-20s is limited to a few cycles, whereas AA6061-20s-T6 at amplitudes with sufficiently large plastic deformation ($\Delta\varepsilon_{total}/2 \geq 0.003$) hardens continually during the total fatigue life (the decrease in the curve at $\Delta\varepsilon_{total}/2 = 0.002$ is caused by growth of a macrocrack). The (secondary) hardening of Al99.85-20s (compare also Ref. [76]) at amplitudes up to $\Delta\varepsilon_{total}/2 = 0.006$ does not occur with the other MMCs.

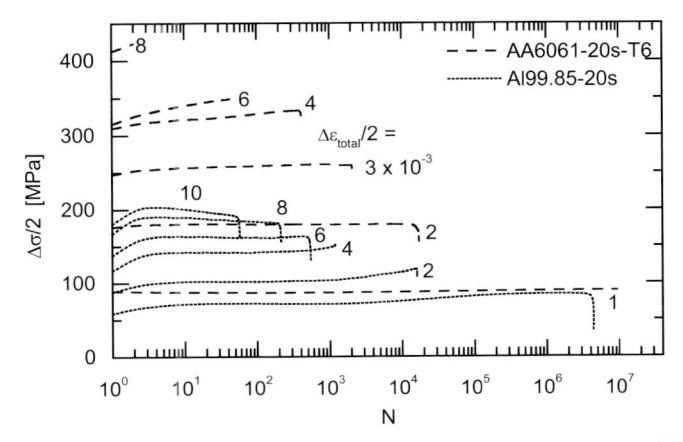

Fig. 7.14 Cyclic deformation curves of AA6061-20s and Al99.85-20s at different total strain amplitudes [2].

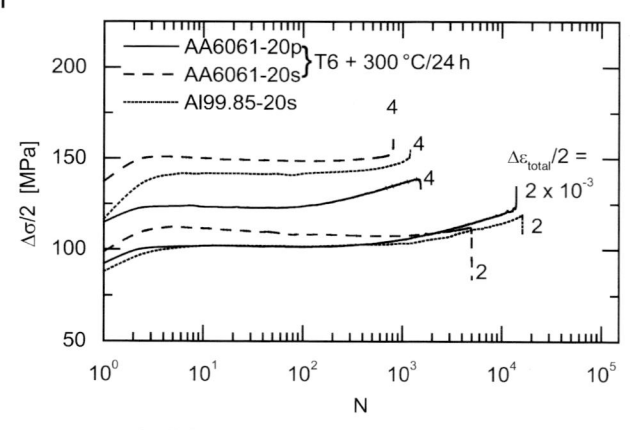

Fig. 7.15 Cyclic deformation curves of overaged MMCs with AA6061 matrix in comparison with Al99.85-20s ($\Delta\varepsilon_{total}/2 = 0.002$ and $\Delta\varepsilon_{total}/2 = 0.004$) [2].

Cyclic softening of Al99.85-20s occurs only partly at high deformation amplitudes of $\Delta\varepsilon_{total}/2 \geq 0.008$ and is caused by damage to the matrix and fibers.

Overageing of MMCs with AA6061-matrix (T6 + 300 °C/24 h) reduces the strength of the matrix and thus the strength of the composite by coarsening of the precipitates. The cyclic deformation curves of AA6061-20p and AA6061-20s (overaged state) presented in Fig. 7.15 are far below the curves of peak-aged T6 treated samples, compare Fig. 7.13 and 7.14. The short-fiber-reinforced MMC shows, contrary to the peak-aged state, in the presented cases of overaged samples a higher strength than the particle-reinforced MMC. Due to the missing interference of dislocation movement, caused by coarsening of precipitates, the present cyclic hardening of the overaged material is comparable with the hardening of Al99.85-20s. Cyclic saturation and secondary hardening are also to be observed. The effect of cyclic hardening can therefore be directly correlated to the ductility of the matrix, which is higher in the composite with a softer matrix [2].

7.4.2
Fatigue Life Behavior

To investigate the relation between the cyclic deformation parameters and the number of cycles to failure, N_f, the corresponding amplitudes of the stress, the total strain and elastic and plastic strain have been taken from the hysteresis loops at half the number of cycles to failure. Figure 7.16 shows the strain values in dependence on the number of load reversals $2N_f$ (one cycle corresponds to two load reversals) of the composite AA6061-20p. The values of the plastic and elastic strain amplitudes can be described after Eq. (5) and (6). The sum of both lines results in a curve, which describes the fatigue life behavior as a function of the total strain amplitude. These curves of all composites are summarized in Fig. 7.17. Furthermore, the results of an investigation of AA6061-15p-T6 are shown (from [78]). Ac-

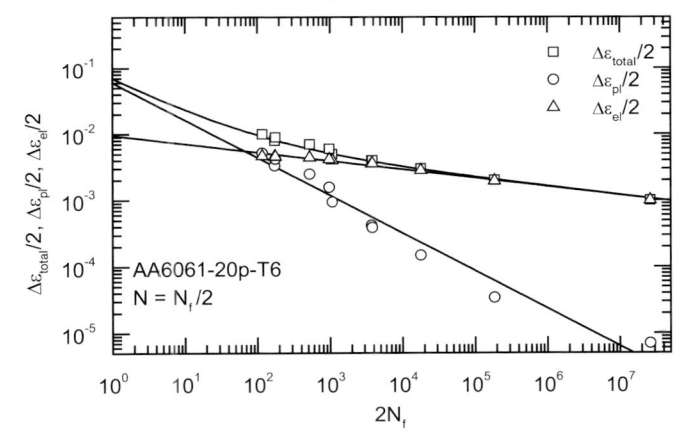

Fig. 7.16 Strain amplitudes at half the number of cycles to failure and their description by the laws after Manson–Coffin and Basquin for AA6061-20p-T6 [2].

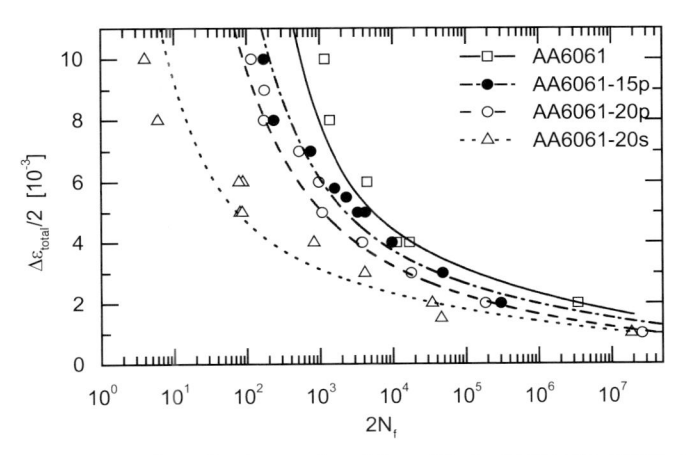

Fig. 7.17 Number of load reversals to failure, $2N_f$, for AA6061-T6, AA6061-15p-T6, AA6061-20p-T6 and AA6061-20s-T6 as a function of the total strain amplitude $\Delta\varepsilon_{total}/2$ [2, 78].

cording to that, the reinforcement by particles and short fibers causes a decrease in the fatigue life in relation to the nonreinforced alloy in total-strain controlled fatigue tests. Furthermore, the comparison of AA6061-15p-T6 with AA6061-20p-T6 shows that the fatigue life reduction is more significant with increasing volume fraction of reinforcements (see also Ref. [74]). The short-fiber-reinforced MMC reaches the lowest fatigue life under all test conditions. All curves can be described with the laws of Basquin and Manson–Coffin; the coefficients and exponents are presented elsewhere [80].

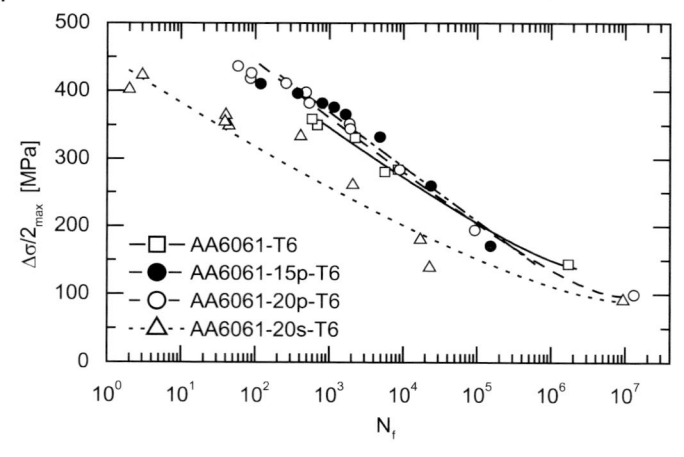

Fig. 7.18 Correlation between number of cycles to failure and maximum stress amplitude for AA6061-T6, AA6061-15p-T6, AA6061-20p-T6 and AA6061-20s-T6 [2].

Figure 7.18 shows the correlation of the maximum occurring stress amplitude of a test on the fatigue life (in the case of a stress-controlled test, Fig. 7.18 would correspond to the Wöhler diagram). The behavior of particle-reinforced composites is identical within the scatter of the investigations. Compared to the nonreinforced alloy, a similar, even tendentially higher fatigue life is reached in the range of large stress amplitudes. In this presentation the fatigue life of the short-fiber reinforced MMC is also lower than the fatigue lives of the other composites with AA6061 matrix.

In an analogous way the influence of the matrix strength on the fatigue life is shown in Fig. 7.19 and 7.20 (from [2]). Figure 7.19 shows the number of load rever-

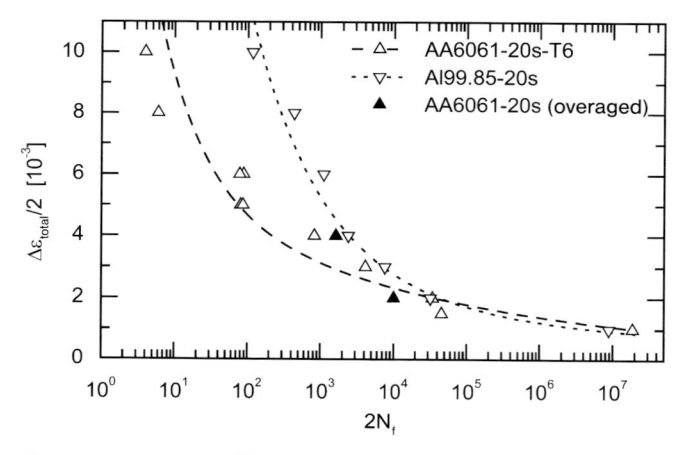

Fig. 7.19 Comparison of fatigue life of Al99.85-20s and AA6061-20s-T6, related to the total strain amplitude (filled symbols indicate the overaged samples) [2].

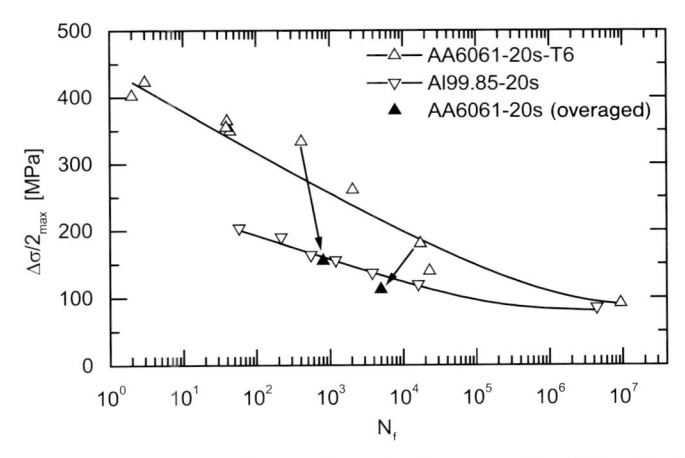

Fig. 7.20 Comparison of fatigue life of Al99.85-20s and AA6061-20s-T6, related to the maximum stress amplitude (filled symbols indicate the overaged samples; the arrows indicate the influence of overageing) [2].

sals until fracture as a function of the total-strain amplitude for both short-fiber-reinforced composites and both tests with overaged AA6061-matrix (compare also Fig. 7.15). The curves differ significantly at higher strain amplitudes. The MMC with a softer matrix reaches a longer fatigue life at amplitudes of $\Delta\varepsilon_{total}/2 > 0.002$ for a given strain amplitude. At a number of cycles to fracture of $N_f \approx 25\,000$ ($\Delta\varepsilon_{total}/2 < 0.002$) the regression curves intersect. The additional heat treatment extends the service life at $\Delta\varepsilon_{total}/2 = 0.004$; however, a reduction in the number of cycles to failure is found at $\Delta\varepsilon_{total}/2 = 0.002$. Thus, a stronger matrix seems to be an advantage at low cyclic loading.

The evaluation of the plastic-strain amplitude has shown an increase by a factor of 7 during the test with $\Delta\varepsilon_{total}/2 = 0.004$ due to overaging, while this has increased by 47 times during the test with smaller strain amplitude. From this point of view a reduction in the fatigue life is expected for both experiments. In the evaluation of the fatigue life in dependence on the matrix strength it has to be distinguished whether it is related to the strain or stress control. Figure 7.20 shows that the MMC AA6061-20s-T6 shows an advantage in fatigue life behavior compared to Al99.85-20s with stress controlled tests for the total range of stress amplitudes. By overageing of the AA6061 matrix, stress amplitudes and fatigue lives are reached, which agree approximately with the regression curve of Al99.85-20s (the effect of overageing is indicated by arrows). Thereby the stress amplitude is much more reduced in tests with a higher strain amplitude than in the test with $\Delta\varepsilon_{total}/2 = 0.002$. The influence of the fatigue life is therefore directly attributed to the strength of the matrix, since the stress in the fibers is strongly influenced by that factor.

Investigations of the fatigue crack paths show that the cracks run through broken fibers or particles with increasing stress amplitude [2]. For small stresses, especially with tests on Al99.85, the reinforcements keep intact to a large extent and therefore in this case the crack runs predominately through the matrix [2].

7.4.3
Damage Evolution

The damage of composites during cyclic deformation has been investigated using stiffness measurements (compare Refs. [2, 14, 77 and 78]). For this the elastic unloadings after load reversals (E_T in tension and E_C in compression, compare Fig. 7.5) have been evaluated as a function of the cycle number for different total-strain amplitudes for the MMCs with AA6061-T6 matrix. The changes in the values E_T or E_C have been divided by the Young's modulus of the samples in the initial state (E_0), whereby characteristic damage parameters D_T and D_C result. Figure 7.21 shows the development of D_T and D_C as a function of the number of cycles for AA6061-20p at different total-strain amplitudes. According to this, the samples' stiffness values decrease during fatigue (this corresponds to an increase in the damage values). This effect is even more significant the higher the total-strain amplitude. Especially at the beginning of the cyclic deformation, a steep increase is found and thus a high number of damage events occur during the first cycles. This could also be proven on the basis of acoustic emission measurements during the cyclic deformation [35, 36]. For amplitudes up to $\Delta\varepsilon_{total}/2 = 0.004$ a regime results in which the damage values increase only slightly with increasing cycle number. At the end of the test the D_T-curves increase again very steeply, which is attributed to the growth of fatigue cracks. The damage curves of the determined damage value in compression D_C are always below the curves of D_T. This is attributed to the fact that damaged areas can carry load in compression again and thus the loss of sample stiffness is not very high. The comparison of the damage curves of AA6061-20p and AA6061-20s does not show significant qualitative differences, see Fig. 7.22.

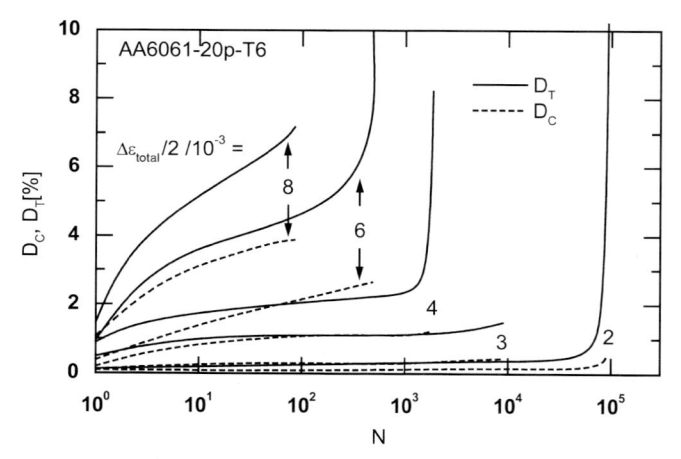

Fig. 7.21 Development of damage parameters D_C and D_T for particle-reinforced composites at different total strain amplitudes [2].

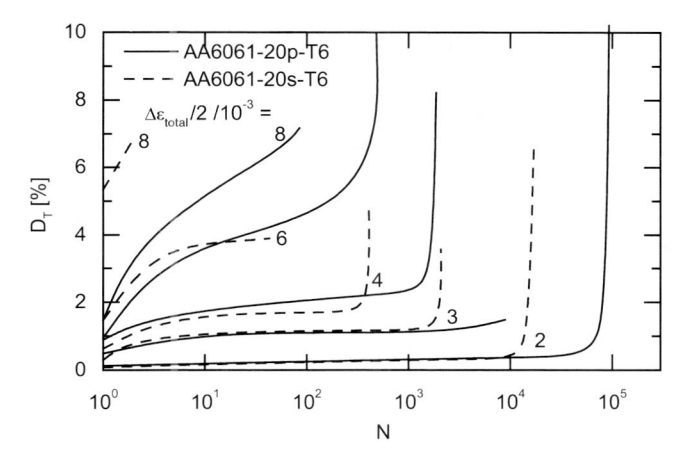

Fig. 7.22 Development of the damage parameter D_T for particle- and short-fiber-reinforced composites with AA6061-T6 matrix at different total strain amplitudes [2].

7.5 Summary

This chapter describes the investigation of the deformation behavior of three different metal-matrix composites (Al_2O_3 particle- and short-fiber-reinforced Al alloy AA6061 as well as short-fiber-reinforced, technically pure Al 99.85), in particular at cyclic load. The fatigue tests were carried out in total-strain control with total-strain amplitudes between $\Delta\varepsilon_{total}/2 = 0.001$ and 0.01 at room temperature. The most important results can be summarized as follows (see also Ref. [2] for more details):

1. The addition of ceramic reinforcements leads to an increase in the Young's modulus and the strength. Due to the inhomogeneous stress distribution in the matrix, plastic (micro) yielding occurs earlier in MMCs compared to the nonreinforced reference material.

2. In the case of cyclic deformation at sufficiently high strain amplitudes the composites show an initial cyclic hardening, which is higher than in the nonreinforced material. The hardening increases with increasing strain amplitude. It is higher in the short-fiber-reinforced MMC (at similar matrix) than in the particle-reinforced MMC. This is due to the fact, that with short-fiber reinforcement more matrix is influenced by the reinforcements than with particle reinforcements; particles are significantly larger on average than fibers and thus the distances between reinforcements is higher for particles than for short fibers.

3. The fatigue life of the materials investigated in this work can be described by the laws of Manson–Coffin and Basquin. With respect to the total-strain amplitude the addition of ceramic reinforcements causes a reduction in fatigue life in the range of the investigated strain amplitudes. In the short-fiber-reinforced MMC

the reduction in the number of cycles to failure using an identical matrix is more significant in the regime of large strain amplitudes compared to particle reinforcement. In the case of short-fiber reinforcement the MMC reaches a higher fatigue life with a softer matrix (Al99.85-20s) than the MMC with an alloyed matrix.

4. If the fatigue life is related to the maximum occurring stress amplitude of a test, a comparable number of cycles to failure is found for the nonreinforced material and for 15 or 20 vol% Al_2O_3 particle-reinforced MMCs. Again, the short-fiber-reinforced MMC shows the shortest fatigue lives.

5. From the stress–strain hysteresis loops a damage parameter has been determined, which evaluates the reduction in the samples´ stiffness. The course of damage with the cycle number increases monotonically. The curves show a strong increase in damage at the beginning of the cyclic deformation, followed by a slight increase and a final increase just before failure (i.e. growth of a macrocrack within the last third of fatigue life). This behavior shows, like additional measurements of acoustic emission, that MMCs already experience the first damage by particle or fiber fracture within the first cycles. The increase at the end of the test is attributed to the growth of macro cracks which can be detected easily. The levels of the damage curves depend on the strain amplitudes.

Acknowledgement

The authors thank Prof. Dr. H. Mughrabi for his interest in this work, Dipl.-Ing. M. Kemnitzer for his cooperation and assistance, and Deutsche Forschungsgemeinschaft for financial support within the Gerhard Hess program (Bi 418/5-1 to 5-3).

References

1 M. D. Skibo, D. M. Schuster, *Mater. Technol.* **1995**, *10*, 243–246.
2 O. Hartmann, Doctorate Thesis, University Erlangen-Nürnberg, **2002**.
3 S. F. Corbin, D. S. Wilkinson, *Acta Metall. Mater.* **1994**, *42*, 1319–1327.
4 H. Biermann, A. Borbély, O. Hartmann, *Verbundwerkstoffe und Werkstoffverbunde*, B. Wielage, G. Leonhardt (Eds.), Wiley-VCH, Weinheim, **2001**, pp. 134–139.
5 P. J. Withers, D. Juul Jensen, H. Lilholt, W. M. Stobbs, in *Proceedings of the 6th International Conference on Composite Materials (ICCM VI)*, F. L. Matthews, N. C. R. Buskell, J. M., Hodgkinso, J. Morton (Eds.), Elsevier Applied Science, London, **1987**, Vol. 2, pp. 255–264.
6 W. Voigt, *Lehrbuch der Kristallphysik*, Teubner-Verlag, Leipzig, **1928**.
7 A. Reuss, *Z. Angew. Math. Mech.* **1929**, *9*, 49–58.
8 D. J. Lloyd, *Acta Metall. Mater.* **1991**, *39*, 59–71.
9 A. Karimi, *Mater. Sci. Eng.* **1984**, *63*, 267–276.
10 A. Mocellin, R. Fougeres, P.F. Gobin, *J. Mater. Sci.* **1993**, *28*, 4855–4861.
11 M. Manoharan, J. J. Lewandowski, *Scr. Metal.* **1989**, *23*, 1801–1804.
12 S. B. Wu, R. J. Arsenault, in *Fundamental Relationships between Microstructure and Mechanical Properties of Metal-Matrix Composites*, TMS, Warrendale, **1990**, pp. 241–253.

13 J.-Y. Buffière, E. Maire, P. Cloetens, G. Lormand, R. Fougéres, *Acta Mater.* **1999**, *47*, 1613–1625.
14 H. Biermann, M. Kemnitzer, O. Hartmann, *Mater. Sci. Eng., A* **2001**, *319–321*; 671–674.
15 S. Elomari, R. Boukhili, *J. Mater. Sci.* **1995**, *30*, 3037–3044.
16 O. Hartmann, H. Biermann, H. Mughrabi, in *Proceedings of the 4th International Conference on Low Cycle Fatigue and Elasto-Plastic Behavior of Materials (LCF-4)*, K.-T. Rie, P.D. Portella (Eds.), Elsevier Science Ltd., Amsterdam, **1998**, pp. 431–436.
17 M. S. Hu, *Scr. Metall. Mater.* **1991**, *25*, 695–700.
18 W. H. Hunt, J. R. Brockenbrough, P. E. Magnusen, *Scr. Metall. Mater.* **1991**, *25*, 15–20.
19 M. Kouzeli, L. Weber, C. San Marchi, A. Mortensen, *Acta Mater.* **2001**, *49*, 497–505.
20 J. Llorca, A. Martin, J. Ruiz, M. Elices, *Metall. Trans. A* **1993**, *24*, 1575–1588.
21 T. Mochida, M. Taya, M. Obata, *JSME Int. J. Series I* **1991**, *34*, 187–193.
22 M. Vedani, E. Gariboldi, *Acta Mater.* **1996**, *44*, 3077–3088.
23 P. M. Singh, J. J. Lewandowski, *Metall. Mater. Trans. A* **1995**, *26*, 2911–2921.
24 J.-Y. Buffière, E. Maire, C. Verdu, P. Cloetens, M. Pateyron, G. Peix, J. Baruchel, *Mater. Sci. Eng., A* **1997**, *234–236*; 633–635.
25 E. Maire, A. Owen, J.-Y. Buffière, P. J. Withers, *Acta Mater.* **2001**, *49*, 153–163.
26 P. M. Mummery, B. Derby, J.C. Elliot, *J. Microsc.* **1995**, *177*, 399–406.
27 A. Borbély, H. Biermann, O. Hartmann, J.-Y. Buffiére, *Comput. Mater. Sci.* **2003**, *26*, 183–188.
28 A. Borbély, F. Csikor, S. Zabler, P. Cloetens, H. Biermann, *Mater. Sci. Eng. A* **2004**, *367*, 40–50.
29 P. Kenesei, H. Biermann, A. Borbély, *Scr. Mater.* **2005**, *53*, 787–791.
30 E. Gariboldi, C. Santulli, F. Stivali, M. Vedani, *Scr. Mater.* **1996**, *35*, 273–277.
31 P. M. Mummery, B. Derby, C.B. Scruby, *Acta Metall. Mater.* **1993**, *41*, 1431–1446.
32 A. Niklas, L. Froyen, M. Wevers, L. Delaey, *Metall. Mater. Trans. A* **1995**, *26*, 3183–3189.
33 A. Rabiei, M. Enoki, T. Kishi, *Mater. Sci. Eng., A* **2000**, *293*, 81–87.
34 H. Suzuki, M. Takemoto, K. Ono, *J. Acoustic Emission*, **1994** *11*, 117–128.
35 H. Biermann, A. Vinogradov, O. Hartmann, *Z. Metallkd.* **2002**, *93*, 719–723.
36 A. Vinogradov, H. Biermann, O. Hartmann, in *Proceedings of the 8th International Fatigue Congress (Fatigue 2002)*, A. F. Blom (Ed.), EMAS, West Midlands, UK, **2002**, Vol. 3, pp. 1939–1948.
37 M. Manoharan, J. J. Lewandowski, *Mater. Sci. Eng., A* **1992**, *150*, 179–186.
38 P. Mummery, B. Derby, *Mater. Sci. Eng., A* **1991**, *135*, 221–224.
39 M. Suéry, G. L'Esperance, *Key Engng. Mater.* **1993**, *79–80*; 33–46.
40 A. R. Vaida, J. J. Lewandowski, *Mater. Sci. Eng., A* **1996**, *220*, 85–92.
41 C. González, J. Llorca, *Scr. Mater.* **1996**, *35*, 91–97.
42 J. Llorca, P. Poza, in *Proceedings of the 6th International Fatigue Congress (Fatigue '96)*, G. Lütjering, H. Nowack (Eds), Elsevier Science Ltd., Oxford, **1996**, pp. 1511–1516.
43 S. S. Manson. Technical report, National Advisory Commission on Aeronautics NACA, Report TN-2933, Lewis Flight Propulsion Laboratory, Cleveland, OH, **1953**.
44 L.F. Coffin, *Trans. ASME* **1954**, *76*, 931–950.
45 O. H. Basquin, *Proc. Am. Soc. Test. Struct.* **1910**, *10*, 625–630.
46 K. K. Chawla, in *Structure and Properties of Composites*, T. W. Chou (Ed.), VCH, Weinheim, **1993**, pp. 121–128.
47 M. Suéry, G. L'Esperance, *Key Engng. Mater.* **1993**, *79–80*; 33–46.
48 C. Li, F. Ellyin, *Metall. Mater. Trans. A* **1995**, *26*, 3177–3182.
49 Z. Wang, J. Zhang, *Mater. Sci. Eng., A* **1993**, *171*, 85–94.
50 M. Levin, B. Karlsson, *Int. J. Fatigue* **1993**, *15*, 377–387.
51 S. Kumai, J. E. King, J. F. Knott, *Fatigue Fract. Engng. Mater. Struct.* **1990**, *13*, 511–524.
52 E. Y. Chen, L. Lawson, M. Meshii, *Scr. Metall. Mater.* **1994**, *30*, 737–742.

53 U. Habel, C. M. Christenson, J. W. Jones, J. E. Allison, in *Proceedings of ICCM-10*, A. Poursartip, K. Street (Eds.), Woodhead Publishing, Cambridge, **1995**, pp. 397–404.

54 D. A. Lukasek, D. A. Koss, *J. Comput. Mater.* **1993**, *24*, 262–269.

55 M. Levin, B. Karlsson, J. Wasen, in *Fundamental Relationships between Microstructure and Mechanical Properties of MMCs*, P. K. Liaw, M. N. Gungor (Eds.), TMS, Warrendale, **1990**, pp. 1–2.

56 A. K. Vasudévan, K. Sadananda, *Scr. Metall. Mater.* **1993**, *28*, 837–842.

57 F. Ellyin, C. Li, *Proceedings of ICCM-10*, A. Poursartip, K. Street (Eds.), Woodhead Publishing, Cambridge, **1995**, pp. 565–573.

58 M. Papakyriacou, H. R. Mayer, S. E. Tschegg-Stanzl, M. Gröschl, *Fatigue Fract. Engng. Mater. Struct.* **1995**, *18*, 477–487.

59 O. Botstein, R. Arone, A. Shpigler, *Mater. Sci. Eng., A* **1990**, *128*, 15–22.

60 W. A. Logsdon, P. K. Liaw, *Engng. Fract. Mech.* **1986**, *24*, 737–751.

61 J. K. Shang, R. O. Ritchie, in *Metal Matrix Composites: Mechanisms and Properties*, R. K. Everett, R. J. Arsenault (Eds.), Academic Press, Boston, **1991**, pp. 255–285.

62 Y. Sugimura, S. Suresh, *Metall. Trans. A* **1992**, *23*, 2231–2242.

63 J. E. Allison, J. W. Jones, in *Proceedings of the 6th International Fatigue Congress (Fatigue '96)*, G. Lütjering, H. Nowack (Eds.), Elsevier Science Ltd., Oxford, **1996**, pp. 1439–1450.

64 J. J. Bonnen, J. E. Allison, J. W. Jones, *Metall. Trans A* **1991**, *22*, 1007–1019.

65 F. Ellyin, L. Chingshen, in *Proceedings of the 6th International Fatigue Congress (Fatigue '96)*, G. Lütjering, H. Nowack (Eds.), Elsevier Science Ltd., Oxford, **1996**, pp. 1475–1480.

66 V. V. Ogarevic, R. I. Stephens, in *Cyclic Deformation, Fracture, and Nondestructive Evaluation of Advanced Materials*, M. R. Mitchell, O. Buck (Eds.), ASTM STP 1184, American Society for Testing and Materials, Philadelphia, **1994**, Vol. 2, pp. 134–155.

67 N. J. Hurd, *Mater. Sci. Technol.* **1988**, *4*, 513–517.

68 K. Toka jr, H. Shiota, K. Kobayashi, *Fatigue Fract. Engng. Mater. Struct.* **1999**, *22*, 281–288.

69 M. Finot, Y.-L. Shen, A. Needleman, S. Suresh, *Metall. Mater. Trans. A* **1994**, *25*, 2403–2420.

70 J. Llorca, S. Suresh. A. Needleman, *Metall. Trans. A* **1992**, *23*, 919–934.

71 G. Meijer, F. Ellyin, Z. Xia, *Composites: Part B* **2000**, *31*, 29–37.

72 A. Borbély, H. Biermann, O. Hartmann, *Mater. Sci. Eng., A* **2001**, *313*, 34–45.

73 J. N. Hall, J. W. Jones, A. K. Sachdev, *Mater. Sci. Eng., A* **1994**, *183*, 69–80.

74 M. Papakyriacou, H. R. Mayer, S .E. Stanzl-Tschegg, M. Gröschl, *Int. J. Fatigue* **1996**, *18*, 475–481.

75 N. L. Han, Z. G. Wang, L. Z. Sun, *Scr. Metall. Mater.* **1995**, *32*, 1739–1745.

76 J. Llorca, *Progr. Mater. Sci.* **2002**, *47*, 283–353.

77 O. Hartmann, K. Herrmann, H. Biermann, *Adv. Eng. Mater.* **2004**, *6*, 477–485.

78 O. Hartmann. Diploma Thesis, University Erlangen-Nürnberg, **1997**.

79 M. Kemnitzer, Diploma Thesis, University Erlangen-Nürnberg, **2000**.

80 O. Hartmann, M. Kemnitzer, H. Biermann, *Int. J. Fatigue* **2002**, *24*, 215–221.

8
Interlayers in Metal Matrix Composites: Characterisation and Relevance for the Material Properties

J. Woltersdorf, A. Feldhoff, and E. Pippel

8.1
Summary

The modification of the interlayer reaction kinetics by process and material parameters is described as a suitable principle to improve the properties of metal matrix composites. The structural and nanochemical peculiarities of the layers down to atomic dimensions will be discussed and correlated with the resulting macroscopic properties measured by *in situ* experiments. It is shown, that control and adjustment of the chemical reactions during the processing route (especially by selection of reaction partners and by suitable precoatings) can be used to tune the desired interface properties and thus the material parameters. It is also demonstrated, in detail, that modern nanoanalytical techniques combined with quantum mechanical calculations can provide valuable information on the reaction kinetics of the interlayers between coatings and fibers down to the atomic scale.

8.2
The Special Role of Interfaces and Interlayers

For all solids, their interfaces and interlayers (phase, grain and twin boundaries as well as special grain boundary phases and reaction layers) are the essential regions of microstructural processes, due to their energetic and structural peculiarities. Therefore the material properties can be controlled by the adjustment of the kind and contribution of these interfaces or interlayers [1], as done in *monolithic* materials for, e.g., the improvement of strength, high temperature behavior, and fracture toughness by the embedding of a polymorph second phase [2, 3]. Notably this statement applies for *composite* materials, as these systems only become compact materials by the effect of the interfaces and interlayers between the composite partners.

The constantly increasing complex demands on high performance composites with metal matrices can only be fulfilled by the development of specially optimised interfaces or interlayers (or even layer systems), which have to fulfil in different ways the suitable parameter combinations. This concerns, e.g., chemical reactivity, diffusion inhibition, and thermomechanical behavior, especially the high temperature strength.

Besides producing interlayers or interlayer systems by precoating of the embedded phase (for example the reinforcement fibers), which can be brought about by the use of pyrolytic techniques from an organic precursor, the intrinsic formation of intermediate phases by solid state chemical reactions, which occur during or after the processing by complex chemical reactions in the nanometer range, can be applied.

In the case of metal matrix composites (MMCs) the aim is to combine the high strengths and Young's moduli of ceramic and graphite fibers with the properties of the metal matrix, where light metals and their alloys are mostly utilised. Magnesium–aluminum alloys are of great interest for applications due to the low densities of their components (each approximately 1.8 g cm^{-3}) and the high strengths (3–4 GPa) and Young's moduli (several 100 GPa) of the graphite fibers. By optimisation of the fiber/matrix interlayers (with simultaneous prevention of fiber degradation) materials can be developed in this system, which are two or three times as strong as the best steels, but have a density of only a third that of steel.

This chapter is structured as follows: After an overview of the experimental techniques used, the microstructural and nanochemical peculiarities of interlayers in MMCs down to atomic dimensions, from which reaction mechanisms and crystal geometry models are derived, are discussed. Furthermore, a correlation of these results with the macroscopic properties measured by *in situ* experiments is performed. The possibilities for adjustment of desired interlayer properties by suitable fiber precoatings are then demonstrated. In this context, the way in which quantum theory can complete the nanoanalytical investigations is evidenced.

8.3
Experimental

The microstructural characterisation of the interlayers was performed by transmission electron microscopy (TEM) using both a high voltage electron microscope (HVEM) JEOL JEM 1000 operating at 1 MV acceleration voltage, and a high resolution/scanning transmission electron microscope (HREM/STEM) Philips CM20 FEG, which is equipped with a thermal supported field emission source (coefficient of spherical aberration of the objective lens $C_s = 2.0$ mm, point resolution 0.24 nm) operating at 200 kV. Besides the methods of diffraction contrast and of selected area diffraction, the high resolution atomic plane imaging technique, which is caused essentially by the phase contrast, is applied [4]. Thereby arrays of atomic planes of especially interesting interlayer regions are made visible, so that this method – parallel to the nanoanalyses – can be used to characterise interlayers on the nanoscale.

To determine the chemical composition in the nanometer range by energy dispersive X-ray spectroscopy (EDXS) and electron energy loss spectroscopy (EELS), the CM20 FEG was used in combination with a light element X-ray detector system Voyager II (Tracor) and an imaging energy filter (GIF 200, Gatan). The chemical composition and the lateral distribution of phenomena occurring by diffusion and chemical reactions at interfaces, with dimensions of often only a few nanometers, cannot be displayed satisfactorily by the X-ray microanalytical method due to a limited lateral resolution by fluorescence phenomena in adjacent areas and, moreover, taking X-ray intensity images is very time consuming. Therefore, the interlayer areas have been investigated in general by EELS (cf., e.g., Refs. [5, 6]). Thereby the analyses of the near-edge fine structures (ELNES, energy loss near-edge structure) are of special importance and are observable at sufficiently good energy resolution (< 1 eV) at the ionisation edges in energy regions of approximately 30 eV (see Refs. [7–10]). These ELNES features are caused by the excitation of electrons of inner shells in unoccupied states above the Fermi energy. The shape, the energy, and the signal intensity of the ELNES details are determined both by chemical bonding and by coordination and distance to the nearest neighbors. Thus, the near-edge fine structures give information on the local partial density of states above the Fermi energy. Furthermore, a chemical shift of the onset of the ionisation energy occurs, amounting to a few eV, depending on the difference in the electronegativities of the binding partners, which also allows statements about the bonding. Altogether, detailed information far beyond the pure chemical composition can be taken from EEL spectra near the ionisation edge. The identification of the chemical bonding state is derived both by comparison with ELNES structures of standard samples (ELNES fingerprinting) and with the help of results of quantum chemical calculations, which are gained with different, problem adjusted approximation methods.

The *in situ* three-point bending tests of notched MMC samples are carried out with an atmospheric scanning electron microscope ESEM-3 (Electro-Scan, Wilmington) with the use of a special microdeformation unit (Raith, Dortmund).

8.4
Interlayer Optimisation in C/Mg–Al Composites by Selection of Reaction Partners

In *metal* matrix composites, in contrast to composites with *ceramic* matrices, the fracture strain of the fibers is lower than that of the ductile matrix, thus the metal matrix deforms in plastic and elastic ways after the first fiber cracks, and the failure behavior is determined by interlayer-controlled secondary microprocesses (e.g. delamination at the fiber/matrix interface, single fiber and bundle pullout, ductile and brittle matrix failure). In the case of graphite fibers in metals, the decisive parameter is the interlayer reactivity, which may lead to carbide formation. Our systematic tests [11–18] using gas pressure melt infiltration techniques [19] on different, manufactured C/Mg–Al composites reveal
1. that the extent of carbide formation reactions is important for the composite properties,
2. that the binary carbide Al_4C_3 does not form, but a ternary compound develops,

3. that it is possible to control the interlayer reactions, and thus the composite properties, by variation of the aluminum content of the matrix and by use of carbon fiber with different surface microstructures.

The last mentioned adjustment of the interlayer reactivity by variation of both composite partners is based on the following properties of the fiber and matrix materials:

The *fiber* reactivity increases with increasing amount of free terminating hexagonal graphite atomic planes serving as docking points for chemical reactions with the metal melt. This means: The transition from high modulus fibers to high strength fiber types increases the reactivity. The reactivity of the *magnesium matrix* is increased by the aluminum content, because only aluminum, and not magnesium, is able to form stable carbides with the fiber carbon. The modified C/Mg–Al metal matrix composites were manufactured at the University Erlangen-Nürnberg at 720 °C with a fiber content of approximately 63 vol%. To control the interlayer adhesion two commercial Mg–Al alloys of 2 and 9 wt% Al (AM20 and AZ91) with carbon fiber M40J (high Young's modulus, surface parallel aligned graphite basal planes, chemically inert) or high strength fiber T300J (free terminating basal planes, chemically reactive) have been combined. The following three composites were investigated, having increasing fiber/matrix reactivity, according to the above statements: M40J/AM20, T300J/AM20 and T300J/AZ91. The investigations show that at too low or too high fiber/matrix reactivity optimal utilisation of the fiber strength is not achieved, but that a moderate reactivity gives the best results.

The *in situ* deformation test of the composite of low interlayer reactivity, prepared by combination of two relatively inert partners, shows, with increasing deformation, an initial linear load increase which flattens due to single fiber damage and then exhibits a continuous load decrease after exceeding the maximum of the tensile strength σ_B of approximately 0.5 GPa (see Fig. 8.1). The fracture surface, also shown in Fig. 8.1, is characterised by single fiber pullout, caused by the low fiber/matrix adhesion. Similarly HVEM and HREM observation does not show any carbide formation or other interaction between matrix and fiber, rather a replica type separation is shown.

At medium interlayer reactivity (provided by combination of high reactive fiber and inert matrix) the strength is doubled to approximately 1 GPa, and the fracture surface shows collective bundle-fracture behavior (see Fig. 8.2). The load decrease reveals characteristic steps, which can be attributed to failure of fiber bundles. Planar fracture surfaces occur within the bundles, consisting of approximately 30 to 80 fibers connected by the matrix metal. HREM images demonstrate that a moderate carbide formation occurs in the fiber/matrix interlayer, which leads to precipitations with a size of 10 to 20 nm.

In the case of high interlayer reactivity, provided by the combination of two reactive partners, brittle fracture occurs, caused by the strong fiber/matrix coupling, and the strength is only approximately 1/6 GPa (see Fig. 8.3). The smooth fracture surface points to brittle fracture, known from monolithic ceramics. Corresponding to the high chemical reactivity, HVEM observations show carbide plates with sizes

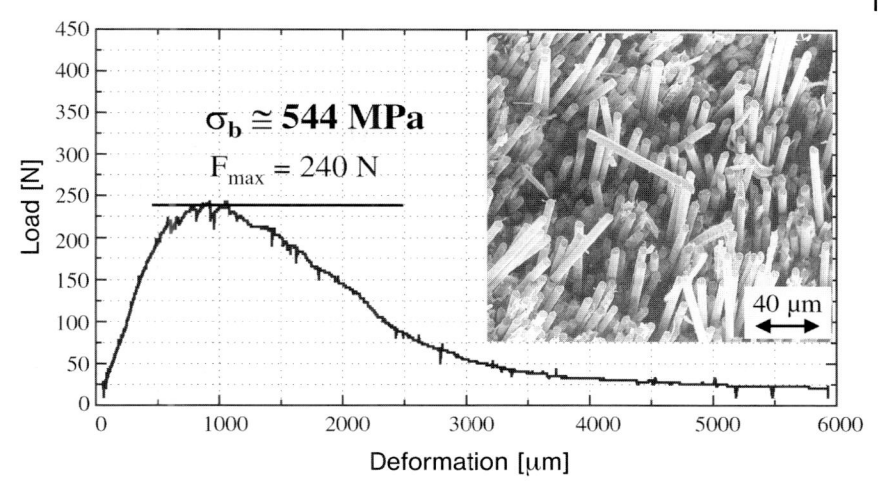

8.4 Interlayer Optimisation in C/Mg–Al Composites by Selection of Reaction Partners | 201

Fig. 8.1 Load–deformation diagram and fracture surface of the composite M40J/AM20 (low interlayer reactivity); $\sigma_B \cong 544$ MPa.

of about 1 μm (see Fig. 8.4), which run through the whole matrix region. High resolution imaging clearly reveals that the carbide precipitations start immediately from the graphite structures of the fiber surface. Thus, fracture-causing notch effects to the fibers can be exerted from these carbide plates at certain load situations.

Figures 8.1 to 8.5 not only show that, as described above, an optimal utilisation of fiber strength is only possible at moderate fiber/matrix reactivity but also that the composite strength with increasing fiber/matrix reactivity is correlated with

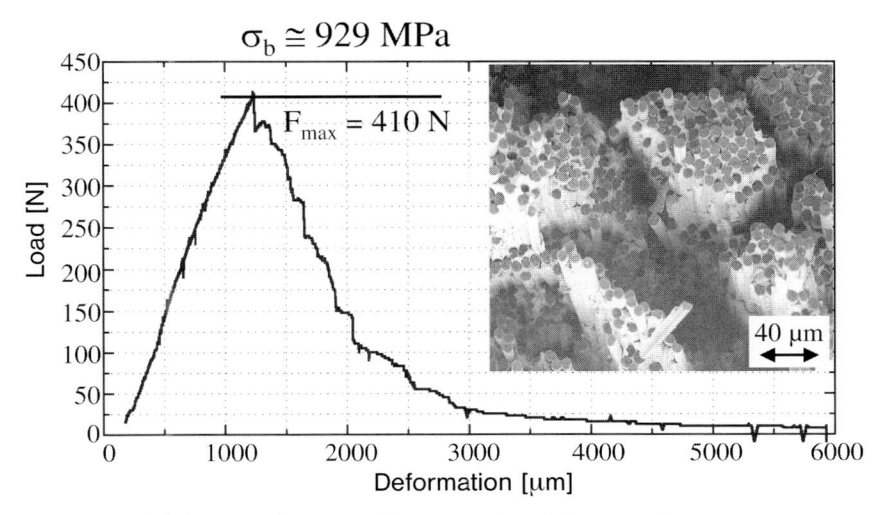

Fig. 8.2 Load–deformation diagram and fracture surface of the composite T300J/AM20 (moderate interlayer reactivity); $\sigma_B \cong 929$ MPa.

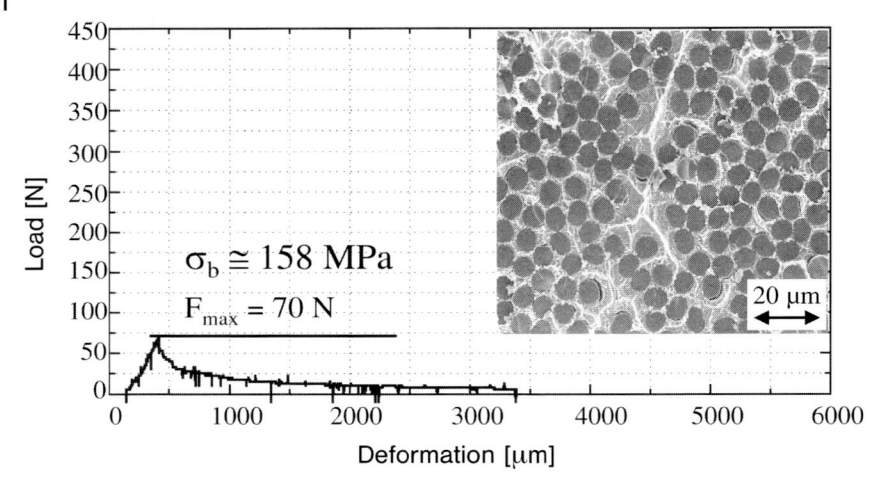

Fig. 8.3 Load–deformation diagram and fracture surface of the composite T300J/AZ91 (high interlayer reactivity); $\sigma_B \cong 158$ MPa.

$\sigma_b \cong 158$ MPa

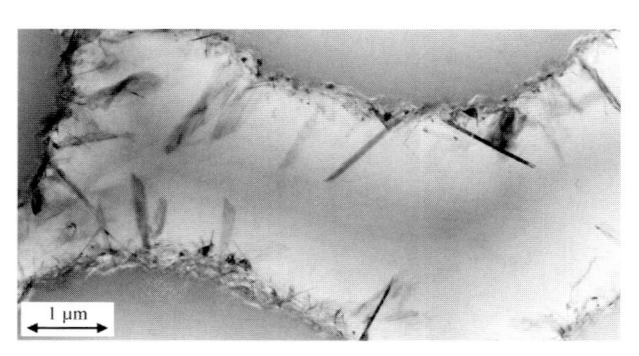

$F_{max} = 70$ N

Fig. 8.4 HVEM image of the plate-shaped carbide precipitations, grown at the fiber/matrix interface, of the highly reactive system T300J/AZ91; middle: Mg-Matrix.

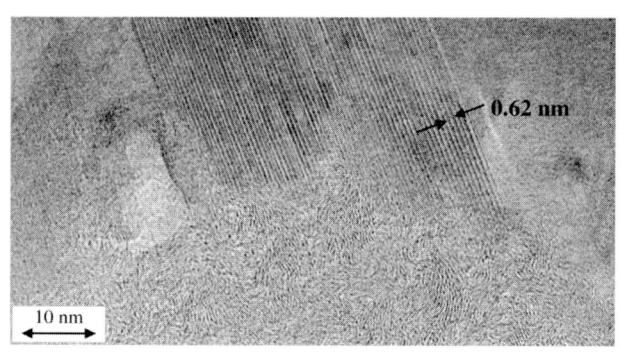

Fig. 8.5 HREM image of the atomic planes of the carbide plates shown in Fig. 8.4, immediately growing at the fiber surface (bottom).

8.4 Interlayer Optimisation in C/Mg–Al Composites by Selection of Reaction Partners

three different failure mechanisms: single fiber pullout, bundle fracture behavior and brittle fracture. The three composites show a different extent of plate-shaped precipitations at the interface, which characterises the special interlayer reactivity and controls the mechanical properties. Therefore it is of general interest for the optimisation of C/Mg–Al composite materials to elucidate the composition, the structure, and the growth kinetics of these precipitations.

EDXS and EELS investigations evidenced that the precipitations contain the three elements Mg, Al and C. However, as described in Section 8.2, additional information concerning details of the specific bonding can be found from EEL spectra near the ionisation edge, because the ELNES features are substantially influenced by bonding states. In Fig. 8.6, the measured fine structures of Mg-L_{23}-, Al-L_{23}- and C-K-ionisation edges of the precipitation phase are compared with different standard substances. From the comparison of the Mg-L_{23}-ELNES curves in Fig. 8.6(a) it is seen that the bonding states of the magnesium atoms in the precipitation differ substantially from those in metallic (bottom) or oxidic magnesium (top). The Al-L_{23}-ELNES (Fig. 8.6(b)) of the Al_4C_3-standard (onset at approximately 73 eV) and the precipitation phase differ significantly from those of the metallic aluminum and aluminum oxide. However, they agree with each other to a large extent. The spectra of the Al_4C_3-standard and the precipitation phase show a steep signal

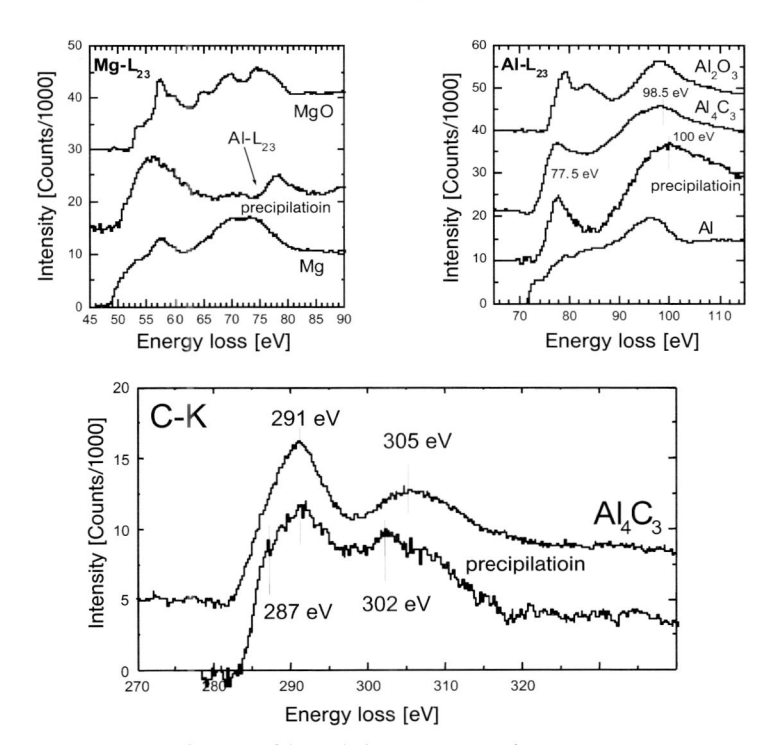

Fig. 8.6 ELNES features of the carbide precipitation phase in comparison to standard substances: (a) Mg-L_{23}-edge, (b) Al-L_{23}-edge, (c) C-K-edge.

increase at the C-K ionisation edge (see Fig. 8.6(c)), which leads to a peak at approximately 291 eV. At the flank of the peak the signal increase flattens slightly in the case of the precipitation at approximately 287 eV, which is caused by an additional interaction of carbon with the 2s electrons of magnesium, which are excited into unoccupied σ^*-orbitals forming sp^3–s–σ-bondings.

These ELNES features indicate that the precipitation phase is an Al–Mg mixed carbide, which is closely related to the binary carbide Al_4C_3. Crystallographic and morphological investigations reveal that the observed precipitations are Al_2MgC_2-mixed carbides with a (0002) lattice plane distance of 0.62 nm. A suggestion for the crystal structure, which describes the ternary carbide Al_2MgC_2 analogous to the well known binary carbide Al_4C_3 as an interstitial mixed carbide (cf. Refs. [12], [14]), is shown in Fig. 8.7, in which the corresponding atomic order in the (1120) crystal plane is presented and compared with the binary carbide (left). Thus, the mixed carbide Al_2MgC_2 consists of a close-packed structure of the metal atoms, in which the Al atoms occur in cubic (c) and the Mg atoms in hexagonal (h) stacking. The carbon atoms fill the gaps in the metal atom host lattice in such a way that they have an octahedral neighborhood between successive aluminum atom layers and a trigonal-bipyramidic one in the height of the magnesium atom layers. The comparison with Al_4C_3 shows that both carbides have one Al_2C structure unit in common.

As already noted in the HVEM overview (Fig. 8.4), the morphology of the C/Mg–Al mixed carbides formed in the composites is plate-shaped. This means that the longitudinal axes of the carbide crystals are always orientated vertical to their [0001] direction. Figure 8.8 shows, in atomic plane resolution, that the front surfaces of the carbide crystals are atomic stepped in relation to the terminating carbide atom planes in the metal matrix, while they show atomic smoothness between the (0001) habit surfaces and the metal matrix.

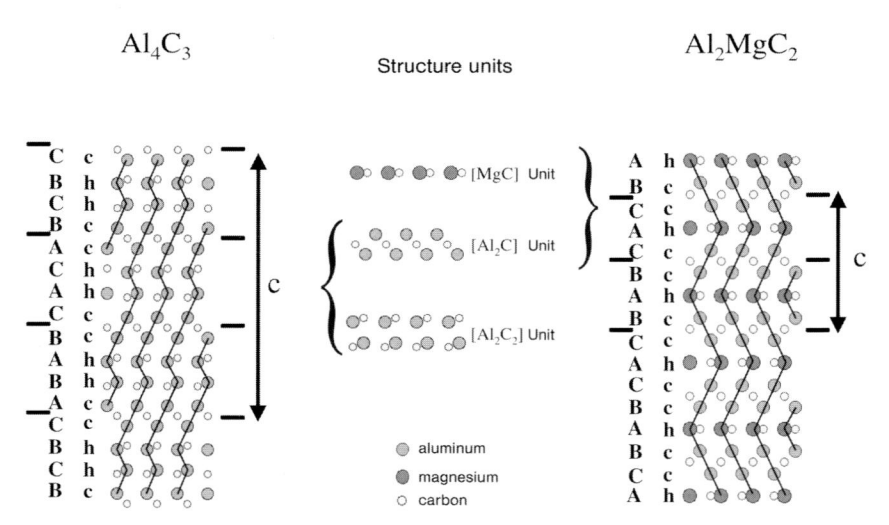

Fig. 8.7 right: atomic configuration in the (1120) plane of the ternary carbide Al_2MgC_2 (proposed structure) and comparison with the binary carbide Al_4C_3 (left).

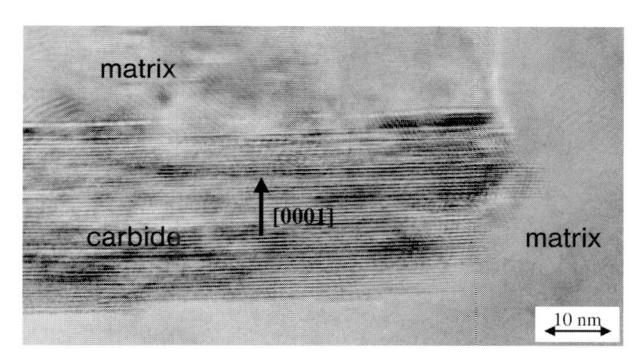

Fig. 8.8 Atomic structure of the two surface types of a mixed carbide-precipitation in a Mg–Al matrix (Composite T300J/AZ91), see text.

This morphological difference of both interface types of the carbide plates is due to two different growth mechanisms, from which the plate-like shape of the precipitations results: The atomic-rough interface moves by a continuous *diffusion*-controlled growth mode, the atomic-smooth interface can only move by slow *interface*-controlled step mechanisms, which require multiple two-dimensional nucleation or a spiral growth mechanism. The diffusion-controlled growth process is, at low driving forces, much faster than the interface-controlled mechanism. Thus the observed plate-shaped morphology of the mixed carbide crystallites is obvious.

8.5
Interlayer Optimisation in C/Mg–Al Composites by Fiber Precoating

A completely different way to modify the interface properties is made possible by a defined fiber precoating, which is especially appropriate for the production of hybrid compounds, in which the fiber reinforcement is only locally embedded in high strength areas, whereas in the less-loaded areas sufficient strength is provided by alloy formation. Therefore it is aimed to apply Mg–Al alloys with a relatively high Al content of approximately 5 wt% or more, which leads, in combination with an uncoated high strength carbon fiber, to extensive carbide formation and thus to the above described brittle fracture behavior. By precoating the fiber with titanium nitride, the carbide formation and resulting embrittlement of the composite can be avoided to a large extent (cf. [20, 21]). The coatings of 10–50 nm thickness produced by chemical vapor deposition from a titanium tetrachloride containing N_2–H_2 atmosphere both serve as a diffusion barrier between fiber carbon and matrix metal, and enable a suitable micromechanical fiber/matrix adhesion. The homogeneous deposition of the fiber surfaces is clearly shown in Fig. 8.9. The atomic plane resolution (Fig. 8.10) reveals that the titanium nitride coating is polycrystalline with grain sizes of the order of layer thickness and that the atomic planes of the Mg matrix terminate at the interlayer surface.

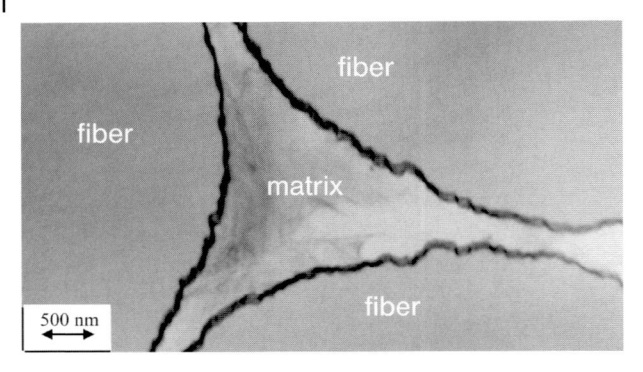

Fig. 8.9 Triple point of three TiN-coated C fibers; middle: pure magnesium matrix; dark: TiN-interlayer.

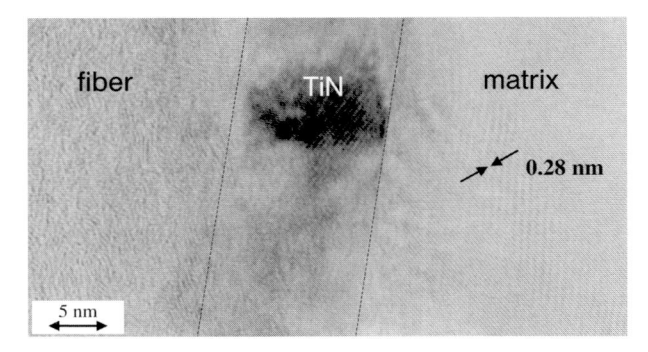

Fig. 8.10 HREM image of a polycrystalline TiN fiber coating (middle), left: turbostratic atomic planes of graphite fibers; right: (1010) atomic planes of the pure magnesium matrix, terminating at the TiN layer.

With a point-to-point distance of 1.3 nm, EEL spectra were taken across the interface at 30 equidistant points. Figure 8.11 shows the C-K, N-K, Ti-L_{23} and O-K ionisation edges with their fine structures in the energy interval between 250 and 750 eV. In the area of the fiber (spectra at the back) only the C-K edge at 284 eV is seen in the exponentially decreasing background. At the fiber/coating interface the Ti-L_{23} edge appears, and the fine structure of the C-K edge changes. Within a gradient of approximately 5 nm the N-K edge emerges (with similar fine structure to the locally adjacent C-K edge) with a continuous gradual decrease of the C-K edge. The Ti-L_{23} ELNES occurs dominantly over the noise background due to the sharp white lines, which appear at the L_{23}-ionisation edges of the transition metals because of the very high density of unoccupied 3d states existing just at or rather above the Fermi edge.

To analyse in detail the bonding states directly at the fiber/coating interface, which are essential for the composite properties, the C-K-, N-K and Ti-L_{23}-fine

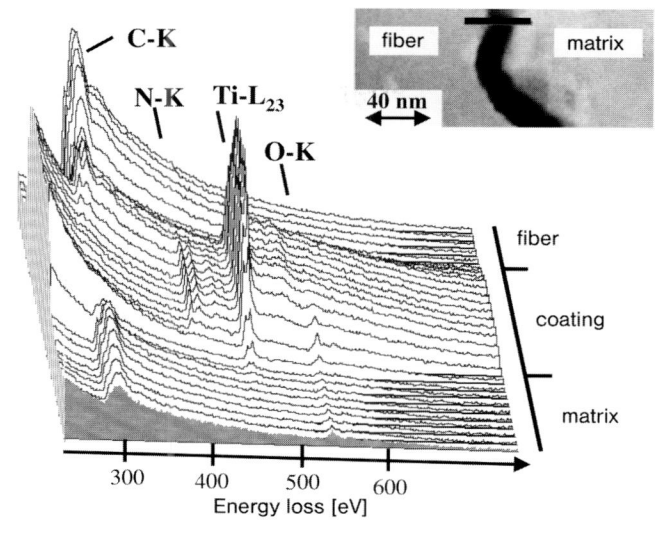

8.5 Interlayer Optimisation in C/Mg–Al Composites by Fiber Precoating

Fig. 8.11 EEL spectra across the fiber/matrix interlayer in the C/TiN/Mg-composite material; measured with 1.3 nm point distance.

structures across the interface were measured additionally with a dispersion of 0.1 eV per channel, and compared with both TiC and TiN standards (cf. Figures 8.12–8.14) as well as electron transition energies [22], which were determined using quantum chemical calculations on the basis of the density functional theory (cf. Fig. 8.15). These results are presented below in more detail as they demonstrate how material science can benefit from the combination of quantum mechanics and modern solid state analytical techniques.

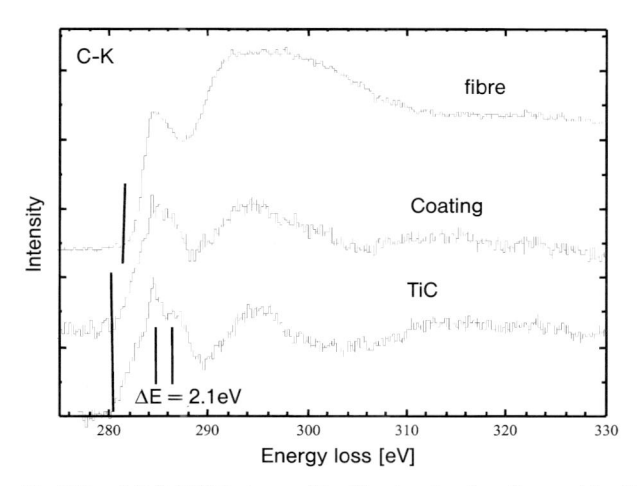

Fig. 8.12 C-K-ELNES features of the fiber/coating interface and the TiC standard (bottom); background-subtracted and scaled.

The measured C-K standard ELNES of TiC (bottom spectrum in Fig. 8.12) starts at 280 eV (onset) and shows three main maxima: (i) a relatively sharp peak at 4.5 eV above the onset, which has at $\Delta E = 2.1$ eV an extra (second) maximum, (ii) a less sharp peak at about 14 eV above the onset, and (iii) a very flat maximum at 34 eV above the onset energy. The top spectrum of Fig. 8.12 shows the ELNES signal of the composite in the outer fiber region immediately at the fiber/coating interface, the middle spectrum is that of the layer region at the interface to the fiber. The

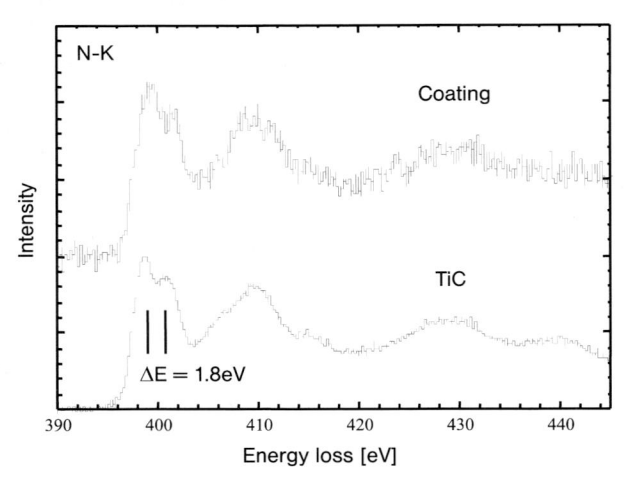

Fig. 8.13 N-K-ELNES features of the TiN interlayer and the TiN standard (bottom), background-subtracted and scaled.

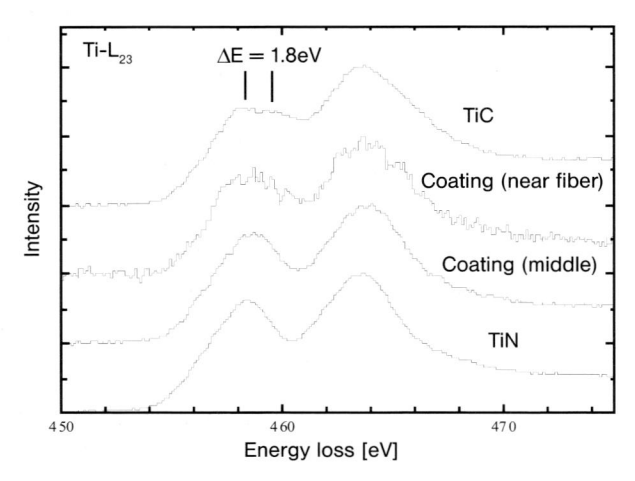

Fig. 8.14 Ti-L_{23}-ELNES features of the fiber/coating interface (middle curves) and the TiC and TiN standards (top and bottom); background-subtracted and scaled.

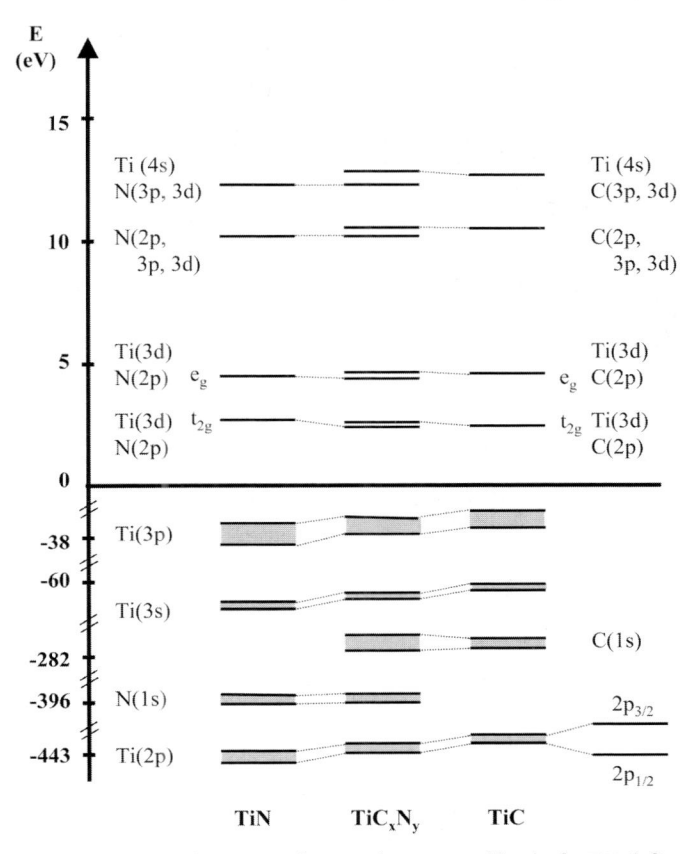

Fig. 8.15 Calculated energies of core and unoccupied levels, for TiN (left), for TiC (right) and for a hypothetical carbonitride TiC$_x$N$_y$ (middle) of octahedrally coordinated titanium; DFT-BP-DN*-calculations (see text).

transition of the fiber to the coating is therefore combined with the decrease in the onset energy by approximately 2 eV, a broadening of the first peak, and a decrease in the second. The observed splitting of the first peak in the TiC standard (bottom curve) cannot be resolved in the layer due to an increase in the noise ratio, caused by the necessary decrease in measurement time.

For the interpretation of the ELNES details of TiC, TiN and a hypothetical carbon nitride TiC$_x$N$_y$ molecular orbital (MO) schemes were derived with the help of density functional theory (DFT) and are presented in Fig. 8.15. These *ab initio* calculations were carried out using the nonlocal self-consistent Becke–Perdew model (BP) (cf. [23], [24]), with a double numerical basis set (DN*), that means one function for the core electrons (1s) and two functions for the valence electrons (2s, 2p for C and N; 3s, 3p, 4s for Ti). Furthermore a set of five 3d functions as an additional valence electron set was treated for Ti, while five additional 3d functions form the polarisation functions. The latter corresponds to the above-mentioned peculiarity of the 3d metals that the Fermi edge just cuts this 3d band. The principle of DFT is introduced in Refs. [25, 26], a complete overview is given in Ref. [27].

The interaction of C 2p electrons with Ti 3d electrons shown in Fig. 8.15 (right) leads thus to two unoccupied MOs with t_{2g} and e_g symmetry, which are 2.3 eV and 4.5 eV above the Fermi edge, respectively. The latter terms originate from the group theory based symmetry fitting of linear combinations of atomic orbitales (AOs) [28]. Thereby the t_{2g} and the e_g orbitals represent MOs with octahedrally co-ordinated central atoms (as Ti with d-like AOs), where the resulting bondings are π-like in the case of t_{2g} MOs and σ-like in the case of e_g MOs. It also follows (i) that the t_{2g} MOs are threefold degenerated orbitals, consisting of d_{xy}, d_{yz} and d_{zx} atomic orbitals of the central atom and p-like atomic orbitals of the bonding partner, and (ii) that the e_g MOs are twofold degenerated, with contributions of d_{x2-y2} and d_{z2} atomic orbitals of the central atom and p-type atomic orbitals of the bonding partner. As both types of MOs contain contributions of p-like AOs (azimuthal quantum number $l = 1$), the excitation of 1s electrons ($l = 0$) in these MOs leads to corresponding peaks in the ELNES of the K edges: The excitation of 1s electrons of carbon in these two unoccupied MOs is the reason for the observation of the relatively sharp first peak above the onset of C-K-ELNES of the TiC standards (cf. Fig. 8.12, bottom curve) and for the splitting by $\Delta E = 2.1$ eV. This measured splitting corresponds very well to the calculated value of 2.2 eV between the t_{2g} and the e_g levels, which is not surprising, because the relative peak locations are better observable than, for example, the absolute location of the ELNES onset and its correlation to the location of the Fermi edge, where specific equipment parameters are of influence (energy resolution of the spectrometer, energy width of the emitted electron radiation, calibration of the energy scale). The next measured peak in the direction of increasing energy values at approximately 14 eV above the onset (less sharp), shown in Fig. 8.15 at the top right, is a result of the excitation of 1s electrons of carbon in unoccupied states with 2p-, 3p-, and 3d-like symmetry, localised at the carbon atom.

Analogous measurements concerning the reaction between titanium and nitrogen are shown in Fig. 8.13, where the bottom spectrum is taken for a TiN standard sample and the top one, which corresponds very well with the standard signal curve, is taken at the middle region of the layer. It can be seen, that the N-K-ELNES of TiN starts at approximately 396 eV (onset) and shows essentially the same features as that of TiC (cf., e.g., Ref. [29]); however, the splitting of the first peak with $\Delta E = 1.8$ eV is slightly smaller than in the case of the carbide. The quantum chemical interpretation of the N-K-ELNES features is enabled again by MO schemes, which are obtained with the help of *ab initio* calculations on the basis of density functional theory (likewise DFT-BP-DN* calculations, see above). Figure 8.15, left, shows the results: The excitation of 1s electrons of nitrogen in unoccupied, energetically adjacent t_{2g} and e_g MOs, arising from the interaction of N 2p- and Ti 3d-like electrons, leads to the first ELNES peak occurring after the onset and to its measured splitting.

The observed ELNES features of the titanium L_{23}-ionisation edges of TiC and TiN standard substances are shown in Fig. 8.14 in the top and bottom spectra. The splitting into two separated peaks L_3 and L_2 caused by the spin–orbit coupling can be seen, where L_2 corresponds to the higher energy due to the stronger bonding to the core. An additional splitting of the L_3 peak in the case of TiC by approximately $\Delta E = 1.8$ eV can be interpreted as crystal field splitting. In principle, this effect should also be observed in other titanium compounds, e.g., TiN, however, in this case, the transition metal–nonmetal hybridisation is described as weaker (cf. Ref. [30]), and therefore the splitting is not found.

The titanium L_{23} ionisation edges of the composite material were also measured across the TiN coating, from the fiber to the matrix with a point-to-point distance of 2.5 nm. Figure 8.14 shows – between the standard spectra of TiC (top) and TiN (below) – two examples of such ELNES profiles, one measured near the fiber (top) and one in the middle of the coating. At first glance, all measured Ti L_{23} features in the layer correspond with those of titanium nitride. However, the spectra taken near to the fiber show additionally a broadening of the L_3 peak, which is typical of carbidic bonding.

Altogether, the ELNES features presented in Fig. 8.11 to 8.14 reveal, that the deposited TiN coatings on the graphite fibers consist of a TiC/TiN mixture of variable composition, and possibly of a titanium carbonitride (TiC_xN_y) region too, whereas the layers have a high amount of carbon near the fiber/coating interface and become increasingly nitrogen-rich towards the middle of the coating. The energy levels of a hypothetic carbonitride, determined again with the help of DFT-BP-DN*-calculations, are presented in Fig. 8.15 (middle), and demonstrate that a distinction between a TiN/TiC mixture (left and right in Fig. 8.15) and a compound in which titanium is bound simultaneously to carbon and nitrogen, cannot be facilitated by ELNES measurements with the recently available energy resolutions: The calculated differences between these binding states result in energy shifting of the ionisation edges of less than 0.1 eV.

On the other hand, the measured and calculated bonding-relevant specific features within the deposited TiN coatings indicate a chemical reaction of the $TiCl_4/N_2/H_2$ reactor gas mixture with the carbon of the fiber surface at the beginning of the CVD process: The Ti 3d electrons interact with both the N 2p electrons of the gas mixture and the C 2p electrons of the graphite fibers, whereas a mixture of p–d–π bondings (which results in t_{2g} MOs), and p–d–σ bondings (which lead to e_g MOs, as described above) results.

Thus it is evidenced that the growth kinetics of the TiN layers during the chemical vapor deposition process is far more complex than described by the usually assumed simple reaction: $TiCl_4 + 1/2\ N_2 + H_2 \rightarrow TiN + 4\ HCl$. Therefore, the deposition of these layers can be described best by the term "reactive chemical vapor deposition" [31]. The involvement of the fiber carbon in the reaction kinetics during the first steps of the layer growth from the precursor gas can also explain the observed good adhesion of the TiN layers to the fibers.

In conclusion, Fig. 8.16 demonstrates the effect of these TiN coatings for the embedding of the fibers in an Mg–Al alloy with high Al content of 5 wt%: While many Al_2MgC_2 carbide plates were observed at the surface of the *uncoated* carbon fibers (cf. Fig. 8.4), which caused an embrittlement of the composite, the *TiN precoating*

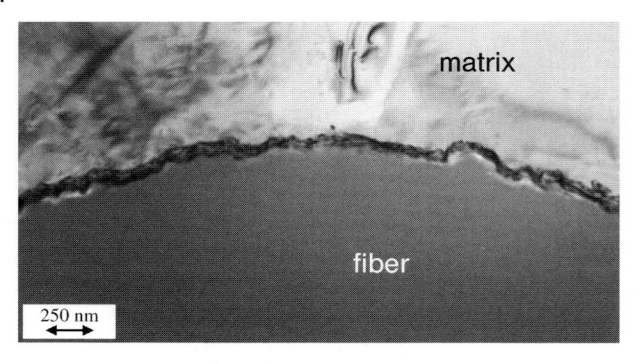

Fig. 8.16 HVEM image of the fiber/matrix-interlayer in C/TiN/Mg + 5 wt% Al composite material.

of the fibers can significantly stop this extensive reaction (cf. Fig. 8.16), and thus effect a considerable increase in strength.

Acknowledgements

We thank our project partner Prof. Dr. R. Singer and his coworkers (University of Erlangen) for the manufacture of the metal matrix composites. For the calculation of energies of electron transitions by density functional theory our co-worker Dr. O. Lichtenberger is thanked. Some of the described investigations are part of the project "Grenzschicht-Reaktionskinetik", supported by the Deutsche Forschungsgemeinschaft.

References

1 J. Woltersdorf, Elektronenmikroskopie von Grenzflächen und Phasentransformations-Phänomenen als Beitrag zum mikrostrukturellen Konstruieren neuer Hochleistungs-werkstoffe, Universität Halle, Habilitationsschrift, 220 pp., 1989.
2 E. Pippel, J. Woltersdorf, High-voltage and high-resolution electron microscopy studies of interfaces in zirconia-toughened alumina, *Philos. Mag. A* **1987**, *56*, 595–613.
3 J. Woltersdorf, E. Pippel, The structure of interface in high-tech ceramics.*Colloq. Phys. C1* **1990**, Suppl. 1, T.51, 947–956.
4 J. Woltersdorf, J. Heydenreich, Hochauflösungselektronenmikroskopie keramischer Werkstoffe, *Materialwiss. Werkstofftech.* **1990**, *21*, 61–72.
5 R. Schneider, J. Woltersdorf, The Microchemistry of Interfaces in Fiber-Reinforced Ceramics and Glasses, *Surf. Interface Anal.* **1994**, *22*, 263–267.
6 R. Schneider, J. Woltersdorf, A. Röder, Characterization of the chemical bonding in inner layers of composite materials, *Fresenius J. Anal. Chem.* **1995**, *353*, 263–267.
7 O. Lichtenberger, R. Schneider, J. Woltersdorf, Analyses of EELS fine structures of different silicon compounds, *Phys. Status Solidi A* **1995**, *150*, 661–672.
8 R. Schneider, J. Woltersdorf, O. Lichtenberger, ELNES across interlayers in SiC (Nicalon) fiber-reinforced Duran glass, *J. Phys. D: Appl. Phys.* **1996**, *29*, 1709–1715.

9 R. Schneider, J. Woltersdorf, A. Röder, Chemical-bond characterization of nanostructures by EELS, *Mikrochim. Acta Suppl.* **1996**, *13*, 545–552.

10 R. Schneider, J. Woltersdorf, A. Röder, EELS nanoanalysis for investigating both chemical composition and bonding of interlayers in composites, *Mikrochim. Acta* **1997**, *125*, 361–365.

11 A. Feldhoff, E. Pippel, J. Woltersdorf, Interface reactions and fracture behavior of fiber reinforced Mg/Al alloys, *J. Microsc.* **1996**, *185*(2), 122–131.

12 A. Feldhoff, E. Pippel, J. Woltersdorf, Structure and composition of ternary carbides in carbon-fiber reinforced Mg-Al alloys, *Philos. Mag. A* **1999**, *79*, 1263–1277.

13 A. Feldhoff, E. Pippel, J. Woltersdorf, Carbon-fiber reinforced magnesium alloys: Nanostructure and chemistry of interlayers and their effect on mechanical properties, *J. Microsc.* **1999**, *196*, 185–193.

14 A. Feldhoff, Beiträge zur Grenzschichtoptimierung im Metall-Matrix-Verbund Carbonfaser/Magnesium, Shaker-Verlag, Aachen, 1998.

15 J. Woltersdorf, E. Pippel, A. Feldhoff, Steuerung des Bruchverhaltens von C/Mg-Verbunden durch Grenzflächenreaktionen, in *Verbundwerkstoffe und Werkstoffverbunde*, K.Friedrich (Ed.), DGM-Verlag, Frankfurt/M. 1997, pp. 567–572.

16 J. Woltersdorf, A. Feldhoff, E. Pippel, Nanoanalyse der Reaktionskinetik von Carbonfasern in Magnesiumschmelzen zur Optimierung von Metallmatrixkompositen, in *Nichtmetalle in Metallen '98*, D. Hirschfeld (Ed.), DGMB-Informationsgesellschaft, Clausthal-Zellerfeld 1998, pp.105–112.

17 J. Woltersdorf, A. Feldhoff, E. Pippel, Formation, composition and effect of carbides in C-fiber reinforced Mg-Al alloys, *Erzmetall* **1998**, *51*, 616–621.

18 J. Woltersdorf, A. Feldhoff, E. Pippel, Bildung und Kristallographie ternärer Carbide in C/Mg-Al-Compositen und ihr Einfluß auf die Verbundeigenschaften, in *Verbundwerkstoffe und Werkstoffverbunde*, K. Schulte, K. U. Kainer (Eds.), Wiley-VCH, Weinheim, 1999, pp. 147–152.

19 O. Öttinger, R. F. Singer, An advanced melt infiltration process for the net shape production of metal matrix composites, *Z. Metallkunde* **1993**, *84*, 827–831.

20 A. Feldhoff, E. Pippel, J. Woltersdorf, TiN coatings in C/Mg composites: microstructure, nanochemistry and function, Philos. Mag. A **2000**, *80* (3), 659–672.

21 J. Woltersdorf, A. Feldhoff, E. Pippel, Titannitrid-Beschichtungen für Carbonfaser-verstärkte Leichtmetallegierungen, *Freiberger Forschungshefte B* **1999**, *297*, 49–56.

22 J. Woltersdorf, A. Feldhoff, O. Lichtenberger, The complex bonding of titanium nitride layers in C/Mg composites revealed by ELNES features, *Cryst. Res. Technol.* **2000**, *35*, 653–661.

23 A. D. Becke, Density-functional exchange-energy approximation with correct asymptotic behavior, *Phys. Rev. A* **1988**, *38*, 3098–3100.

24 J. P. Perdew, Density-functional approximation for the correlation energy of the inhomogeneous electron gas, *Phys. Rev. B* **1986**, *33*, 8822–8825.

25 P. Hohenberg, W. Kohn, Inhomogeneous electron gas, *Phys. Rev. B* **1964**, *136*, 864–868.

26 W. Kohn, L. J. Sham, Self-consistent equations including exchange and correlation effects, *Phys. Rev. A* **1965**, *140*, 1133–1137.

27 R. G. Parr, W.Yang, Density Functional Theory of Atoms and Molecules, Oxford University Press, Oxford, 1989.

28 D. F. Shriver, P. W. Atkins, H. L. Cooper, *Inorganic Chemistry*, Oxford University Press, Oxford, 1990.

29 J. Fink, J. Pflüger, Th. Müller-Heinzerling, in *Earlier and Recent Aspects of Superconductivity*, J. G. Bednarz, K. A. Müller (Eds.), Springer, Berlin, 1990, pp. 377–406.

30 J. Hosoi, T. Oikawa, Y. Bando, Study of titanium compounds by electron energy loss spectroscopy, *J. Electron Microsc.* **1986**, *35* (2), 129–131.

31 H. Vincent, C. Vincent, J. P. Scharff, H. Mourichoux, J. Bouix, Thermodynamic and experimental conditions for the fabrication of a boron carbide layer on high-modulus carbon fiber surfaces by RCVD, *Carbon* **1992**, *30* (3), 495–505.

9
Metallic Composite Materials for Cylinder Surfaces of Combustion Engines and Their Finishing by Honing

J. Schmid

9.1
Introduction

Improving fuel consumption and reducing emission of combustion engines have become relevant competition factors in the automotive industry. The series standards have more and more extra features which sometimes counteract the required weight and fuel consumption reduction. In the light of increasing fuel costs and diminishing resources, modern, efficient, economical and light engines are needed for the future. This requires developments in construction and also increasing application of new materials. Both must go hand in hand, many improvements in construction are only realizable by application of new materials. Modern light metal composites combine several positive properties:

- Weight saving
- Higher load resitivity in service and therefore better efficiency
- Tribologically most convenient surfaces, that is, for reduction of emissions by lower oil and fuel consumption.

9.2
Composites Based on Light Metals

9.2.1
Manufacturing Possibilities

In comparison to common gray cast iron, light metals offer a great variety, not only of properties but also of manufacturing possibilities.

9.2.1.1 Casting of Over-eutectic Alloys

The classical and currently used "composite" of aluminum and precipitated silicon crystals from the melt, is manufactured in a relatively expensive low-pressure die-

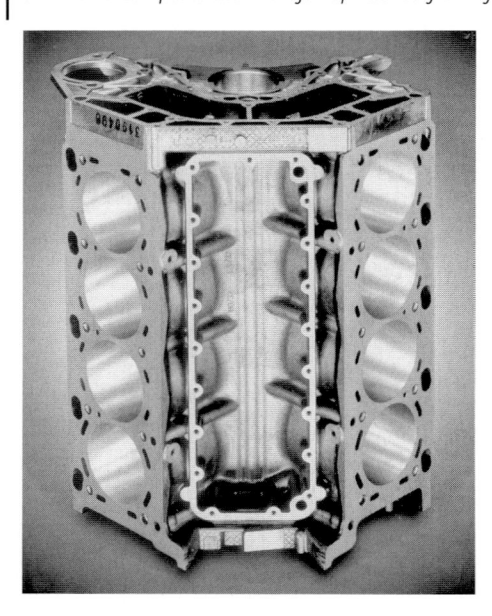

Fig. 9.1 Monolithic V8- engine block of AlSi$_{17}$.

casting technique. Both surfaces and engine blocks are produced, although the latter are restricted so far to relatively small eight- and twelve cylinder productions, Fig. 9.1.

9.2.1.2 Infiltration
Particles or fibers of different materials are combined in an open porous framework (preform cylinder) and afterwards under slowly increasing pressure (squeeze casting) infiltrated with a light metal melt, Fig. 9.2. By the use of fibers, increases in strength are possible. Especially with magnesium matrix alloys, infiltration can be the solution, where the precipitation of hard phases from the melt is a problem, Fig. 9.3.

9.2.1.3 Sintering
Nonmetal particles and light metal powders are cold pre-pressed and sintered, or hot pressed under a protective gas . Relatively good reproducible, regular structures can be processed by sintering.

9.2.1.4 Stirring of Hard Particles into the Melt
Corundum or SiC particles are stirred into the melt by the special Duralcan®-procedure and distributed free of segregations. Afterwards the blank is transformed to liners by extrusion. Co-extrusion is an interesting variation, Fig. 9.4. Here an inner layer of wear resistant, particle-containing material with a particle-free easy to process cladding is manufactured in one working step.

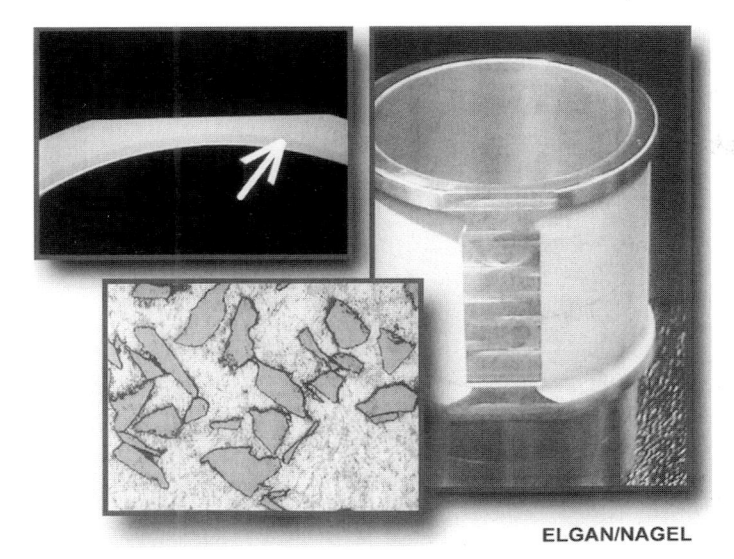

Fig. 9.2 Infiltration material Locasil from Kolbenschmidt: detailed image of the preform area and the final honed surface

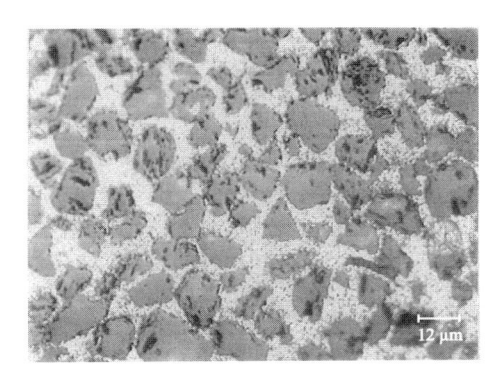

Fig. 9.3 Particle content approximately 60 vol%

Fig. 9.4 Cut of the co-extruded liner (LKR/Ranshofen).

9.2.1.5 Spray Forming

Fine atomized aluminum melt and a particle jet (alternatively an overheated, overeutectic Al–Si melt) are aligned on a rotating carrier and formed into a massive raw material. Cylinder liners are then extruded from this mass. This procedure is very suitable for the manufacture of homogeneous, finely dispersed materials. To reduce the porosity further compression has to take place in a subsequent process (forging, rolling or hammering).

9.2.1.6 Addition of Reactive Components into the Melt

By reaction with the melt, hard phases (for example boride or nitride) can be precipitated.

9.2.1.7 Thermal Coating

A layer of a light metal alloy and ceramic particles (for example Si, Al_2O_3) is sprayed onto the cylinder surface by laser or plasma gun. The actual engine block can be manufactured by the cheap die-cast technique. There, as with coating procedures, a defined maximum size of the cast pores may not be exceeded. However, nowadays plasma coatings based on iron are used exclusively. These are described in Section 9.3.

9.2.1.8 Laser Alloying

In comparison to laser coating only Si from an external source is brought into a locally made molten bath. The Si powder is very quickly resorbed in the strongly overheated melt spot of the surface of the cylinder drilling and, due to the fast cooling, precipitated as extremely fine (2–10 μm diameter) primary crystals with irregular shape. The main advantage is that the intermetallic compound is bound to the base material with outstanding adhesion and optimal heat transmission, Fig. 9.5.

All these variants have in common the light metal matrix in the surface. For this reason the possible occurrence of problems like poor resistance to special fuels,

Fig. 9.5 Laser alloyed Al–Si surface (approximately 20% Si) in low pressure casting.

poor heat transmission, crack formation and deformation due to different heat expansion coefficients can be avoided, or at least reduced. The materials described here are called metal matrix composites (MMC). Figure 9.6 gives an overview of the possibility of designing surfaces of light metal engines.

1. Grey-cast iron liners (in general centrifugal casting with relatively high P content)
2. Hypereutectic "AlSi$_{17}$" alloy in low pressure/gravity die casting, e.g. Alusil, Silumal
3. Infiltration materials with ceramic particles and/or fibres, e.g. Lokasil (squeeze casting, reaction infiltration)
4. Addition of reactive components in the molten metal, e.g in situ formation of TiB$_2$
5. Mixing in inert hard materials (Al$_2$O$_3$, SiC into the light metal melt, e.g. Duralcan)
6. Powder-metallurgical materials (sintered) from LM powders and various ceramic powders (Si, SiC, etc.)
7. Spray compacting
   a) hypereutectic alloys (AlSi$_{25}$) e.g. "Silitec"
   b) Al alloys with Al$_2$O$_3$/SiC particles
8. Fine-grained, hypereutectic AlSi alloys by suppressing the Si crystal growth and then heat treatment
9. Laser alloying of Al blocks by adding Si powders (or similar material)
10. Laser depositing of AlSi alloys
11. Thermal coatings (plasma, HVOF, wire arc spraying
    a) pure metals (predominantly iron-based)
    b) Al-Si alloys
    c) MMCs
    d) pure ceramic layers
12. Plasma-chemical Al$_2$O$_3$ coatings/MAO
13. Thin layers (PVD/CVD)
14. Electro plating of e.g. Ni-SiC (Nicasil)

Fig. 9.6 Possible materials for cylinder surfaces for light metal engines.

9.2.2
Selection Criteria

9.2.2.1 **Strength**
The addition or the precipitation of hard materials does not lead inevitably to higher strength. The strength can even be reduced with larger particles by the inner notch effect. Essentially strength increase, especially for surface properties, can only be reached when using hard phases when the particle size changes from the micron to the submicron range. This is not the case for the materials used so far, thus only moderate strength increases are reached in this way. Significant strength increase could be achieved by:

- Dispersion solidification of the matrix, for example in powder metallurgy (spray forming, sintering), but also during casting, the finest precipitations of intermetallic phases contribute to higher strength. Strength at higher temperatures is most important for materials that are to be used for the wall of a combustion area. Thus intermetallic compounds which dissolve at higher temperatures (during casting), or recrystallize to larger grains, have to be avoided.

- The application of fibers with as large as possible length/diameter ratio. Correspondingly, with endless fibers the highest stability increase can be reached. In that case their use in high compression engines is also possible. A further advantage is the good high temperature strength.

9.2.2.2 Tribology
Here a wide field for systematic research opens up. So far only little is known of the influence of the kind, content and size of hard phases on the in-engine service on real existing friction relationships. Referring to adhesion of the oil film, the direct material properties are less decisive than the processable surface structure.

9.2.2.3 Flexibility
With regard to kind, shape and size of hard materials, preform solutions, like Locasil®, seem very adaptable. For the application of fibers, their convenient orientation is easy to control.

Controllable laser or plasma coating is a very interesting solution, not only in small serial manufacture with changing drilling diameter, but also the coating process can be integrated into the engine manufacturer's production line. Further flexibility is given concerning the coating composition. However, coating with light metal alloys is relatively complex due to the required vacuum and protective gas unit. Therefore atmospheric sprayable alloys based on steel are increasingly applied (see Section 9.3). When comparing the considerably lower density of magnesium (1.74 g cm^{-3}) to that of aluminum (2.70 g cm$^-$_), the application of the possibilities described here will also play a role for magnesium in the future. For example, silicon as a hard phase will no longer be used due to the reactivity and decrease in the mechanical properties.

9.2.2.4 Design Criteria
To save weight, the web width between the cylinder boreholes is often very small, so that an application of liners is no longer possible. For these cases monolithic blocks of over-eutectic alloys complete one metal matrix and hard materials or local material changes by preform infiltration, or applying thin coatings (galvanic, plasma and laser alloying) are to be considered.

9.2.2.5 Processing and Machining

Coarse hard phases tend to break increasingly during processing (extrusion). Thus this technology is reserved for the relatively fine grained MMCs, like Dispal® or Duralcan®. At least during machining with geometrically defined cutting edges (drilling, turning and milling) high service times are reached with decreasing particle size. In general, apart from honing of cylinder drillings, tool service times during other processing of monolithic MMC blocks are only a fraction of those of processing particle-free die-cast crankcases. That is why, at least for large series production, local material solutions are more suitable.

9.2.2.6 Strength of the Material Composite

Contrary to widespread opinion, even surfaces of aluminum alloys included in aluminum crankcases are not combined "in one piece" with the block. The reason for this is found in both the oxide layer of the surface and also in the temperature difference to that of the hot melt. Neither coating nor pre-heating of the liner could effect a complete melting. In the top area of the drilling, especially, where the highest thermal and mechanical strains apply, the connection is still a problem due to the small wall thickness and thus the short melting time. This observation applies to all surface types. For this reason almost all liners are mechanically clamped. From this point of view monolithic blocks or infiltration solutions are of advantage. With the adhesion of plasma sprayed surfaces there have now been positive experiences with continuous operation, at least in small series.

9.2.2.7 Heat Transmission Ability and Heat Expansion

In comparison to pure aluminum, addition of corundum and silicon carbide hard material lead to a considerable reduction in the heat capacity whereas with addition of pure silicon there is a slight reduction. The heat capacity reduces with increasing content of such materials while the heat expansion behaves inversely. At least with the use of cast-in and pressed-in liners, attention has to be paid to reach low deviations (tolerances) to the basic material.

9.2.3
Fine Processing

The best properties inside the material are useless if unsuitable surface conditions lead to poor friction and wear behavior. Materials of friction partners are therefore only as good as their surfaces. Consequently, honing as the final process step in processing functional surfaces is of crucial importance. Especially for combined composites of hard and soft components, previous attempts at finishing have been cost intensive, due to the small area between wear resistant and poor conditions, and they are not suitable for series production. It is shown in the following that by the development of modern honing technology, cost convenient advantageous materials can be reproducibly processed in large piece numbers.

9.2.3.1 Processing Before Honing

In general, for the materials described here, MCD (monocrystalline diamond) or PCD (polycrystalline diamond) are used for fine drilling or turning. The content of abrasive particles, like silicon or corundum, nowadays does not allow a meaningful alternative.

Despite the use of diamond, the geometrically defined cutting edge of the tool becomes rounded off after a few drillings and free surface wear occurs. The consequence for the tool is the appearance of deep cold deformation, and spalling of the embedded hard phases for reinforcement. Even particles underneath the surface show crack formation, see Fig. 9.7(a) and (b).

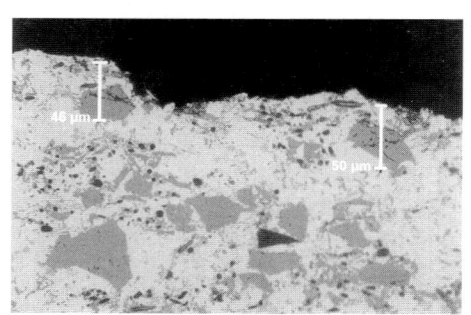

Fig. 9.7 (a) Cross-section of a fine turned cylinder liner. Due to the waviness of the fine drilled surface, material of at least 100 μm (per side) has to be removed. (b) Due to strain hardening within the material, cracks also occur in crystals below the surface.

9.2.3.2 Honing Step 1

For pre-honing the following development steps are desired:

1. High wear rate due to deep damage in the pre-treatment.
2. Good cutting ability at low pressing pressure to reach an optimal borehole geometry.
3. Possibly low damage of the embedded hard materials.
4. High tool service time.

These aims are reached in the following ways:

During scrub honing material wear of more than 0.3 mm is already reached at 60 s honing time (Alusil®, Locasil® und Silitec®). Even for the difficult to machine composites of aluminum and corundum or silicon carbide (for example Duralcan®) comparable wear rates can be reached during honing in the first honing step.

The big advantage of honing compared to any other process with defined cutting edges is the permanent self-sharpening of the honing ledges. With proper arrangement of cutting grains and binding these never get blunt during processing of

MMCs. Therefore the tools can also be used for fine drilling for longer times because even deeper deformations and damage are worn off in a relatively short time. With appropriate machining and tool equipment during honing, rectangularity between cylinder and main surface drilling can be reached. Even corrections of the drilling center are possible. Thus fine drilling can be completely replaced by the economical honing, at least for critical materials with reinforcements of Al_2O_3 or SiC. Due to the significantly higher tool service times for honing, clear advantages in process costs can result. Additionally, constant and low deforming depths are reached due to the good self-sharpening of newly developed honing ledges. Process times for subsequent operations can thus be shortened. For more easy to machine materials, for example $AlSi_{17}$, the material additions during pre-honing are between 60 and 120 µm.

Almost all new engines allow only a small tool overflow due to the weight saving building construction. To reach a good cylinder form, the selection of the right cutting edge and the application of suitable honing tools is important. Here improvements are possible which save, completely or partly, repeated grinding of the tools during the blind hole process.

By the use of diamond as cutting grains, tool life quantities from a few thousand to over ten thousand drillings have been reached. For the processing of common materials, an over proportional dependence of the related material removal rate exists.

9.2.3.3 Honing Step 2

Both the improved drilling geometry and the lower damage depth of crystals require only small wear thanks to the improved pre-honing process. The main emphasis can be set mostly on the areas of surface quality and tool service time during the second honing step. Thus the degree of damage of the silicon crystals of Alusil® in the series can be reduced from over 40% to below 10%. Surface roughness down to Ra = 0.003 µm is possible. The surface quality is almost comparable with a metallographical polish, even phase differences within the aluminum matrix become significant. Furthermore the relatively high waste ratio caused by groove formation could be reduced to zero, see Fig. 9.8 and 9.9. By the exchange of

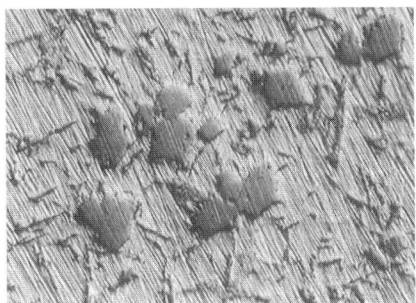

Fig. 9.8 Alusil after the second honing step.

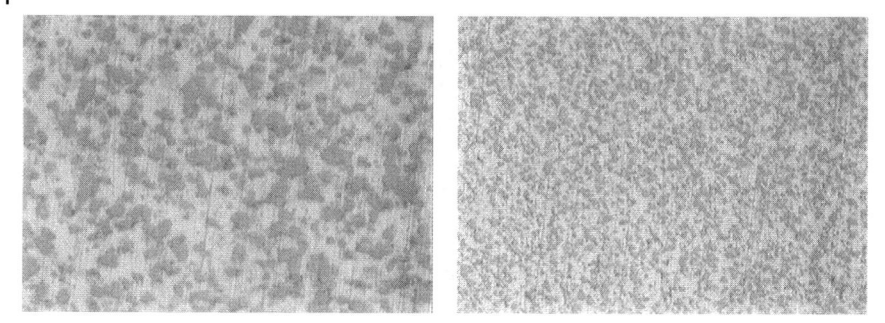

Fig. 9.9 Spray formed aluminum–silicon alloy after the second honing step.

ceramic cutting grains with diamond, tool life quantities from 10 000 to 20 000 drillings in series could be reached. This is an important requirement for more consistency in the working processes.

9.2.3.4 Honing Step 3

Especially with cylinder surfaces of engines, hard phases, which partly stick out of the metal matrix, are required for functionality reasons. This is possible so far only with very cost intensive etching techniques. Its reproducibility (problems with clearing of engine blocks, different reactivity during etching) has also proved to be problematic. Therefore the third honing step has been developed with the aim being to replace etching by a more cost convenient, reliable mechanical technique. Material independent uncover depths to 4 µm have been reached. The required honing times are between 30 and 90 s. The reachable uncover depth within a determined time is not the same for all materials; composition and structure have a strong influence, so that an early cooperation between material and process development is important to reach an optimal final result, see Fig. 9.10 to 9.12.

In principle, stripping of hard phases can be reached in the second honing step with newly developed cutting materials. However, this requires some concessions of cycle time and tool lifetime, whereas three-step honing is profitable, at least during large series runs. As Fig. 9.10 and 9.13 show with the example of Duralcan®, the possible extreme uncover depths are also possible for outside processing. The reachable surface is optically comparable with etched materials. In addition to the enormous cost saving, a further advantage has arisen for this honing technique. With re- initialization of the light metal matrix by etching, crystals with relatively sharp edges are uncovered. In comparison, edges are slightly rounded off by mechanical stripping during the third honing step. Damage to the piston ring, which leads to high oil consumption and bad emission values, can be already avoided in the run-in period. Comparing the relatively regular shaped, precipitated Si crystals from a melt, for example Alusil®, with manufactured hard phases Duralcan® or Locasil® (by cutting up) with sharp edges, makes the advantages of the mechanical stripping of the latter group clearly visible, see Fig. 9.14.

9.2 Composites Based on Light Metals | 225

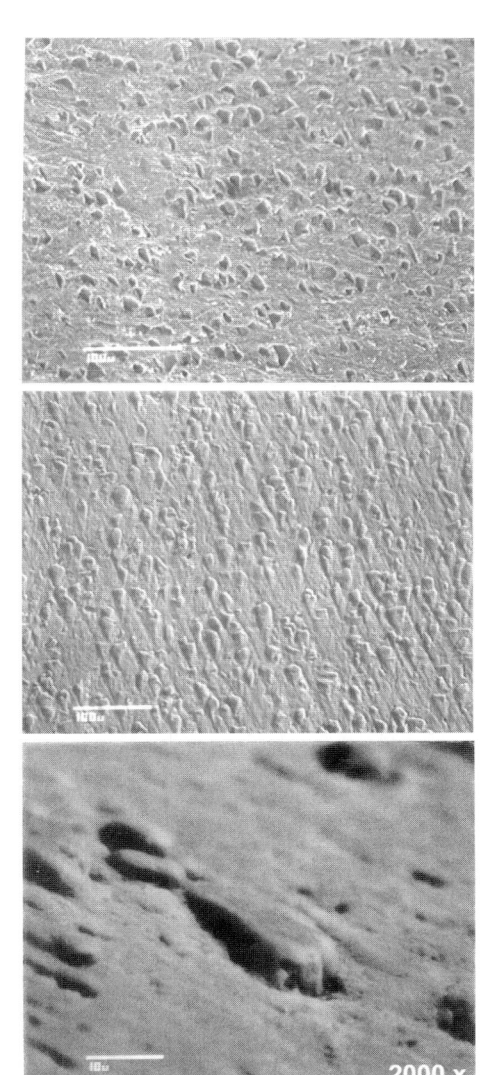

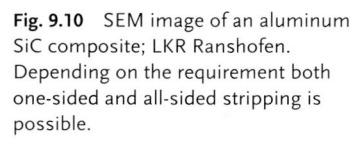

Fig. 9.10 SEM image of an aluminum SiC composite; LKR Ranshofen. Depending on the requirement both one-sided and all-sided stripping is possible.

Fig. 9.11 Liner made from spray formed material, surface after mechanical stripping.

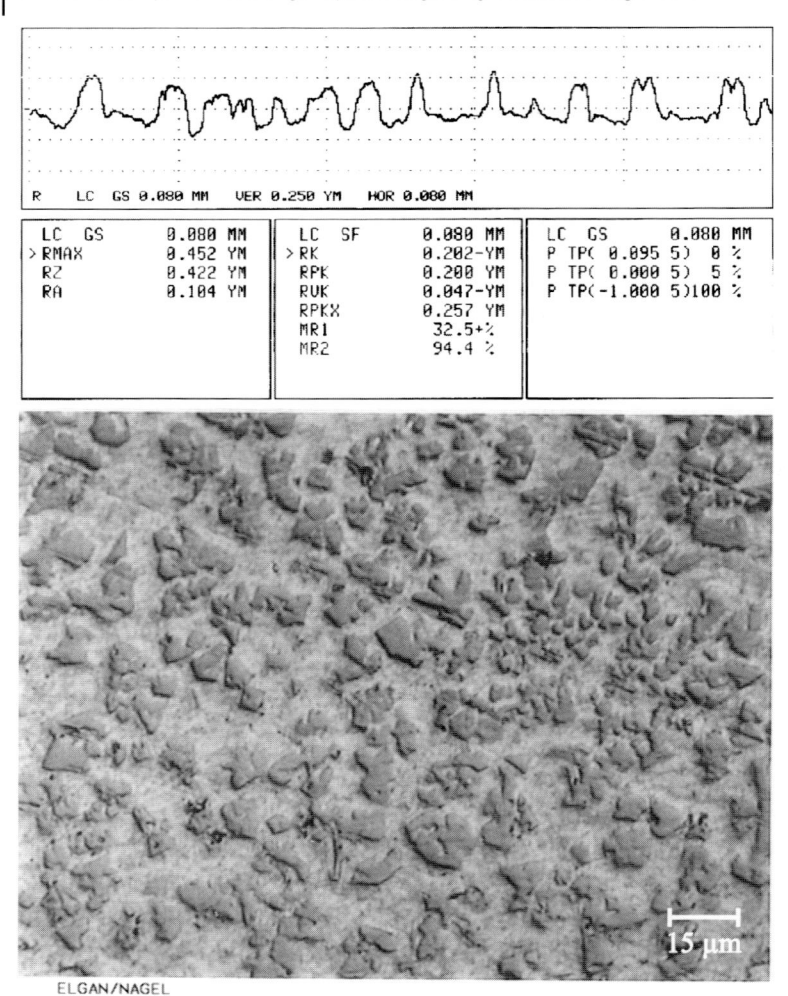

LC GS	0.080 MM	LC SF	0.080 MM	LC GS	0.080 MM
>RMAX	0.452 YM	>RK	0.202-YM	P TP(0.095 5) 0 %
RZ	0.422 YM	RPK	0.200 YM	P TP(0.000 5) 5 %
RA	0.104 YM	RVK	0.047-YM	P TP(-1.000 5)100 %
		RPKX	0.257 YM		
		MR1	32.5+%		
		MR2	94.4 %		

Fig. 9.12 Laser alloyed Al–Si surface after honing.

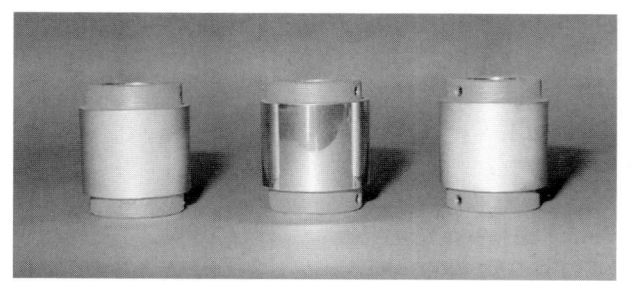

Fig. 9.13 Left: Duralcan (with SiC particles) after the third honing step; middle: Duralcan (with SiC particles) after the second honing step, right: Duralcan (with Al_2O_3 particles) after the third honing step.

Si crystals after
mechanical stripping

Si crystals after
chemical etching

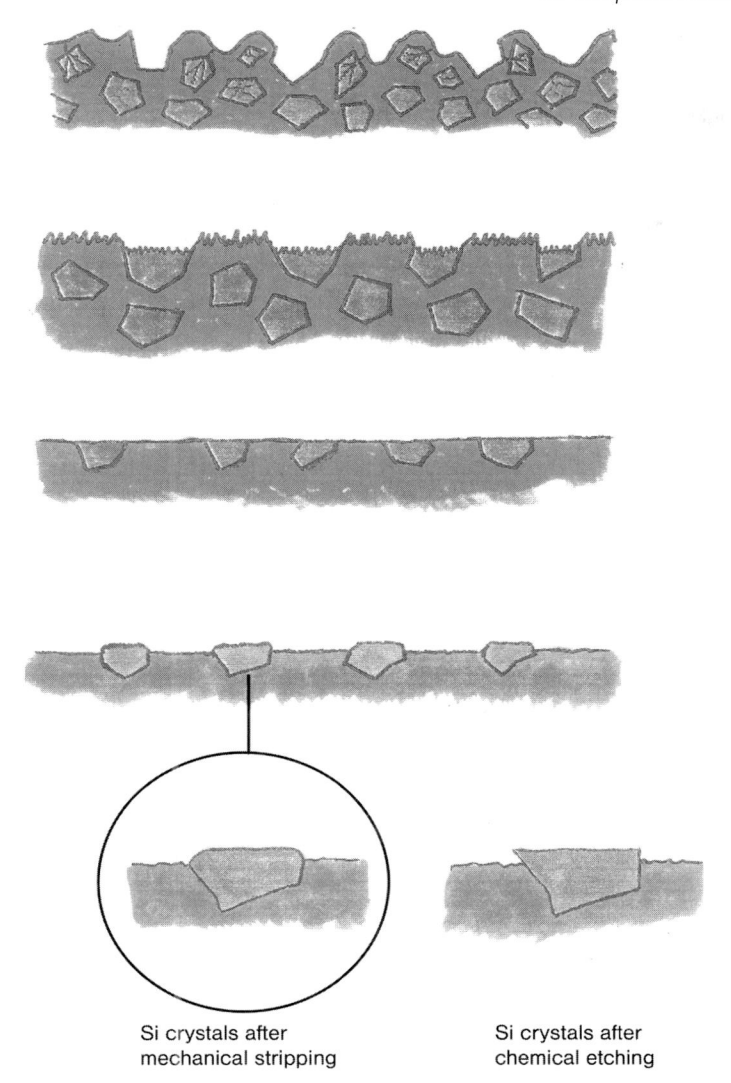

Fig. 9.14 Machining sequence during honing of MMCs. The figure shows from top to bottom the initial state and honing steps 1 to 3.

However, for the processing of fiber-reinforced materials this is an important aspect. On or just under the surface, longitudinal aligned fibers can rise up after etching, like needles out of the matrix.

Additional structuring of the surface is an interesting new development, especially for fine grained materials. During the last process step, a variable structure can be processed, using a double honing tool in a few seconds at a low cost, which improves the adhesion of the oil film by coarsening the specific surface in the break-in phase, see Fig. 9.15 and 9.16.

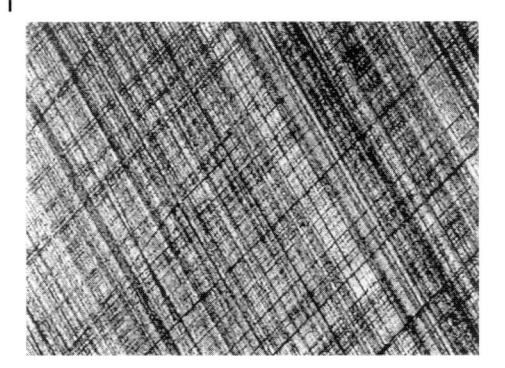

Fig. 9.15 Example of surface structuring of a spray formed material.

Fig. 9.16 Detailed image of the surface structure of a spray formed material: additional structuring does not cause damage of the silicon crystals.

9.2.4
Marginal Conditions

9.2.4.1 Expanding systems

The honing operation described has been carried out successfully both with hydraulic and with mechanical delivery. During pre-honing a mechanical delivery is preferable in general; during the intermediate honing step both expanding systems can be used. Regarding optimally reproducible surfaces, a hydraulic expanding system during the third honing step is recommended due to the constant acting pressure force.

9.2.4.2 Cooling Lubricants

Of the previously tested cooling lubricants the following have proven to be practical:

- a thin liquid, mineral product
- a thick liquid, mineral product
- medium viscosity, to esterify vegetable- or mineral-based honing oils.

Beside the cutting medium, the cooling lubricant has an influence on the wear ratio and remaining condition of the hard phases. However, the main point here should be the ability to infiltrate since the cleaning of cooling lubricants is technically very demanding during light metal processing. Thus it has to be done without the usual magnetic cutter during cast processing.

In a highly concentrated, watery emulsion (more than 10% mineral oil content) lubricating of the honing ledges can be avoided by good cutting behavior with respect to material wear. Nevertheless the application is only recommended for low material removal rate volume, otherwise the crystal damage by the aggressive cutting behavior is relatively high and the tool lifetime is significantly lower. Due to the resulting unsafe process guidance the application of water-containing cooling lubricants is not advised.

9.2.4.3 Cutting Speeds

Investigations have shown that an optimization of cutting speed gives advantages for each single honing step. The right alignment is strongly dependent on the material, the delivery, the cooling lubricant and the required surface. Thus no general recommendation can be given here. During the use of double honing tools, a machine with adjustable, variable rotary rate is suggested.

9.2.5 Summary

In the engine industry the honing process has been optimized in respect of new light metal composites. With that the following could be achieved:

1. Relief or replacement of fine drilling by honing with high wear rate. The self-sharpening system during honing makes a high tool lifetime possible. In connection with an already proven successful machining concept which guarantees both the drilling center and the rectangular arrangement to the main surface drilling, this is an economical interesting solution, especially for materials with strongly wearing hard materials such as Al_2O_3.

2. The use of special diamond honing ledges with high tool lifetime within the second honing step. With that, extremely smooth surfaces with nondamaged hard phases could be achieved for all aluminum materials. This is also an optimal basis for eventual steps like the re-initialization of the metal matrix.

3. The replacement of chemical etching by honing. Therefore by the development of suitable, selective cutting tools, a further contribution to the economical processing of modern composites is achieved.

9.3
Plasma Coatings

9.3.1
General

The high heat energy density available to melt powder in plasma, connected with the possibility to build plasma burners with short spraying distances for specific applications, has widened the application areas of plasma spraying. All materials which show a stable liquid phase and are available in the right powder size range and shape can be coated onto metallic substrates. This allows a great degree of freedom for selecting the coating material.

In the special case of the coating of cylinder drillings, the transmission of heat must be held within limits during the coating process. The obtained temperatures should be below the maximum service temperature of the engine, so that unwanted changes in the microstructure and mechanical properties of the substrate do not take place. Plasma spraying allows the coating of a large range of layer types without disadvantages concerning the structure and deformation of engine blocks due to the temperature of the interface between coating and base material, which is approximately 100 °C, Fig. 9.17.

However, the selection and use of this great variety of coating materials is only possible if suitable process solutions are available. In the area of cylinder surfaces, honing is the final process step so far for all materials, and for a few materials the only meaningful one. To exclude the limitation "ability to hone" for the material selection, the manufacturer of the equipment responsible for the final process (finishing), has a duty to work directly with the material producer on a process solution at a very early stage.

Fig. 9.17 Plasma spraying allows the application of various coating types.

9.3.1.1 Coating Materials

Different materials are used as coating materials, especially iron–carbon alloys of different chemical composition. Because the coating usually takes place by the APS technique (atmospheric plasma-spraying), a certain degree of oxidation of the coating results, which acts positively due to the formation of correct oxide types with appropriate optimization of the oxygen content.

The oxides FeO (Wüstit) and Fe_3O_4 (Magnetite) can be considered as solid lubricants and they increase the tribological properties, as well as the machinability, Fig. 9.18.

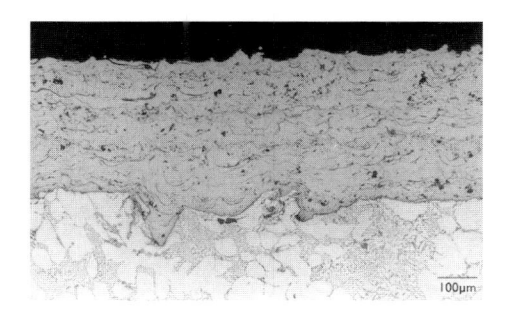

Fig. 9.18 Solid lubricants improve the tribological properties of coatings.

The application of corrosion resistant coatings to iron based alloys is also possible by the addition of chromium and molybdenum. These coatings are resistant to sulfuric acid and formic acid during their application in combustion engine machines. These materials can be reinforced by addition of fine particles of a tribologically functionable ceramic. Thus an increase in the compressive strength, abrasion resistance and a reduction of galling ability are achieved. These MMC-coatings, which are highly resistant to strain, have already been tested and are especially suitable for petrol and diesel engines.

Further composites, consisting of many metallic alloys or components are also used. The drillings are prepared by corundum spraying before coating. The obtained roughness makes a stable mechanical clamping of the coatings possible.

9.3.1.1.1 Layer Characteristics and Tribological Properties

The adhesion strength of the coating (in the case of coatings based on Al–Si alloys with approximately 7 to 10% silicon) is in the range 40 to 50 MPa, see Fig. 9.19. Experience shows, that a minimum value of 30 MPa has to be reached to avoid problems in combustion engine machines during service.

The microhardness is selected in such a way that the necessary comprehensive strength (at least $HV_{0.3} = 350$) is reached and the ability to machine by honing is still kept at a reasonable level to limit the process times (max. $HV_{0.3} = 650$). The remaining porosity of the layers (1–3 vol%) fine stochastic contributed, allows convenient tribologic properties.

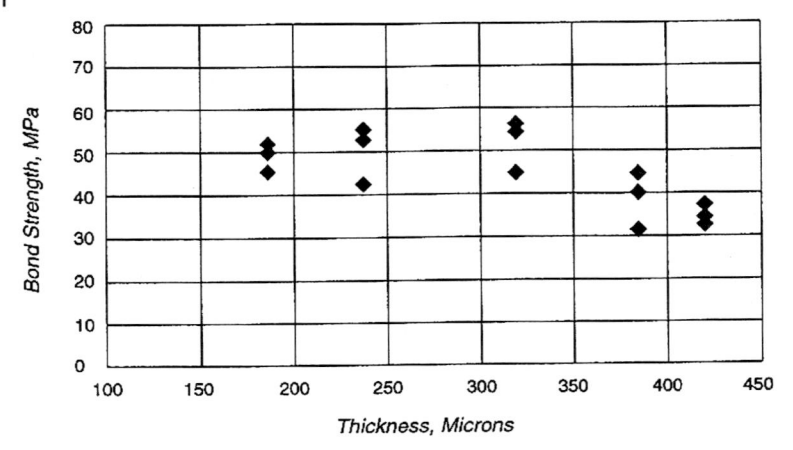

Fig. 9.19 Adhesion strength of plasma coatings of cylinder bores in dependence on the coating thickness (material: low alloyed steel; surface treatment: corundum spraying, Ra = 10 µm; substrate: AlSi-cast alloy; inspection: DIN EN 582.

The pores of diameters of a few micrometers become filled with lubricating oil and their fine contribution allows a secure lubrication of the system, whereas the oil consumption can be further reduced in comparison to cast iron.

9.3.1.1.2 Application Potential

The technology of inner plasma spraying of cylinder surfaces can be applied to both petrol and diesel engines. This technology has already been introduced to series production in Europe. The relation between costs and benefit can be considered as very attractive for the automotive industry.

The considerable technical advantages are the reduction of the friction coefficient between piston ring and cylinder surface, the potential reduction of the oil consumption, and also the significant increase in the wear resistance. In addition, the distance between the cylinders can be reduced, so that weight and space can be reduced by the compact design. At larger volumes (10 000 drillings a day) the coating costs (depending on the powder type) are more economical than using gray cast liners.

9.3.1.2 Comparison of Honing with Other Processing Technologies

In polishing, drilling or turning, the wear process is concentrated on a very narrow contact area. In this quasi-linear-shaped process zone relatively high working forces occur, which lead to crack formation or even to spalling of whole coating areas with increasing hardness and embrittlement of the coating.

In contrast, wear forces during honing are distributed over the relatively large area of honing ledges appearing during the process. Additionally, honing is a self-

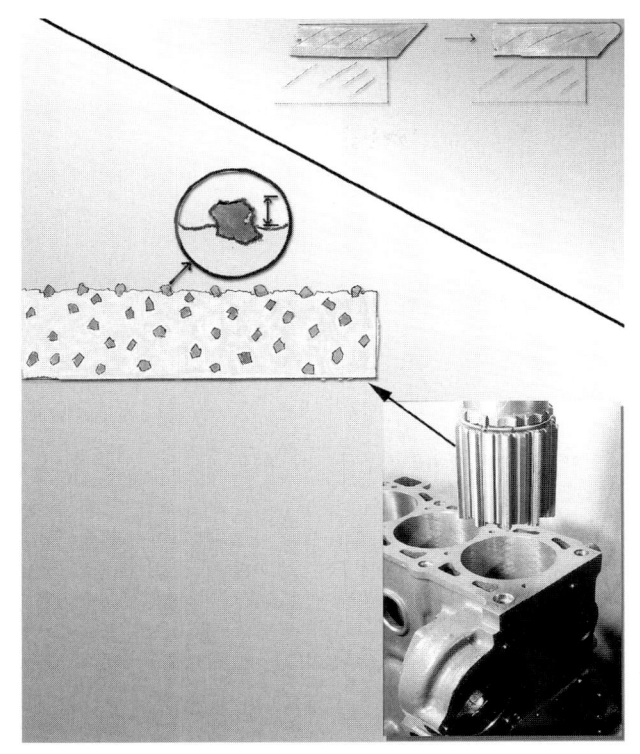

Fig. 9.20 Comparison of processes between geometrically defined cutting edge (top) and honing (bottom).

sharpening system and therefore, compared to processing with geometrically defined cutting, guarantees a constant treatment process with constant low specific wear until the cutting coating is used up, Fig. 9.20.

During the processing of plasma-sprayed coatings, the following advantages arise during honing:

- Coating damage (cracks, spalling) can be avoided as well as lubrication within the pores (important as oil pockets).
- A very large variety of coating types is applicable for series production. Thanks to the self-sharpening process during honing, very hard ceramic or metal–ceramic (MMC) coatings can be processed with high tool lifetimes. Such coatings lead otherwise, even with application of diamond, to the rounding off of the geometrically defined cutting edge.

9.3.2
Definition of the Process Task

9.3.2.1 Process Adding, Geometry

The maximum possible coating thickness is determined by the heat expansion coefficient and the mechanical properties of the coating. If the heat expansion coefficient of the coating is very different from that of the basic material, or if the elasticity, or rather ductility, is too low, then detachment can occur with too thick coatings due to residual strain. On the other hand a remaining coating thickness of 100 to 180 µm after honing is required to safely cover casting pores in the basic material. Therefore it is aimed to keep the necessary honing wear as low as possible above the required minimum coating thickness. However, it has to be large enough to correct the deformations which occur during coating.

A run-in, stable coating process shows process adding between 100 and 150 µm drilling diameter. With single coating procedures significantly higher honing adding can be demanded at first.

9.3.2.2 Requirements of the Surface Processing

The first honing tests on plasma coatings were carried out with a specification analogous to cast processing. The result was large spalling at the surface. Subsequent engine tests gave negative results regarding wear and oil consumption. To produce a functional surface, material investigations have been carried out by the equipment producer himself. Metallographic grindings, both conventionally crosswise to the layer and also parallel to the coating surface (that means to the surface of the piston ring) were carried out. Both investigations resulted in important information for optimization of the honing procedure. Figure 9.21 shows the typical structure of thermal sprayed coatings using a crosswise Schliff at strong magnitude. By the high impact speed, the melting drops are deformed, flat, or rather lens type. In a honing operation, with cast iron, the single lenses are torn off completely. This also explains why with such honing – within a certain grain spectrum – the roughness of a coating is independent of the cutting grain size used. These irregular shaped spalls not only increase the oil consumption, they also decrease the

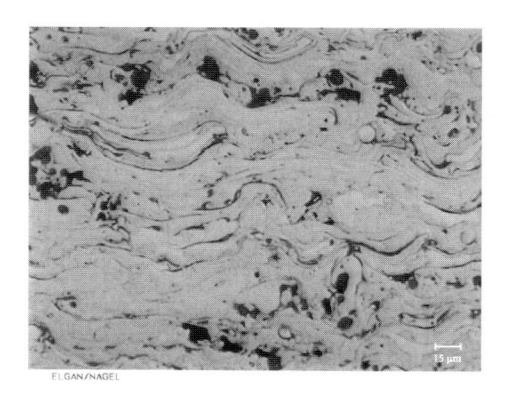

Fig. 9.21 Metallographic cross-section of a plasma sprayed coating (magnified section).

creep resistance of the layer in the form of notch areas. As a consequence, cutting performances had to be developed, at least for the last honing operation, which cut through the deformed melting drops without weakening their laminated structure.

Independent from this, it was necessary to clarify which surface topography is actually meaningful for the plasma coating, from the tribological viewpoint. To do this a polished grinding parallel to the surface was investigated. In comparison with the cross-section grinding, a completely different picture is revealed, Fig. 9.22. It is shown that a sufficient amount of pores, i.e. oil pockets, already occur in the plasma coating. Therefore the machining of additional honing grooves could be avoided for further tests. Anyhow only very fine honing structures, which do not damage the coating, are possible and these disappear after a certain running-in period. Therefore the honing angle makes no difference from the tribological viewpoint.

An advantage of an even plateau structure consists in anticipating the running-in behavior and thus the increased starting oil consumption. This has already been impressively proven by the slide-honing of gray cast cylinder surfaces [4, 5].

For safe running behavior it is therefore important that all pores are opened by honing. With pores removed by lubrication not only the volume of oil contained is decreased, Fig. 9.23. In the course of service these spalls can break off. This leads

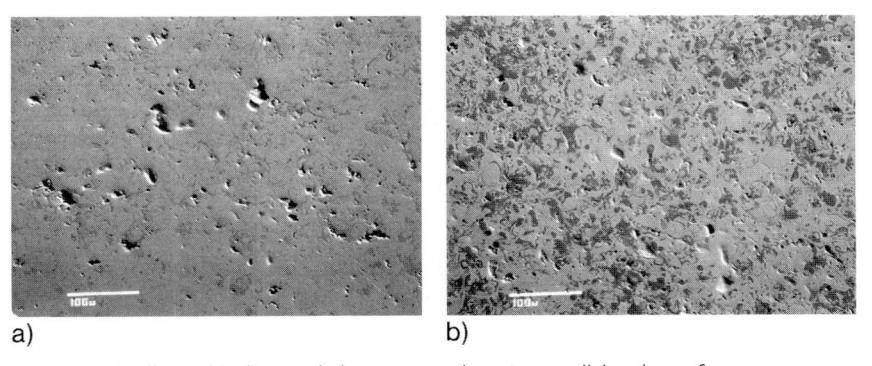

Fig. 9.22 Metallographic dissected plasma sprayed coating parallel to the surface: pure steel coating (a), particle reinforced coating (b).

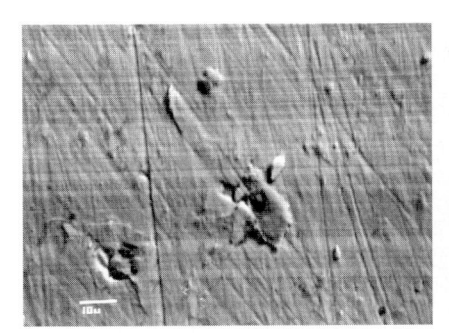

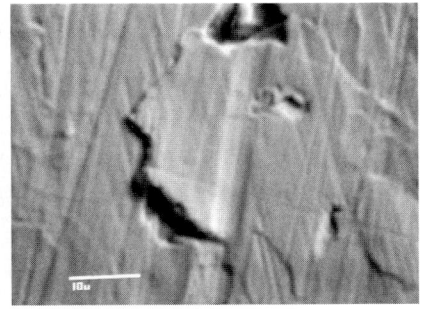

Fig. 9.23 Spall formation of plasma sprayed coatings at nonoptimal processing.

correspondingly to piston movement in the axial grooves, whereby there is an increase in the oil loss in the combustion chamber and also blow-by in the crankcase. For these reasons the optimal functionality of a surface of the honed plasma coating cannot be fulfilled by the global demand for a low Rz value or rather "as smooth as possible". At this set-up of specifications, a coating's own porosity is always to be considered.

9.3.3
Results of Honing Tests

9.3.3.1 Investigation of Adhesion
The investigations concerning this topic have been carried out by the equipment producer considering those points which are relevant to honing.

9.3.3.1.1 Adhesion of the Plasma Coating at High Machining Rates (Rough Honing)
During application of cutting mediums, which are not suitable for plasma coating, large coating spalling can occur even at conventional wear rates. The reason for this is the high pressing pressure of the honing ledges, where the flexible plasma coating is pressed into the soft aluminum basic material. This leads to strains which remove the coating. On the other hand, the coating can cut in with overlarge cutting grains and thus become torn off. That is why special cutting ledges have been developed for these coatings, which, even at small grain sizes, have a high self-sharpening ability, and thus make high material wear possible at low specific delivery forces. With these cutting mediums a wear rate up to 200 µm (diameter) in 30 s honing time could be reached without crack formation or spalling of the substrate.

9.3.3.1.2 Adhesion Strength During Honing of Thin Coatings
In further tests it was found that, below a certain minimum coating thickness, spalling occurs during honing processing. Thus a plasma coating in 15 µm steps

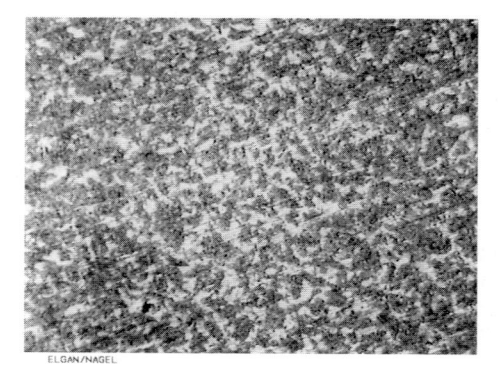

Fig. 9.24 Testing of the adhesion of honing. After the end of the test there is no spalling on the plasma coating in the sand blasting scratches.

becomes worn. Due to the low content of metal the adhesion is seen as more critical and for this test a coating with ceramic particle reinforcement was chosen. Surprisingly, at very low coating thickness no spalling could be seen at a wear rate of 15 µm in 5 s. Figure 9.24 shows the good adhesion at the end of the test: even the plasma coating still remains completely in the processing grooves existing before the coating procedure (sand spraying).

9.3.3.2 Surface Qualities, Removal of Spalling
In these tests it was aimed to minimize spalling of the pores already in the honing step before the final one. By doing this the main focus of the process end step could be on achieving the most even plateau formation possible in economical cycle times. This aim could also be reached by the application of good cutting honing ledges, which already work effectively at a low specific pressure. Furthermore the composition of the cutting medium has been changed so that the adhesion between plasma coating and cutting medium could be almost completely avoided. Figure 9.25 shows crosswise grindings of the pores of a processed plasma coating after the second honing step. Only by the costly and time consuming preparation of the sample with a nickel protection coating on the surface, applied after honing, can the finest tongues or spalling be seen in the plasma coating. These are orientated according to the direction of rotation. By a simple change in the direction of rotation, as is possible in the honing process, these can be easily removed by the final honing step. The result is quasi-run-in cylinder surfaces with open pores and smooth, nonspalled plateaus. Figure 9.26 shows this for the example coating of XPT 512. With especially tough and ductile coatings, for example stainless steel, in some cases finest spalling on the surface can remain, even after a three-step honing process. This spalling can be removed by a subsequent process step, which only lasts a few seconds. "Re-scaling" occurs by application of elastic cutting ledges, which have been specially developed. Figure 9.27 shows the comparison using electron microscopic images. The surface print technique (Fax-film) usually applied for documentation of gray cast processing is therefore less suitable; such structures are only incompletely reproduced.

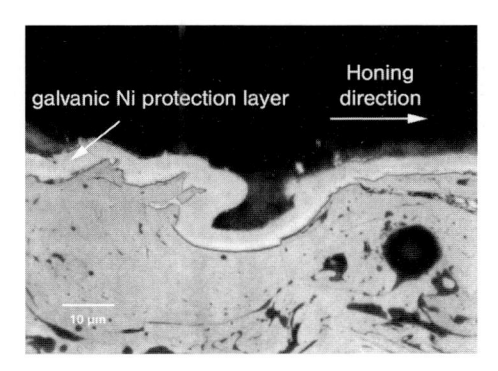

Fig. 9.25 Minimum tongue formation after intermediate honing in dependence of the honing direction.

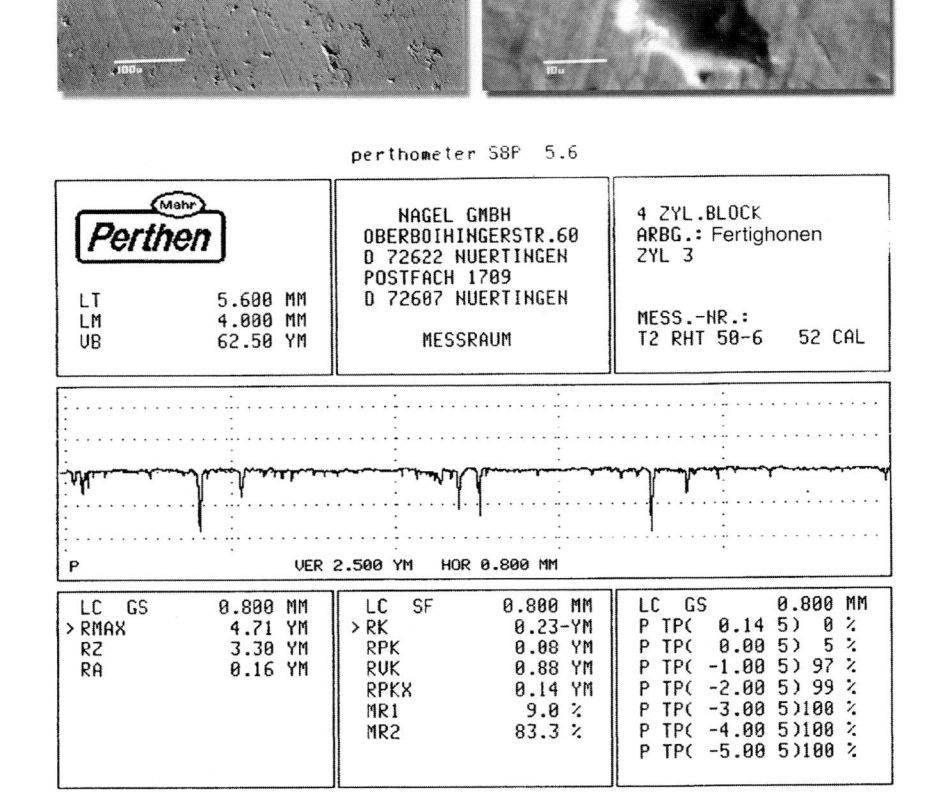

Fig. 9.26 Final honed surface (SEM image and roughness print) of the coating XPT 512.

Further development of an even more gentle honing procedure has not been undertaken. The experience of honing with a variety of materials (not only plasma coatings) has shown that functionable surfaces can be manufactured for very sensitive and labile coatings at appropriate cost. However, in the final engine check it has been shown that few of these coatings, for example where fatigue or adhesion strength is concerned, can withstand the real strains in engine service. This means interval orientated honing tests are also an aptitude test for surface coatings. The honing process could even be interpreted as part of the quality control; poorly adhering coatings tear off during processing and can be discovered before fitting to

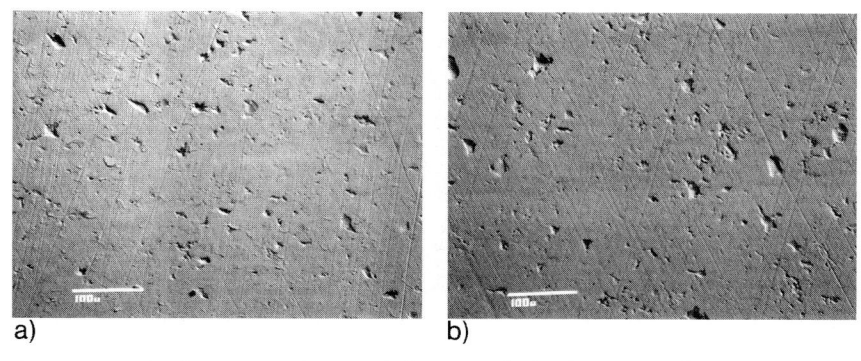

Fig. 9.27 Ductile plasma sprayed coating before (a) and after (b) removal of spalls.

the crankcase (for example by using reflectometric, ultrascan or eddy current detectors).

9.3.3.3 Reachable Form Accuracies

The embedded metal oxides and the porosity of the plasma coatings lead to short honing swarfs and act positively towards the machining ability. Together with the free cutting tools, which are developed for these coatings, very good shape tolerances can be obtained. After honing of four-cylinder aluminum crankcases with pure metallic plasma spray coating under series production conditions, values of 3.5 μm are typical for cylindricity. At these outstanding drilling tolerances, the necessity for a rough surface for "running-in" of the piston rings is unnecessary. The combination of high drilling tolerances with even plateau surfaces (which can be best characterized by the low R_{pk}-values) is the ideal prerequisite for low oil consumption, from the beginning of engine service.

9.3.3.4 Processing of Metal–Ceramic Coatings

When processing MMC materials (metal matrix composites) an additional requirement to those already known for pure metallic spraying coatings arises: the indestructive reproduction of the embedded ceramic particle on the honed surface. Due to the breaking off of these particles, the porosity of the coating, and therefore the oil consumption of the engine, would increase and the partially broken off ceramic particles would lead to abrasion of the piston rings, due to their hardness and sharp edges.

Previous experience with diamond honing of over-eutectic aluminium–silicon alloys forms a good basis to optimize the honing process for this plasma coating [5]. By the development of suitable cutting mediums, it has been shown that after the second honing step ceramic particle damage could be avoided. This is shown in Fig. 9.28 on a crosswise grinding of a sample with a nickel surface. All together surfaces from three honing steps can be processed, as seen in Fig. 9.29. The matrix

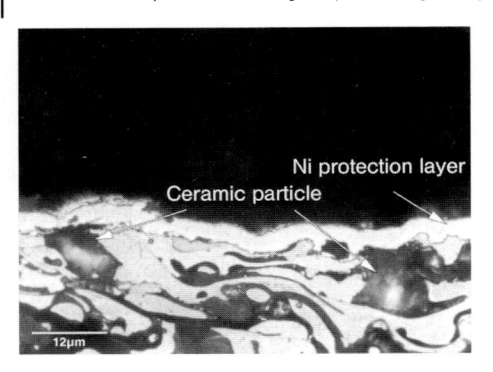

Fig. 9.28 Condition of ceramic reinforcement particle after the second honing step. Plasma coating with ferrite matrix.

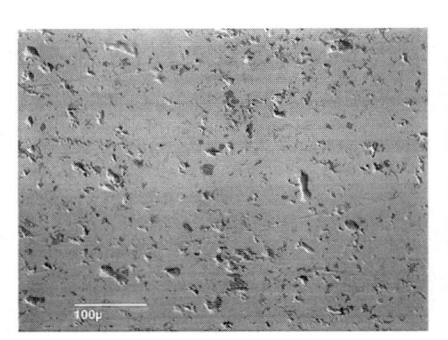

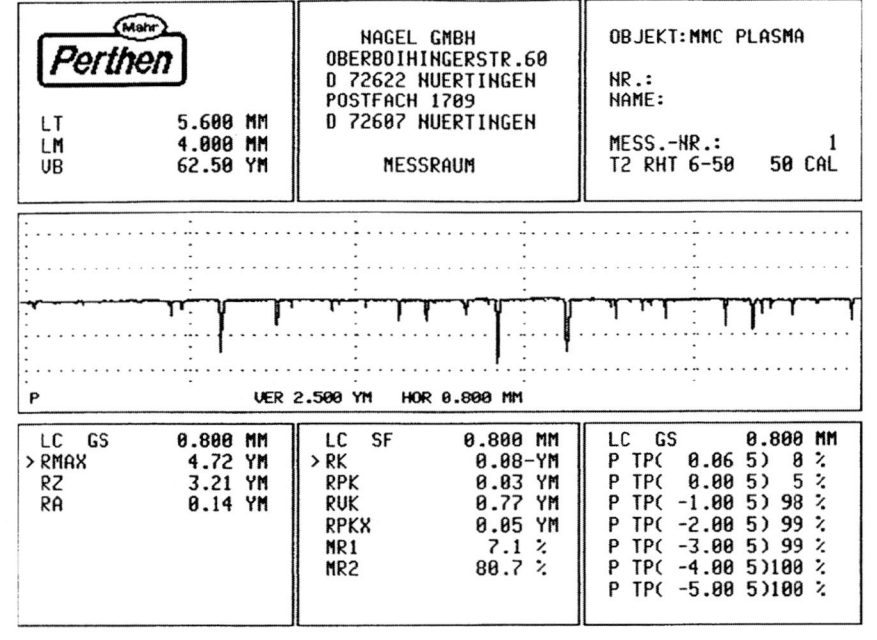

Fig. 9.29 Surface of a three-step honed MMC coating.

of the honed MMC coatings is low alloy steel. In comparison to composite materials with an aluminum matrix, in this case uncovering of the ceramic particles by etching or uncover honing could be avoided. Test bench experiments with engines having these honed surface coatings were positive, also with regard to galling ability and abrasion of the piston rings.

9.3.3.5 Cooling Lubricants

9.3.3.5.1 Honing of Pure Metallic Plasma Coatings

Pure metallic plasma coatings can be processed well both with honing oils and also with aqueous cooling lubricant materials (emulsions or solutions). Little advantage is shown with the use of honing oil during surface smoothing. An indisputable advantage is the better wetting behavior of honing oils. Thus less material of the wear honing sticks in the coating pores. In the previous engine tests a subsequent cleaning was sufficient to rinse the pores. Alternatively the application of watery cooling lubricants allows problem free cleaning of the pores by pressure blasting of the cylinder drillings.

9.3.3.5.2 Metal Composites

For large patch production the application of honing oil is suggested for MMC coatings with an aluminum matrix; with the application of emulsion there often exists the risk that the worn aluminum particles will lubricate (close) the cutting ledges. For MMC materials with a matrix based on iron alloy the use of honing oil is also important. Recent tests have shown that the lowest damage and also the best plateau formation of the embedded ceramic particle can be guaranteed with the use of honing oil. However, at low requirements the use of special emulsion is possible for these coatings; a higher content of lubricant and mineral oil is then desirable.

Plasma coatings based on ferrite are of advantage with regards to cleaning the cooling lubricant. In these, the worn-off material, analogous to gray cast processing can be mostly removed cost effectively from the cooling lubricant cycle by an upstream magnet cutter. During the series production of large numbers the cost advantage is of significant importance.

9.3.4 Summary

With reference to cost, plasma coating is a very attractive technique for the production of functional surfaces for light metal engines and a highly flexible selection of materials is available. By the development of suitable treatment processes this flexibility is fully usable. Different honing variants can be chosen within a coating group to realize different requirements regarding surface quality, structure and processing costs. The process interpretation for series production will still be very customer orientated. The test results will form the basis for the optimal price: benefit ratio, corresponding to the motor properties.

References

1 G. Barbezat, S. Keller, K. H. Wegner, Innenbeschichtung für Zylinderkurbelgehäuse. Thermisch Spritzkonferenz TS 96, Essen, 1996.
2 G. Barbezat, The state of the art of the internal plasma spraying on cylinder bores, in AlSi cast alloy, FISITA congress, Seoul, June 12–15, 2000.
3 M. Winterkorn, P. Bohne, L. Spiegel, G. Söhlke, Der Lupo FSI von Volkswagen – so sparsam ist sportlich, in *ATZ* **2000**, *102* (10).
4 G. Haasis, U.-P. Weigmann, Neues Honverfahren für umweltfreundliche Verbrennungsmotoren, *WB Werkstatt Betrieb Jahrg.* **1999**, *132*, 3.
5 A. Robota, F. Zwein, Einfluss der Zylinderlaufflächentopografie auf den Ölverbrauch und die Partikelemissionen eines DI-Dieselmotors, *MTZ* **1999**, *60*, 4.
6 J. Schmid, Moderne Leichtmetallwerkstoffe für den Motorenbau und deren 'Endbearbeitung durch Honen, MTZ **1998**, *59*, 4.

10
Powder Metallurgically Manufactured Metal Matrix Composites

Norbert Hort and Karl Ulrich Kainer

10.1
Summary

Powder metallurgically manufactured metal matrix composites (PM-MMC) offer economical solutions for the production of high performance materials. A multiplicity of material combinations can be manufactured, which can be optimally adapted to their respective application purposes. At the same time the current procedures for the production of the source powders for both the metallic matrix as well as the selected reinforcement components offer further shapeable parameters, for which the characteristics of materials and construction units can be optimally conceived to their application purpose. With manufacturing and processing methods available, construction units can be produced close to the final contour, which have economically optimal characteristic combinations and can also be produced in long production runs. At the same time with PM-MMC many disadvantages, associated with the fusion metallurgical production of composite materials with a metallic matrix (MMC) can be avoided.

10.2
Introduction

Metal matrix composites (MMC) represent an alternative to conventional materials for the production of high performance materials. The materials can be manufactured in such a way as to exhibit a combination of the characteristics of the metallic matrix and the reinforcement phase. The characteristic profile thus developed can be adapted to the respective requirements of application and an optimal utilization of the characteristics of matrix and reinforcement component is made possible. Matrix and reinforcement components have shared interfaces, which are absolutely necessary for the fulfillment of the tasks undertaken by the MMCs.

There is a set of possibilities for the definition of metal matrix composites, which can differ strongly. While it is agreed that the matrix has proportionally the largest

Metal Matrix Composites. Custom-made Materials for Automotive and Aerospace Engineering.
Edited by Karl U. Kainer.
Copyright © 2006 WILEY-VCH Verlag GmbH & Co. KGaA, Weinheim
ISBN: 3-527-31360-5

part of the total composition, there is at present hardly any clarity about materials called reinforcement components or the volume content percentage required in order to call a material a reinforcement component. However, it has been generally accepted to speak of reinforcement as second phases at a volume content starting from approximately 5 vol% although this definition ignores most ODS (oxides dispersion strengthened) materials, which clearly contain smaller contents of oxidic components [1]. Nevertheless, these ODS alloys will be described in the following, since they are manufactured with common powder metallurgy techniques and also contain certain ceramic portions as a reinforcement component.

Generally it is assumed that the reinforcement components are of a ceramic nature. With this definition some composite materials are neglected, for example some superconductors, where both phases are of metallic nature or high temperature materials [2]. A third point for discussion in the definition of composite materials results from the fact that the formation of intermetallic phases or ceramic components can occur during in situ manufacturing, and these develop a composite material due to their property profile when combined with the matrix. Also this point can be interpreted by some definitions in such a way that a composite material is not concerned [1].

However, in the present report on the topic of powder metallurgically manufactured metal matrix composites (PM-MMC) the definitions specified above and the associated problems are as far as possible ignored. As soon as two or more phases are already present in the starting condition or as a consequence of the production or further processing, the final product will be called PM-MMC. Substantially, there is only a clear difference in the physical and/or mechanical characteristics between the phases.

Procedures exist for the production of composites. The procedures, which use melted liquid metal, fusion infiltration procedures, became generally accepted as squeeze casting, gas pressure infiltration and so-called compocasting (often also called melt stirring) [1, 3–14]. In squeeze casting and also in the gas pressure infiltration method so-called preforms are used, i.e. molded parts which consist of the selected reinforcement components [15–18]. Both short fiber preforms and particle preforms are used, and also hybrid preforms, where short fibers and particles are both used. In the following process the preforms are infiltrated with an overheated melt. In the case of compocasting the particles or fibers serving the reinforcement are stirred directly into a molten bath and then cast into a semi-finished material or fully finished construction unit. Examples of fusion metallurgically manufactured composites can be found in Ref. [1].

The manufacture of composites by the powder metallurgical process is based on the fact that it is often impossible to use fusion metallurgical procedures. Several reasons exist. One lies in the high melting point of the metallic partner involved. During the fusion liquid production of MMCs it can be assumed that serious problems will arise when the metallic matrix exceeds a melting temperature of approximately 1200 °C. Whilst Cu, Al and Mg as well as their alloys have been successfully applied as molten baths for the production of MMCs, there are restrictions with other metals and alloys having higher melting points. Besides the more difficult handling of the melt there is the influence of the temperature itself. At higher temperatures reactions can occur, where the reinforcement components are so strongly damaged during the production that they no longer serve their purpose [19]. Furthermore some metals and alloys are so reactive that it is practically impossible to use them in fusion procedures for the production of MMCs. Examples of this are especially titanium and its alloys. The high reactivity of titanium causes unwanted reactions between the melt and the reinforcement component [20] and as a consequence the composite generally shows worse properties than expected in theory. The remedy here can be the use of powder metallurgy methods. Both elementary metallic powders and alloy powders can be applied. In the first case mixtures of the required content are processed, for example, by mechanical alloying, in order to get a composite material together with the reinforcement component. The use of element powders and reinforcement components and processing methods like hot isostatic pressing can also lead to the production of composite materials with suitable processing parameters. In this way long fiber reinforcements as well as discontinuous reinforcements can be utilized. However, long-fiber-reinforced titanium MMCs are extremely costly, particularly because of the special long fibers and the coatings of the fibers usually needed in order to avoid fiber damage of the titanium basic fiber.

The use of powder shaped source materials represents an economical procedure for the production of high-stressable MMCs for a broad application field. Among others, traffic-, energy-, information-, processing technology and plant construction use these techniques. These applications are concerned with both structure and functional materials. PM-MMCs possess great importance, mostly in wear or corrosion protection, where they are used as a spraying powder. Due to the broadly varied applications, practically unlimited combination options are possible between the metallic powders and reinforcement materials, as well as an almost unlimited range of variation in their powder form. Large requirements exist on the manufacturing processes for such powders. Depending on the application area and the processing method, the resulting characteristic profile of the materials or composite layers must be ensured. This depends primarily on a homogeneous, adapted distribution of the components, a high reliability of the powder manufacturing process, stable compositions of the materials and, above all, on high economic efficiency. However, due to the large variety of composite powders used as spraying powders, as well as the various procedures applicable for it, this topic is not described in greater detail here.

10.3
Source Materials

10.3.1
Metallic Powders

Several manufacturing processes are applicable for the production of powder metallurgical composites with a metallic matrix, however, the basis is the same: pure metallic powders and/or alloy powders. For the production of metallic powders a number of chemical processes are relevant, as well as the mechanical crushing of monolithic metals or alloys. However, mechanical processes, machining for example, are only suitable under certain conditions to produce a fine source powder [21]. Spark erosion represents a further possibility for the production of powders [22, 23] but the output usually is not very high and impurities in the powders have to be expected. However, very fine powders can be produced with this procedure.

During chemical deposition only pure metallic powders of very irregular shape can be produced in general, although an advantage is the high powder fineness. The production of iron or nickel powders by the carbonyl procedure is probably the most well-known procedure for the production of highly pure powders [21]. Besides the carbonyl procedure, electrolytic procedures, or the reduction of metallic bonding permit the production of a broad range of pure, very fine metallic powders [21]. Even finer metallic powders can be manufactured by evaporation and then deposition of metal vapor onto cold surfaces. In order to produce such powders, where their grain size is in the range of some nanometers, laser or electron beams are often used [24–29]. However, this succeeds also by an increase in the vapor pressure and the employment of resistance or inductive heating mechanisms for the production of molten baths [21]. Nanoscale metallic powders hardly have any commercial importance at present. However, they have been intensively investigated [24–32].

A further possibility for the production of metallic powders lies in different atomization procedures, with the use of liquid or gases as atomization media [21, 33–37]. In the case of the atomization of molten baths with gases or liquids several advantages result [21, 36]:

- homogeneous particle size distribution
- homogeneous distribution of alloying elements
- supersaturation at alloying elements also in the manufactured powder
- isotropic material properties
- cooling rates within the range of 10^{-2}–10^{-4} K s^{-1}
- production of large quantities
- high economic efficiency

Atomization procedures are wide ranging, since they are economically very interesting, and permit the manufacture of large quantities of metal and alloy powders with accurately defined compositions. This concerns the use of water or gases such as argon, nitrogen and possibly air as the atomization medium. Impurities also are hardly a problem. Such powders can undergo, with suitable modifications, the usual procedures for subsequent treatment, as described in the following. Small modifications enable such plants to be used in the area of spray forming (Fig. 10.1). The modifications concern mainly a substrate carrier plate, onto which the sputtered metal particles land. The particles are thus either already solid, in the part-liquid condition or still completely liquid. By suitable process control and adjustment of the atomizing conditions to the melt, semi-finished material in the form of bands, pipes or pins can be manufactured (Fig. 10.2). These then, for example, can be processed by extrusion, forging, hot or cold isostatic pressing to the final product. The procedure has already found application for the production of PM-MMC based on Mg, Al, and Cu alloys, and is also used for the production of high-speed steels.

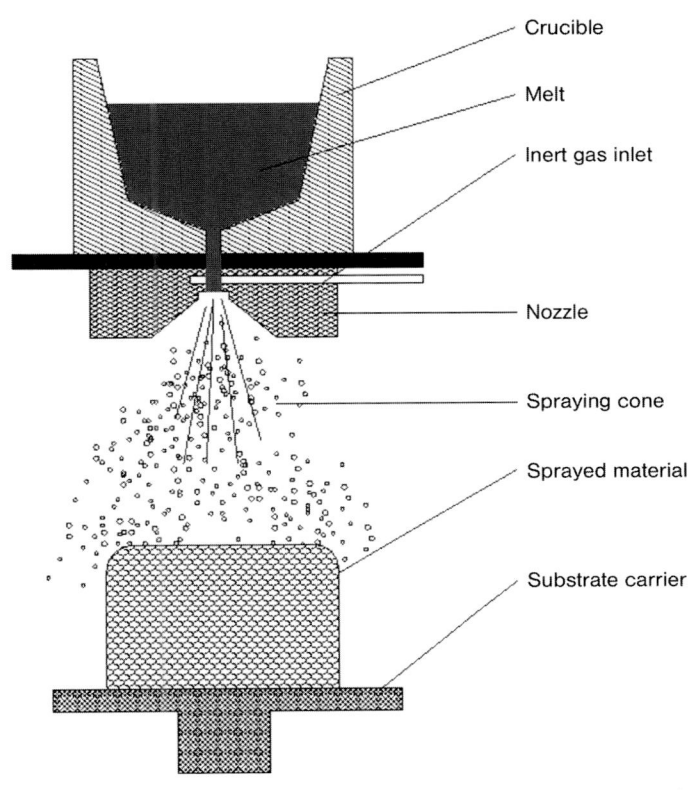

Fig. 10.1 Schematic presentation of the gas atomization of metallic melts to produce metallic powder and semi-finished product, produced by spray forming.

Fig. 10.2 Spray formed pins (QE 22 + 15 vol.% SiC) [105].

Beside the procedures already mentioned, particularly for titanium and its alloys, special procedures exist for the production of titanium powder and powder from titanium alloys, especially if a spherical shape of the powders is of importance [38–41]. However, a set of procedures were developed, one being the PREP (plasma rotating electrode process), which have become generally accepted, and which are also used commercially. However, for the production of titanium materials, pure element powders are often used, which can be produced by the different powder manufacturing processes already mentioned. Additional manufacturing methods are essentially based on mechanical alloying, described later, or the *in situ* route.

10.3.2
Ceramic Reinforcement Components

A whole class of ceramic powders occur naturally. These can be used directly for the production of composite materials after appropriate cleaning and classifying to the desired grain size. However, a lot of the desired particle shaped reinforcement components are considered as hard materials and must be artificially produced. For these a number of thermochemical procedures for the production of hard materials exist [21, 42]. They are predominantly the carburetion or nitrating of metals or metallic oxides, similarly a set of procedures exist, for the preparation of pure oxidic ceramic particles, where the synthetic final products are manufactured by reactions with acids and bases during precipitation processes [21].

Practically all current common reinforcement components can be introduced into the manufacturing procedure of the powder metallurgically manufactured MMCs, also known as other composite materials with metallic matrix. Both long and short fibers are used, as well as whiskers and particles. The particular reinforcement components are differentiated essentially on the basis of their shape, and for fiber reinforcement components the relationship of length and thickness (aspect ratio: l) plays a role in the distinction, particularly with the differentiation of short fibers and whiskers (l > 20: whisker) [43]. Long fiber use is usually called continuous reinforcement, and other reinforcement components are all considered as discontinuous reinforcements [15].

Ceramic and carbon fibers are both in use as long and short fibers. The different ceramic long fibers can be extracted either directly from a melt, or an appropriate starting material is first spun and then transformed by a subsequent thermal treatment into the desired fiber. The thermal treatment often takes place in two steps, first the material is dried, and then the fiber crystallizes at high temperatures [15, 43]. Short fibers of desired length can be made from long fibers in further processing by suitable milling or cutting procedures [43, 44]. For whisker production procedures exist which work with thermal decomposition of the source material, or with deposition from the gas phase. Details of production and the characteristics of whiskers can be found in Refs. [15, 43]. Carbon fibers can be differentiated with respect to the source material, pitch or polyacrylonitrile (PAN). Fibers extracted from pitch or PAN fibers are thermally treated in further steps (carbonizing, graphitizing). Depending on the application planned a surface treatment can take place [45]. Depending on the kind of thermal treatment C fibers can be produced in such a way that they can exhibit major differences in Young's modulus and strength in some parts. [15, 43, 45].

Due to their shape, the use of particles in powder metallurgically manufactured composite materials is naturally much easier, and with them materials with a higher homogeneity or more isotropic in nature can be manufactured. When using long fibers, short fibers or whiskers an anisotropy results due to either the shape of the reinforcement component or the selected manufacturing and processing procedure.

10.4 Manufacture of MMCs

Mixing and mixing following consolidation of mixtures is a common procedure after the production of the output powders. Sometimes a certain degree of consolidation can already take place during the production of the source material by such as spray forming, which will be further described. However, the most common method is to manufacture a PM-MMC by first mixing the existing powders, matrix metal and reinforcement with each other. This can happen in the dry condition, with the help of dispersion agents and/or controlled atmospheres. For mixing, normal devices can be used, for example: asymmetric moved mixers, mills or attrition grinders, which provide additional energy and therefore affect the material. In the following the mechanical alloying, coated powders and *in situ* composite materials will be described. The representation follows normal powder metallurgy routes (mixing and consolidating) as well as spray forming as a procedure, which combines the powder and the semi-finished material production in one step. Figure 10.3 shows schematically the operational sequence of the powder metallurgy production.

The different powders for the production of sprayed coatings etc. will not be described in further detail. These powders consist of different components and are usually mixed and milled, so that very fine powder particles form, which are not manageable under normal conditions and are poorly processable. Here are meant the group of powders for metal-bound hard materials such as WC/Co, Cr_3C_2/NiCr, WC–Cr_3C_2/Ni, WC–VC–Co [46–49] and metal-bound diamond [50]. Structure images of these powders are shown in Fig. 10.4 to 10.6.

For such powders a multitude of different manufacturing possibilities are available, e.g. melting and breaking, mixing, agglomerating, sintering, spheroidizing, precursor technology (sol–gel) and cladding. The techniques ensure the even distribution of the hard materials in the binder phase whereby a very fine grain size of the hard materials (1–5 µm) can be produced. However, the powders must at the same time be agglomerated in such a way that they are capable of flow (Fig. 10.4). The application of the powders is by thermal spraying for the production of wear-protection layers, sintering of tungsten carbides and diamond tools. A further

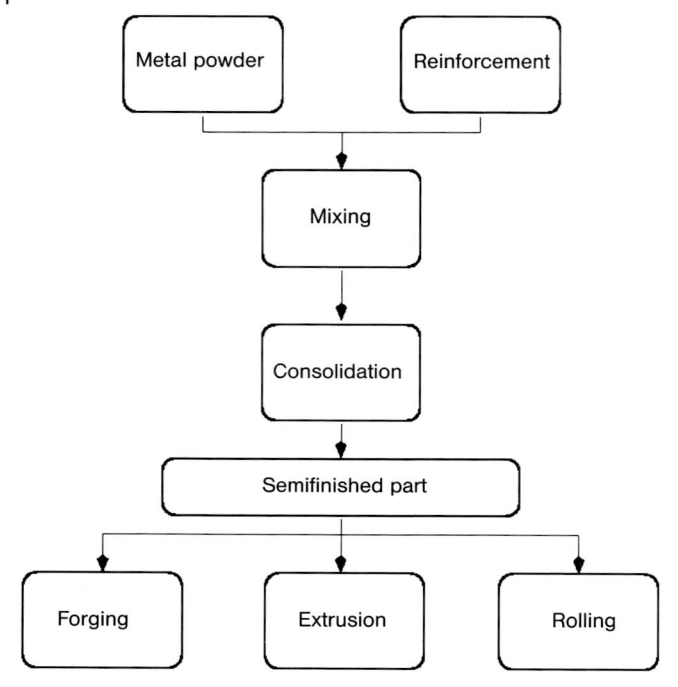

10 Powder Metallurgically Manufactured Metal Matrix Composites

Fig. 10.3 Schematic presentation of the manufacturing process for PM-MMCs.

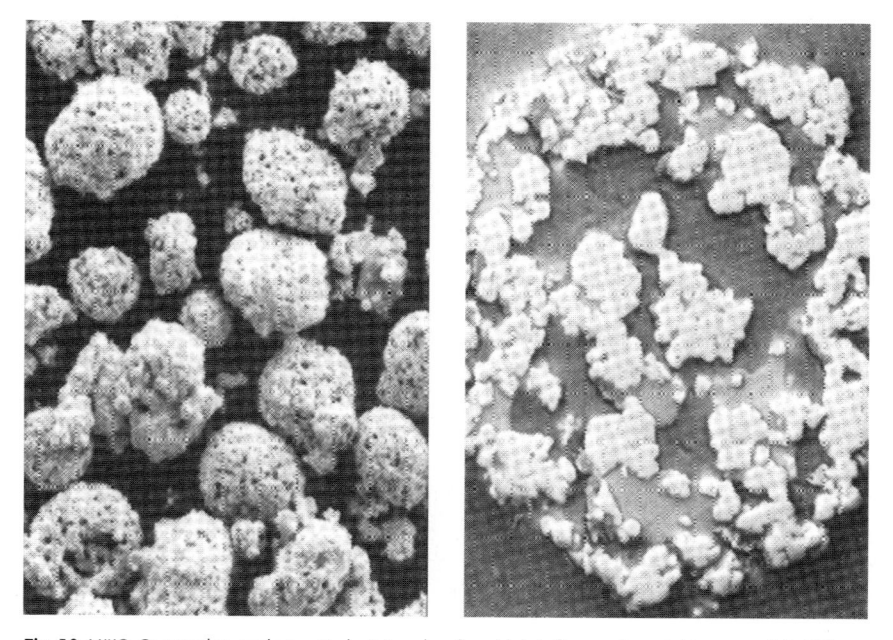

Fig. 10.4 WC–Co powder, agglomerated, sintered, spheroidal. Left: overview, right: cut particle [48].

10.4 Manufacture of MMCs | 251

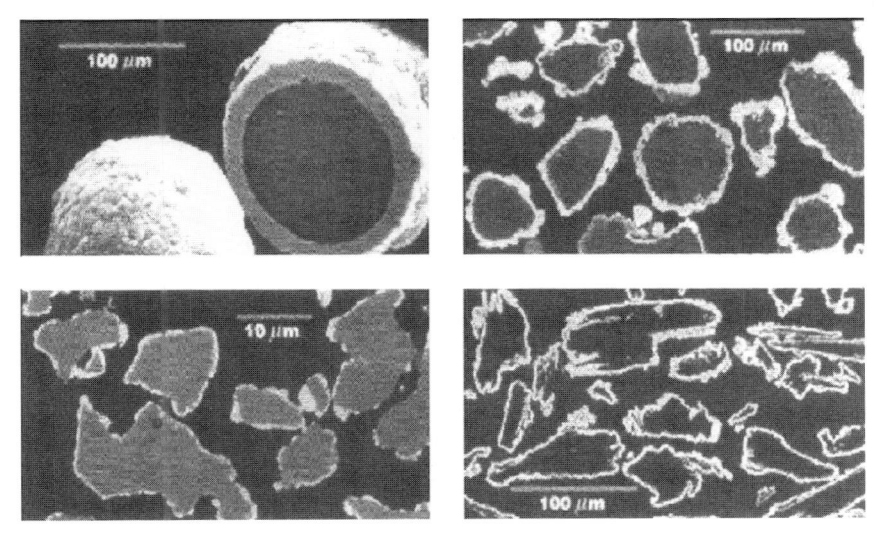

Fig. 10.5 Structure of different nickel cladded powders: (a) Ni–Al, (b) NiCrAl-Betonit, (c) Ni metal carbide, (d) Ni–C [119].

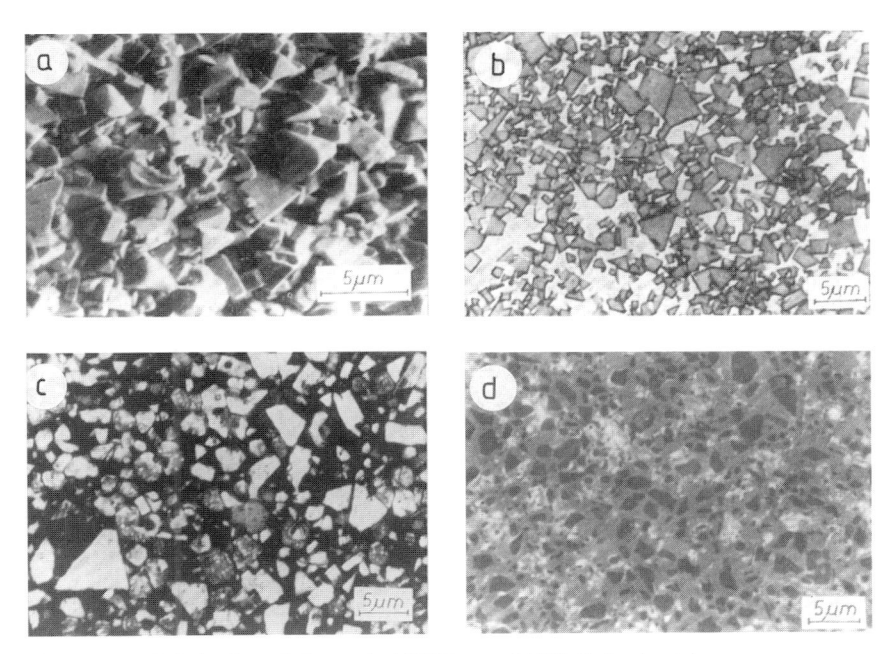

Fig. 10.6 (a) WC–Co hard metal, deep etched SEM image, (b) WC–Co hard metal, WC: dark, Co: bright, (c) WC–TiC–TaC–Co hard metal, Co: black, (W, Ti, Ta)C: gray, WC: bright, (d) TiC–Mo2C–Ni hard metal, TiC: black, (Ti, Mo)C: gray, Ni: white.

group of procedures is the cladding of powders by galvanic coating, CVD coating and plastic coating. Examples of material systems, processing and application can be seen in Table 10.1. Figure 10.5 shows the composition of different cladded powders. In Fig. 10.7 different structures of materials which were manufactured by

Tab. 10.1 Overview of composite powders produced by coating, their processing and application [106, 110, 113–116, 119].

System	Powder manufacture	Processing, consolidation	Application
C/Ni WTiC$_2$/Ni Cr$_3$C$_2$/NiCr NiCrAl/Bentonit	electrolytic deposition hydrometallurgical coating	thermal spraying	wear protection coating turbine blade coating bearing applications
Al/Cu, Cu/SiC, W/Cu	electrolytic deposition, from Cu on Al, W or SiC	cold forming	carrier for electronic components
Fe-polymer (soft magnetic powder)	coating	cold pressing, annealing	electric engines, injection system, transformation

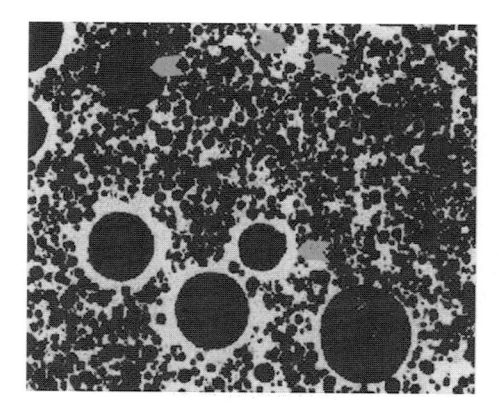

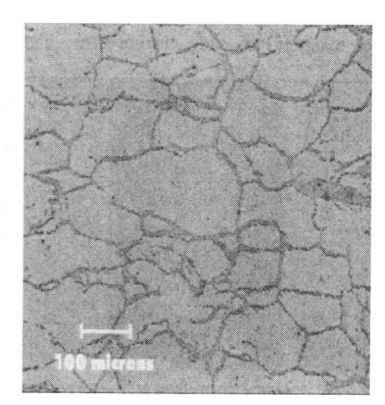

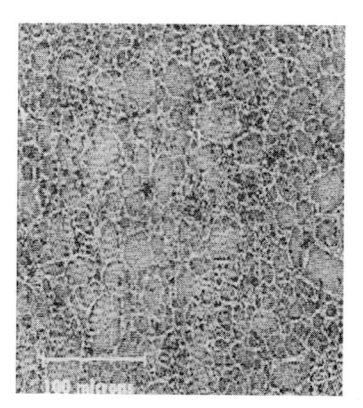

Fig. 10.7 Structure of cold pressed, coated powder. Right top: Cu–W, right bottom: Cu–Al, left: Cu–SiC [106, 110, 113].

10.4 Manufacture of MMCs

Fig. 10.8 Carrier of electronic components, made of cold compressed, coated metal powders [106].

cold pressing of such powders are shown. Heat sink carriers for electronic construction units are an application example (Fig. 10.8).

10.4.1
Mechanical Alloying

Mechanical alloying (MA) is a widespread technique for the manufacture of high performance materials, including PM-MMC. Figure 10.9 gives an overview of applications, where mechanical alloying has been used. Table 10.2 gives an overview of MMCs which are produced by mechanical alloying as well as their processing and applications.

With mechanical alloying the metallic powders are often mixed with each other or in combination with nonmetallic powders. The mixing relationship corresponds

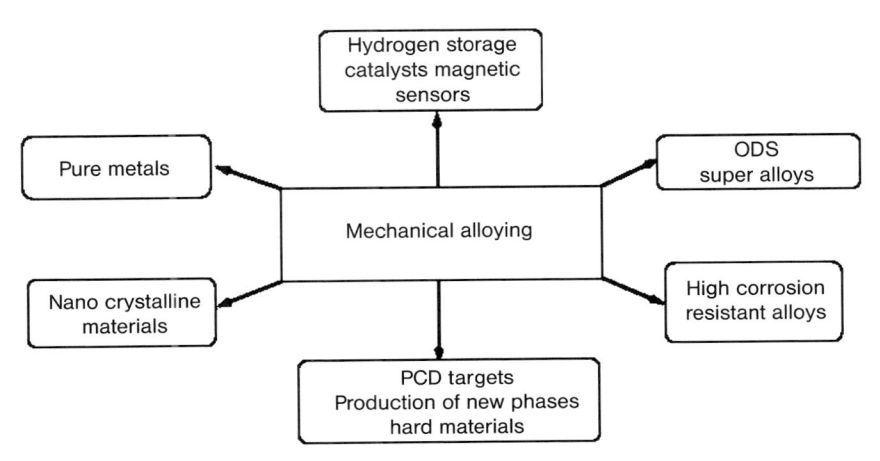

Fig. 10.9 Typical present and potential applications of mechanical alloying [117].

Tab. 10.2 Overview of MMC powders produced by mechanical alloying (MA), their processing and application [2, 51–60].

System	Manufacture	Processing	Application
ODS Ni alloys	MA	HIP, extrusion, hot and cold forming, heat treatment	high temperature area, engine industry, glass processing, air and aerospace industry, chemical plant engineering
dispersion hardened Al (Al/Al_2O_3, Al/Al_4C_3)	MA	CIP, heat treatment (vacuum), extrusion, hot and cold forming	engine industry, high absorbing materials, optical devices, high temperature batteries, compressors, fans
dispersion hardened Cu alloys	MA inner oxidation	axial pressing, CIP, sintering extrusion, hot and cold forming, heat treatment	contact materials, electro-mechanical components, springs
Cu–W	MA	axial pressing, CIP, sintering, extrusion, MIM	spot welding electrodes, contact materials, carrier for electronic components
dispersion hardened Ti alloys	MA	CIP, HIP, extrusion, hot- and cold forming	air and aerospace industry, medical technology

in general to the composition of the final alloy or composite material. A high-energy milling process follows the mixing process, where ball mills or related systems are used, for example attritor mills. Depending on the energy supply during grinding the final product can show different characteristics. However, only the following systems are described, which produce dispersion-solidified materials or fine dispersive metal–metal combinations [2, 51–60]. Metastable systems, which are produced by high-energy milling, are not considered here [61]. Yttrium-oxide dispersion-solidified nickel and iron-based alloys are examples of the group of dispersion-solidified metals, see Table 10.3 [2, 51]. The operational sequence for the production of these high temperature materials is represented in Fig. 10.10; an overview of their advantages and areas of application is shown in Table 10.4. Besides nickel- and iron-based materials, dispersion-solidified copper and titanium alloys are also used [52].

By mechanical alloying, oxide particles are inserted in the element powder particles of the alloy powders, simultaneously cut and distributed evenly. After the milling procedure the yttrium oxide particles are isolated within the mixture, evenly fine dispersive distributed, with a size in the range of some nanometers. This is an optimal dispersoid size for a high temperature material. Figure 10.11 gives an impression of the procedures during mechanical alloying. From the mechanical alloyed powders, semi-finished materials and construction units are made by hot-isostatic pressing, extrusion, warm or cold-rolling and/or forging.

Tab. 10.3 Chemical composition, in wt%, of selected iron- and nickel-based ODS super alloys, [2, 51 107–109, 117].

	Fe	Cr	Al	Ti	W	Mo	Ta	Zr	Y₂O₃	Ni
Inconel MA 6000	15.0	4.5	0.5	4.0	2.0	–	2.0	–	1.1	rem.
Inconel MA 760	20.0	6.0	–	3.5	2.0	–	–	–	0.95	rem.
Inconel MA 758	30.0	0.3	0.5	–	–	–	–	–	0.6	rem.
Inconel MA 757	16.0	4.0	0.5	–	–	–	–	–	0.6	rem.
Inconel MA 754	20.0	0.3	0.5	–	–	–	–	–	0.6	rem.
Inconel MA 957	rem.	14.0	–	1.0	–	0.3	–	–	0.25	–
Inconel MA 956	rem.	20.0	4.5	0.5	–	–	–	–	0.5	–
PM 1000	3.0	20	0.3	0.5	–	–	–	–	0.6	rem.
PM 1500	3.0	30	0.3	0.5	–	–	–	–	0.6	rem.
PM 2000	rem.	20	5.5	0.5	–	–	–	–	0.5	–

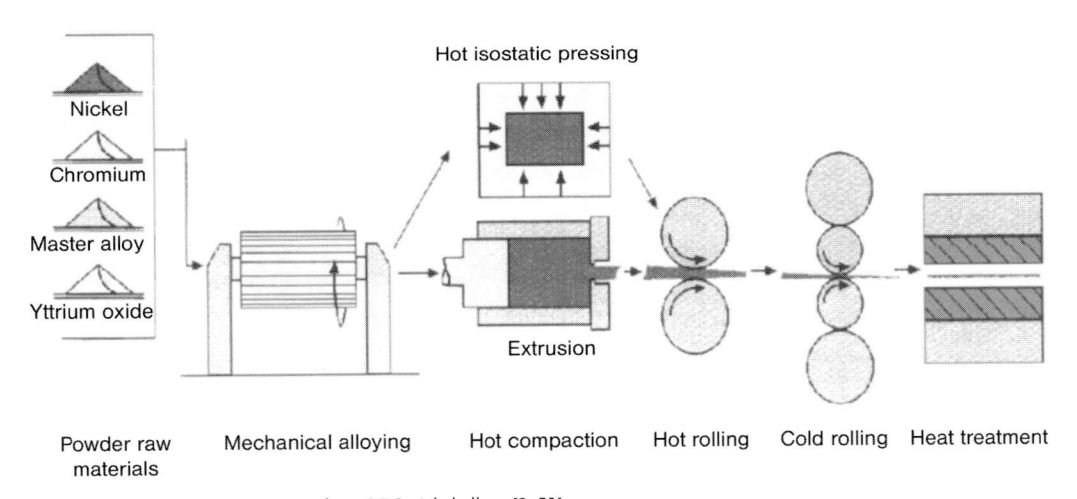

Fig. 10.10 Process sequence to produce ODS nickel alloys [2, 51].

Figures 10.12 to 10.14 show examples of construction units of high-temperature resistant nickel alloys. These were manufactured by mechanical alloying and powder metallurgy by the operational sequence shown in Fig. 10.3. The basis for the production of composite powders is the high availability of most different metallic powders as element powders, semi-alloyed or alloyed powders, which can be manufactured by a wide range of procedures. These metallic powders usually represent the matrix in the final metallic composite materials or the composite layers and determine, in dependence on their composition, the matrix characteristics. In the following the best known techniques to manufacture powder metallurgically produced MMCs are described.

Tab. 10.4 Applications and advantages of ODS super alloys [2, 51, 107–109, 118].

Application	Advantage
outer skin in aerospace industries	excellent creep strength at high temperatures
centrifuge rings in glass and ceramic manufacture	outstanding fatigue properties, also at high temperatures
batch production frames in high temperature furnace fabrication	optimal emission properties for aerospace applications at high temperatures
clamping units in the high temperature test engineering	excellent hot gas corrosion resistance
injection nozzles in engines	outstanding resistance for high temperature furnace fabrication
stirring tools for homogenization and to divide liquid glass	outstanding creep resistance up to 1300 °C
batch production units in high temperature furnace fabrication	high temperature strength
burning chamber in engines	outstanding oxidation resistance
stirring tools in glass manufacture	high stability in high speed flow up to 1100 °C
high temperature furnace fabrication	
honeycomb components in aerospace industries	
internals in chemical plant engineering and construction	
components in gas turbines	
heat exchanging device	
linings in vacuum furnaces	
power train components for aerospace industries	
internals in nuclear power stations	

10.4 *Manufacture of MMCs* | 257

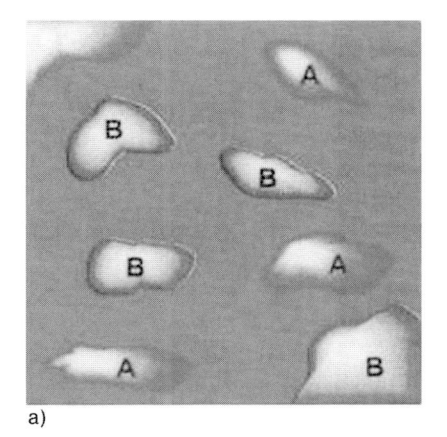

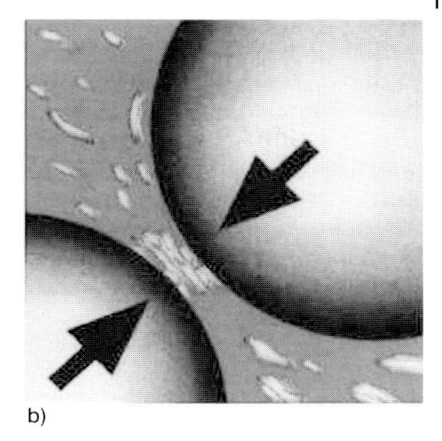

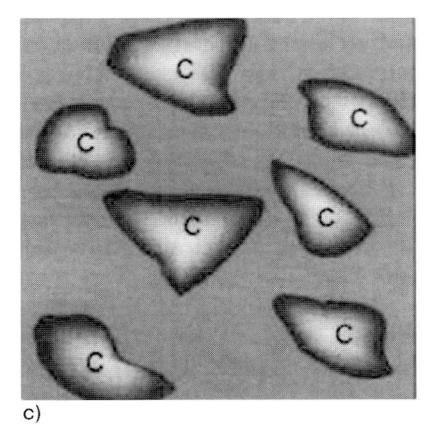

Fig. 10.11 Processes during mechanical alloying of lead powder A and B to homogeneous powder particle of the composition C [52].

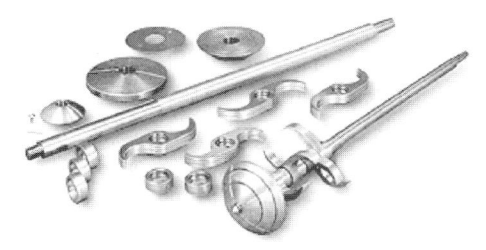

Fig. 10.12 Glass melt stirrer of dispersion hardened nickel made of PM 2000 [2, 51].

Fig. 10.13 Spark protection made of PM 2000 [2, 51].

Fig. 10.14 Diesel engine pre-chamber made of PM 2000 (pin and burner base) [2, 51].

10.4.2
In situ Composite Materials

In situ composite material is a material class which uses the reactions of different components during the production to produce new components. Here source materials are the different metal or alloy powders as well as other powder shaped components. The reaction between components can thus happen both during mixing the powders and in the subsequent processing. However, a minimum of energy must first be introduced to get the reaction going. This can take place, for example, with the thermal energy of hot isostatic pressing, with a subsequent thermal treatment after extrusion, or by high kinetic energy in the case of the mechanical alloying. In the latter case the process is known by the abbreviation MSR (mechanically induced self-propagating reaction). Takacs [63] gives an overview regarding the procedures applied and the possible reaction partners etc. Table 10.5 shows a composition of possible reaction systems. These reactions are usually exothermic reactions, which run independently after activation. The components react completely

Tab. 10.5 Examples of reaction systems of in situ reactions during powder atomization in the manufacture of composite powders [114].

Gas–liquid reaction	Liquid–solid reaction
Cu[Al] + $N_{2/O2}$ → Cu[Al] + Al_2O_3	Ti + SiC → Ti[Si] + Ti[C]
Fe[Al] + N → Fe[Al] + AlN	g Fe[Ti] + Xr_{xN} → g Fe + TiN
Fe[Al] + $N_{2/O2}$ → Fe[Al] + Al_2O_3	Cu[Al] + CuO → Cu + Al_2O_3
Fe [Ti] + Fe [C] → Fe + TiC	
Fe [Ti] + Fe[B] → Fe + TiB_2	
Liquid–solid reaction	
Cu[Ti] + Cu[B] → Cu + TiB_2	

with each other in relation to the percentage by volume and in a matrix, specified before; a new phase develops with a high percentage by volume, which then takes over the reinforcing function. However, with the production of *in situ* composite materials attention has to be paid to the fact that the percentages by volume of the developing phases sometimes clearly differ from each other. Systems whose manufacturing process offers the possibility to adjust internal strain should therefore be selected.

10.4.3
Mixing

The different manufacturing processes specified above result in very different shapes of both metallic source materials and selected reinforcement components. This is important if further consolidation pressing procedures are used. However, during manufacture by mechanical alloying, or *in situ* reactions, the shape of the components hardly makes a difference. In each case all initial components can be mixed. However, the size relationship between metallic powders and the reinforcement has to be considered. Investigations here have shown that, dependent on the size relationship, separations can occur, where, for example, particles segregate to the grain boundaries (Fig. 10.15). Thus the formation of agglomerates, which have an unfavorable effect on the properties of the composite material, can be found [8, 62, 64].

Also the kind of mixing process and the separation of mixing parameters were determined, dependent on the shape of the components to be mixed, the procedure during mixing and the equipment used. It is also important whether the mixing process is undertaken in a gaseous or liquid environment. In the latter case the mixing process can also be affected by suitable additives [8, 62]. However, spherical particles are usually assumed for these investigations. Regarding the behavior

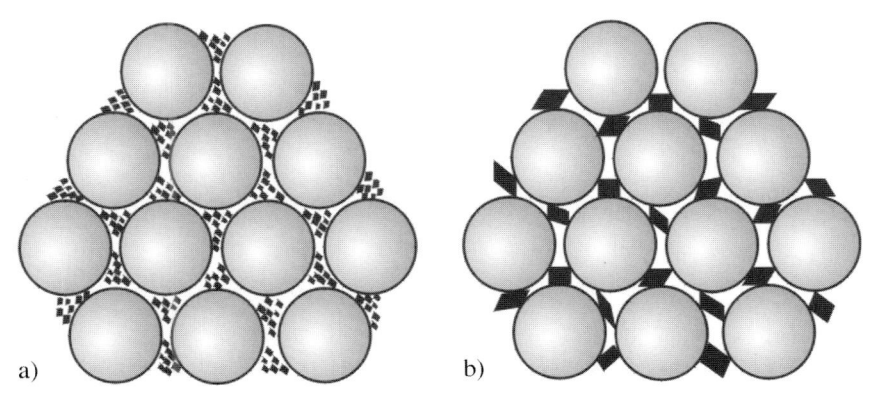

Fig. 10.15 Schematic presentation of separation of metal particle and particle shaped reinforcements during mixing of metallic powders, (a) agglomerate, (b) homogeneous contribution.

of very irregular particles, as are manufactured for example by water atomization, practically no information is present in the literature. However, it can be assumed that here also clear separation procedures occur, especially if the added component shows clearly smaller dimensions than the metallic powders. Beside segregations at the grain boundaries, and separations, which are caused by the mixing parameters, local segregations of particles in some areas of particles can occur, which show hollow structures or voids. In each case an isotropic distribution of all components involved and thus a homogeneous property profile is impossible.

10.4.4
Consolidation

In the classical powder metallurgical production technique, consolidation follows the manufacture of the powder and powder mixtures. There a variety of techniques have been developed, which are applied alone or in combination. The following techniques are most important:

- Sintering [9, 21, 65–75]
- Extrusion [9, 21, 27, 30, 40, 64, 70, 72, 76–81]
- Hot-pressing [9, 21, 27, 64, 72, 73, 81–86]
- Hot isostatic pressing (HIP) [9, 21, 40, 66, 72, 79, 85, 87–93]
- Cold isostatic pressing (CIP) [9, 21, 64, 72, 78, 86, 94]
- Forging [9, 21, 70, 72, 78]
- Rolling [9, 21, 70, 72, 89]
- Metal injection molding [71]
- Production of coatings [46–48]

Sometimes the respective consolidation procedures are combined and have some advantages over the respective procedures. In general, for example, compressing the powders, the subsequent evacuating, sintering and extruding procedures or also forging are applied to reach a density, which corresponds completely, or closely, to the theoretical density. With procedures such as forging or extrusion, reinforcement components can also be aligned to obtain direction-controlled properties. Figure 10.3 shows schematically the normal operational sequence for the production of PM-MMC.

Regarding subsequent processing, both irregular and regular, spherically shaped particles show advantages and disadvantages. Advantages of irregularly shaped particles are the high green strength of the blanks, which are produced mainly by a cold pressing procedure. However, the irregular shape also causes a comparatively small green density, and the miscibility of such metallic powders with reinforcement components is not satisfactory. In contrast spherical powders can be mixed well with other particles and the green parts produced from them by cold pressing have a high green density. However, this decreases the green strength, which is clearly smaller than in the case of irregularly formed particles.

After mixing reinforcement components and the metallic matrix, powders are pressed and by a suitable procedure manufactured to construction units, which can have a density close to the theoretical density. During pressing or the subsequent treatment, helping materials are sometimes added, as for example combined procedures in powder injection molding, where the pressing and the shaping are accompanied by such materials. These helping materials are mainly waxes or other organic materials, which facilitate the shaping by decreasing the internal friction between the powder particles and, in addition, give the greenling a mechanical strength suitable for further processing. However, the helping materials for pressing have to be removed in the subsequent processing or if possible transferred as reinforcement components. While the last step is so far not possible, these materials are normally removed by burning. This step happens directly before sintering and causes the materials to begin with a certain porosity, which is reduced as far as possible by a sinter process up to a value close to the theoretical density.

An alignment of fibers during the further manufacturing steps can occur for short fibers and, furthermore, damage of the fiber can occur, depending on the subsequent treatment steps. With the extrusion of metallic powders and short fibers as reinforcement it could be proven that the short fibers were both aligned parallel to the extruding direction, and were broken by extrusion. While broken fibers prevent the full reinforcement effect of the short fibers, composite materials with extremely anisotropic fiber distribution show very different mechanical properties parallel and perpendicular to the fiber orientation, which sometimes limits their applicableness. However, processes like hot-pressing or hot-isostatic pressing are suitable for the manufacture of short-fiber-reinforced or even long-fiber-reinforced MMCs.

Sintering is often the only possibility to produce construction units, and it is mainly used with a high-melting partner in the composite material, in the case of powder injection molding. By suitable sinter atmospheres the possibility of influencing the properties of the composite material arises. By suitable temperature control, the interface between the matrix and the reinforcement component can be optimized. With sintering the porous bodies desired can be manufactured with a density close to the theoretical density. However, a value of approximately 97 % cannot be exceeded. To achieve the full theoretical density a further compression is necessary by, for example, forging, extrusion, hot or cold isostatic pressing. With the selected procedure the property profile is also affected.

Without pressing helping materials semi-finished materials or near the final contour construction units can be made of the powder substances directly by the procedures or a combination of procedures already specified. Hot and cold isostatic pressing (HIP, CIP) are here mentioned. In both cases powder substances are filled into caps; these caps are evacuated and then exposed to an all-round isostatic pressure. Cold isostatic pressing (CIP) works at room temperature or only slightly above; hot isostatic pressing (HIP) uses temperatures close to sinter temperatures. There is the advantage of directly producing a dense, close to the final contour construction unit, which can be applied directlyor with only slight processing. In contrast CIP is seen generally as a preliminary stage, after which the semi-finished material is then consolidated by extrusion, forging, rolling etc.

10.4.5
Spray Forming

Spray forming is an advancement of conventional gas atomization and was developed to save procedure steps and thus to result directly in a semi-finished material. During the production of conventional MMC a metal or alloy powder is manufactured, the desired grain fraction is separated by sieving and mixed with the particle shaped reinforcement component. The mixture is filled into caps afterwards and completely consolidated by a suitable process. Sighting and/or sieving, the subsequent mixing, and the filling into caps are all comparatively complex processes, which sometimes result in impurities or segregations. Thus the risk exists to produce an inhomogeneous material which cannot fulfil the demands. The different process steps are also cost-intensive. Spray forming fulfils these demands better. During the production of the matrix powders, these are sprayed in the gas atomization facility onto a substrate carrier. The metallic particles are thus still liquid, part-liquid or already solidified, depending on the size and selected manufacture parameters. Depending on the application, differently dimensioned pins, bands or pipes can be manufactured directly in one working step. However, there is still another compression needed; due to the semi-finished material produced having a density of approximately 90–95 % of the theoretical density. However, no contamination of the semi-finished material, for example by atmospheric components such as oxygen or nitrogen, has to be expected.

A further advantage of spray forming is that, even during the atomization procedure, reinforcement particles can be brought into the atomization beam. Thus a particle-strengthened semi-finished material can be manufactured in only one working step. This production is economical and shows in addition a good, reproducible homogeneity. However, a disadvantage of this procedure is that only particle-strengthened MMCs can be manufactured practically. Attempts to use short fibers as reinforcement components have failed so far because the supply of the particles in the atomization beam could not be realized. Furthermore it is likely that the short fibers in the final compression step get broken and therefore their full reinforcement effect is not produced. Spray formed composite materials have already been realized with different metallic matrices such as copper, aluminum, magnesium and their alloys.

10.4.6
Subsequent Processing

The reason for powder metallurgical manufacture route is that the manufactured article is near to the final contour. Thus as many treatment steps as possible can be saved, making possible economical manufacture of a long production run. Nevertheless it can occur that a semi-finished material, or an already finished work piece, must be further worked on. In this case it has been realized that conventional tools are practically unsuitable to work on MMCs in a reasonable way. Diamond tools should be used to ensure sufficient surface qualities and to hold the wear of tools within limits.

10.5 Materials

In the following section an incomplete selection of PM-MMCs is given, discussed according to the respective matrix alloys. Besides these matrix materials and alloys there is still another area of metallic matrix materials, which are applicable for composite materials. In particular noble metals and alloys should be mentioned here but they will be reviewed separately in Ref. [95].

10.5.1 Magnesium-based MMCs

Magnesium and its alloys are at present the lightest construction materials. By appropriate procedures and alloy development it could be possible to produce mechanical properties which are, in principle, comparable with aluminum and its alloys. However, this applies essentially only from room temperature up to approximately 150 °C. Above these temperatures only the relatively expensive Mg alloys of the QE series (Mg–Ag–RE) or WE series (Mg–Y–RE) show sufficient mechanical properties and creep resistance to be able to compete with aluminum alloys. Here a set of improvements can be reached by the application of magnesium-based MMC. This applies to the creep behavior, the mechanical properties and the wear behavior, heat expansion coefficient and thermal conductivity. At present fusion infiltration, mainly squeeze casting and gas pressure infiltration are the procedures selected most frequently to manufacture Mg-MMC [96]. However, attempts have also been made to manufacture Mg-MMC by powder metallurgy methods, particularly by mixing Mg-alloy powders with hard materials (particles, whiskers and short fibers). Examples and characteristic values of powder metallurgically manufactured Mg-MMCs are shown in Table 10.6.

The metallic powders used are manufactured mainly by the gas atomization of melts, where commercial hard materials are used as reinforcement. A clear improvement in the mechanical properties has been shown compared to the non-strengthened basic alloys. Depending on the production process, an adjustment of whiskers and fibers could be detected, as well as the destruction of short fibers as

Tab. 10.6 Change in mechanical properties for the example of gas atomized and extruded magnesium alloy ZK 60 and their composites, the specifications of the reinforcement are stated in vol. % [97].

	Density [g cm^{-3}]	YS [MPa]	UTS [MPa]	El. in %	Young's modulus [GPa]
ZK60A	1.83	220	303	15.2	42
ZK60A + 20% SiC$_w$	2.1	517	611	1.3	97
ZK60A + 20% SiC$_p$	2.1	409	502	2.0	85
ZK60A + 20% B$_4$C	1.97	404	492	1.7	83

soon as extrusion was used as a consolidation procedure. A further consolidation investigated is spray forming of Mg-MMC, where SiC particles were brought into the atomization beam. However, these investigations are still in their infancy.

10.5.2
Aluminum-based MMCs

Aluminum and its alloys and a set of Al-MMCs have found wide application within practically all areas of traffic engineering. Special requirements must be fulfilled for air and space travel applications. Here a low density is of especial interest. With sufficiently good mechanical properties aluminum has been one of the first candidates for MMCs at slightly increased temperatures, good corrosion properties and a good processing. However, primarily fusion metallurgical procedures are used to produce composite materials. Nevertheless, a set of composite materials has also been manufactured by powder metallurgy methods. Tables 10.7 to 10.9 show the properties of some different aluminum-based MMCs. Thus it has to be pointed out that the powder metallurgy manufacturing route of MMCs is ideally suitable for inserting high percentages by volume of reinforcement components into the material. Values up to approximately 50 vol% are obtainable. This differentiates such systems from particle-strengthened MMCs, which can be manufactured in fusion metallurgical ways; where typically about 20 vol% of reinforcement components can be brought into the matrix. Higher percentages by volume are likewise reported; however, process engineering limits are reached. A further advantage of the powder metallurgy manufacturing of MMCs is the homogeneous structure. Figure 10.16 shows the example of a particle-strengthened Al-PM-MMC. Besides mixing source materials with subsequent consolidation, mechanical alloying is used or *in situ* reaction for aluminum -based PM-MMC. Examples of mechanically alloyed

Tab. 10.7 Properties of some aluminum -based MMCs, the specifications of the reinforcement are stated in vol.%. a, cast; b, gas atomized and extruded; c, spray formed and extruded [112].

	CTE [10^{-6} K]	YS [MPa]	UTS [MPa]	El. [%]	Young's modulus [GPa]	Comment
2124 + 20% SiC[b]	–	379	517	5.3	105	T6
2124 + 30% SiC[b]	–	434	621	2.8	121	T6
2124 + 40% SiC[b]	–	414	586	1.5	134	T6
2618 + 13% SiC[c]	19.0	333	450	–	75	T6
6061 + 20% SiC[b]	15.3	397	448	4.1	103	T6
6061 + 30% SiC[b]	13.8	407	496	3.0	121	T6
6061 + 40% SiC[b]	11.1	431	538	1.9	138	T6
7090 + 20% SiC[b]	–	621	690	2.5	107	T6
7090 + 30% SiC[b]	–	676	759	1.2	124	T6
8090 + 12% SiC[b]	19.3	486	529	–	100	T6
A356 + 20% SiC[a]	–	297	317	0.6	85	T6

Tab. 10.8 Properties of selected aluminum alloys, the specifications of the reinforcement are stated in vol.%. a, cast; b, gas atomized and extruded; c, spray formed and extruded [97, 112].

	CTE $[10^{-6} K]$	YS [MPa]	UTS [MPa]	El. [%]	Young's modulus [GPa]	Comment
2014[a]	25.4	414	483	13.0	73	T6
2124[b]	–	345	462	8.0	71	T6
2618[c]	23.0	320	400	–	75	T6
6061[a]	26.1	240	310	20.0	69	T6
6061[b]	23.0	276	310	15.0	69	T6
7090[b]	–	586	627	10.0	74	T6
8090[c]	22.9	480	550	–	80	T6
A356[a]	–	200	255	4	75	T6

Tab. 10.9 Properties of selected in situ MMC with a pure aluminum matrix, the specifications of the reinforcement components are stated in vol%. a, *ex situ* composites [84].

	YS [MPa]	UTS [MPa]	El. [%]	Young's modulus [GPa]
pure Al	64	90	21	70
10.5% Al_2O_3 + 23.7% Al_3Ti	110	145	5	
10.5% Al_2O_3 + 6.3% TiB_2 + 7.9% Al_3Ti	271	311	5	
10.5% Al_2O_3 + 7.9% TiB_2 + 4.0% Al_3Ti	301	328	5	
10.5% Al_2O_3 + 9.5% TiB_2	320	353	6	
11% Al_2O_3 + 9.0% TiB_2 + 3.2 wt% Cu	427	478	2	
11.4% Al_2O_3 + 8.6% TiB_2 + 6.0 wt% Cu	588	618	2	
20% SiC_p [a]	117	200	10	
20% SiC_w [a]	176	278	11	
20% Ti_2	235	334	7	131
20% Ti_2 [a]	121	166	16	96

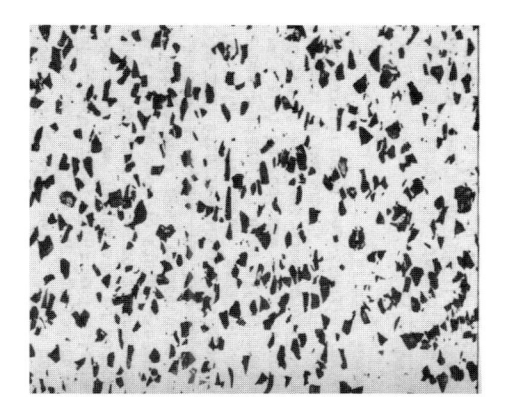

Fig. 10.16 Structure of a spray formed, extruded particle composite with a aluminum matrix (6081 + 20 vol% SiC) [111].

powders are the systems for dispersion-solidified aluminum (Al/Al$_2$O$_3$, Al/Al$_4$C$_3$) [53, 54]. Further materials manufactured by mechanical alloying are aluminum alloys, where carbon is ground in an oxygen-containing ball mill atmosphere, which leads to the formation of Al$_2$O$_3$ and Al$_4$C$_3$.

10.5.3
Titanium-based MMCs

Titanium and its alloys are used in chemical apparatus engineering, medical technology and especially in air and space travel. In air and space travel, weight saving, the combined properties of high stiffness, high temperature properties and good strength are of interest. Compared to aluminum, titanium alloys show generally better high temperature properties, very good corrosion resistance and also a higher Young's modulus. This made titanium and its alloys early candidates for the production of MMCs. However, above all, titanium melts are very reactive in relation to practically all well-known crucible materials, as well as the applicable ceramic reinforcement components. This made the fusion metallurgical production of discontinuous reinforced MMCs impossible at an early stage [97]. A further difficulty is the high reactivity of titanium in the solid state. The application at higher temperatures causes damage to the reinforcement components and brittle phases form at the interface between the matrix and reinforcement component. Also additional strain arises as a result of large differences in thermal coefficients of expansion, and thus is a further criterion for an early failure of the composite material.

A number of different procedures to manufacture titanium MMCs have been examined, including the powder metallurgy procedure and *in situ* reactions [98, 99]. In the case of the powder metallurgically manufactured titanium-MMC, titanium powders and reinforcement components are mixed and then usually hot pressed. Both uniaxial and hot isostatic pressing have been used successfully for the production of MMCs. Titanium and titanium alloy powders were used to manufacture foils for the production of layer composite materials. For *in situ* composite materials TiB$_2$ is the preferred compound, which reacts during sintering or hot-pressing with titanium from the matrix and forms TiB. Table 10.10 shows the properties of selected Ti-MMCs. Ti-MMC, which contains higher portions of TiB, is a promising candidate for Ti-based, powder metallurgically manufactured composite materials [78, 94, 98, 99]. A composite material (CermeTi) [87, 90–101], made of elementary powders (Ti, Al, V) and SiC particles, is already commercially available. Since such a composite material is not producible by the fusion metallurgical production, the advantages of powder metallurgy production become effective during the production of this material. CermeTi is made by a combined procedure of cold and hot isostatic pressing. This composite material has been developed for applications in air and space travel, as well as for engines (Fig. 10.17).

However, a commercial success for the titanium composite materials investigated so far remains elusive. The reason is the high reactivity of titanium as matrix material, which makes it necessary to coat the selected reinforcement components in a complex way. The reactions between matrix and reinforcement, and the costs

Tab. 10.10 Properties of selected in situ MMC with a pure titanium matrix, the specifications of the reinforcement components are stated in vol.% [84].

	YS [MPa]	UTS [MPa]	El. [%]	Young's modulus [GPa]
pure Ti	393	467	20.7	109
10% TiC	651	697	3.7	
5% TiB	639	787	12.5	121
10% TiB	706	902	5.6	131
15% TiB	842	903	0.4	139
15% (TiB + Ti$_2$C)	690	757	2	
22.5% (TiB + Ti$_2$C)	635	680	<0.2	
25% (TiB + Ti$_2$C)	471	635	1.2	

Fig. 10.17 Valve, piston rod and pin, made of CermeTi composite [101].

connected with the coating of the reinforcement, as well as the costs in the production, cause problems at present, which makes the application of powder metallurgically manufactured titanium composite materials appear uncertain [20].

10.5.4
Copper-based MMCs

A major consideration for the production of MMCs based on Cu and its alloys is their thermal and electrical conductivity [85, 92]. It mainly depends on obtaining good characteristic values and increasing wear resistance significantly at the same time. These demands are contradictory goals. However, the problem can be solved by the application of composite materials, better than with the alloy technique. Spray forming gains increasing importance compared to procedures such as mixing and subsequent sintering, extrusion or other consolidation procedures. Here attempts have already been made to bring hard materials and graphite combined into a copper matrix to keep electrical conductivity as high as possible and to achieve an improved wear resistance at the same time. By imbedding lighter reinforcement components, the density of the composite material has been lowered, making such materials interesting for air and space travel [85, 92]. However, because copper possesses a very good ductility due to its lattice structure, the conventional processing by mixing with subsequent cold pressing or extrusion has at present the greater importance. Figure 10.7 shows the structure images of cold pressed Cu-MMC. Such PM-MMCs are, for example, applied as carriers for electronic construction units (Fig. 10.8) and also for electrical contacts (Fig. 10.18). Besides these alloys dispersion-solidified copper alloys are also used [55–57].

However, some composite materials, based on copper, can be classified in the group of tungsten carbide-based composite materials. Here copper is the matrix and ensures that the operating temperature of the work piece remains within the desired parameters due to its good heat conductivity. At high operating temperatures, like those existing at electrodes for spot welding; the copper matrix also fulfils another task: During service copper evaporates and thereby takes a quite significant part of the energy [2]. Figure 10.19 shows the structure thereby developed;

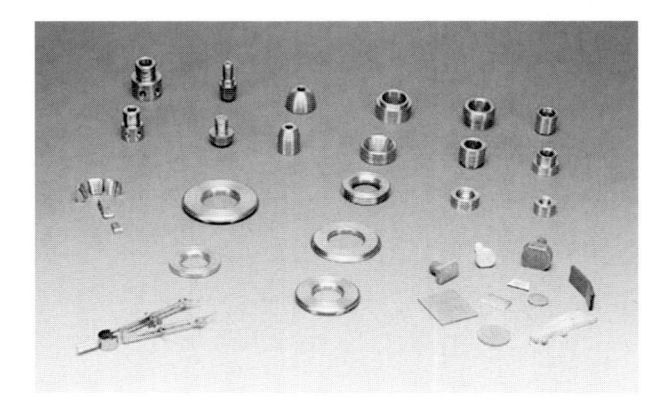

Fig. 10.18 Electronic contacts of Cu–W, Ag–W, Ag–WC, Ag–CdO, Ag–Ni–C [104].

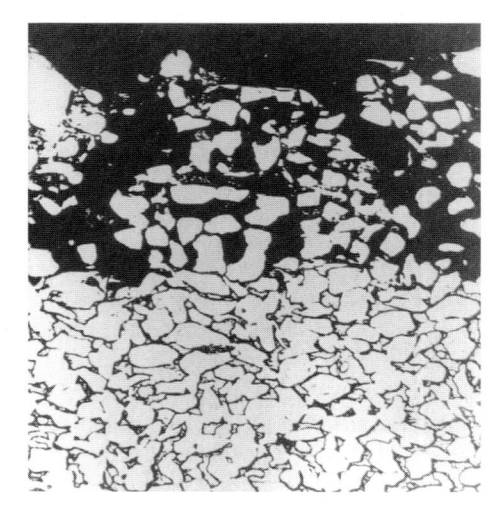

Fig. 10.19 Partial evaporation of copper matrix with a W/Cu contact material [104].

Fig. 10.20 Burn down contacts of W–Cu [2, 51].

Fig. 10.21 Heat sink of a W/Cu contact material [2, 51].

Fig. 10.20 shows the corresponding components. The evaporation of the copper causes a higher operating life of the electrode. An application here is the heat sink shown in Fig. 10.21.

10.5.5
Iron-based MMCs

Iron is the basis for all steel materials, which are of major importance in the technical area. Due to the relatively high Young's modulus and the extremely high range regarding the possible, adjustable tensile strength and yield strengths, considerations regarding the density reduction are only of minor importance if, for example, lightweight construction is concerned. However, the wear behavior of steel is of major interest for tools. By alloying or by thermal mechanical processes the carbides produced or the martensitic parts are often not sufficient to fulfil the demanded properties. Also there are restrictions on fusion metallurgical techniques regarding the size and the distribution of the hard materials in the iron matrix. In order to get a homogeneous material containing a high proportion of hard phases with optimal distribution and particle size, the PM-MMC route is chosen. In such a way materials are produced which are clearly able to exceed the already outstanding properties of tool steels. An example for this is Ferrotitanit, a material from the company Edelstahl Witten Krefeld GmbH [102]. In Fig. 10.22 some tools can be seen, which consist of Ferrotitanit. Such tools show extremely high wear resistance, even at higher temperatures. For example, they are most suitable for use as a tool in friction stir welding.

A further example of Fe-based PM-MMC is the material PM2000 from the company Plansee [2, 51]. Its composition is shown in Table 10.3. Figures 10.12 to 10.14 and 10.23 to 10.26 show some applications of this material. These includes: Glass melt stirrer (Fig. 10.12), engine components (Fig. 10.13 and 10.14), burner nozzles for gas turbines (Fig. 10.23), charger frames for the ceramic industry (Fig. 10.24), honeycomb seals for high performance turbines (Fig. 10.25) as well as clamping devices for test engineering at higher temperatures (Fig. 10.26).

Besides the production of iron-based MMCs by the common powder metallurgical path, i.e. mixing the basic materials, followed by consolidation, *in situ* iron-

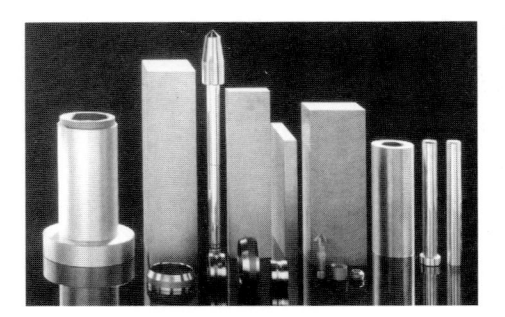

Fig. 10.22 Tool made of Ferrotitanit [102].

Fig. 10.23 Burner nozzles for gas turbines of PM 2000 [2, 51].

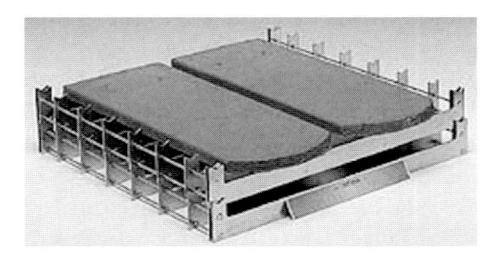

Fig. 10.24 Charger frame of PM 2000 to burn roof tiles [2, 51].

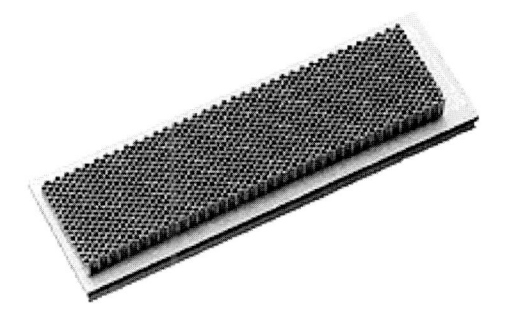

Fig. 10.25 Honeycomb seal of PM 2000 for high performance turbine engines [2, 51].

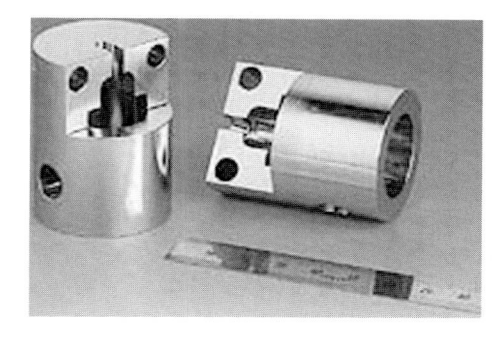

Fig. 10.26 Clamping device for test engineering at high temperatures of PM 2000 [2, 51].

based composite materials can be manufactured [103]. In other *in situ* MMC, one of the phases involved reacts with other existing components of the mixture and the desired phase develops during for example hot isostatic pressing. The materials produced in such a way are comparable with other tool steels and show somewhat better wear properties. So the costs of production can simultaneously be significantly lowered [103].

10.5.6
Nickel-based MMCs

Nickel and its alloys are very often used for chemical plants, since they show good corrosion resistance at high temperatures. However, arrangements have to be made in order to achieve sufficiently high mechanical properties at these temperatures. This can take place either by suitable alloy-technical arrangements or by the employment of powder metallurgy. The latter applies particularly to the ODS super alloys. Table 10.2 shows typical compositions of some alloys manufactured by mechanical alloying. Figure 10.27 shows one example of applications. However, examples of the alloy PM of 2000 can be also manufactured with nickel-based ODS alloys, since the application ranges of iron- and nickel-based ODS alloys overlap within wide ranges.

Fig. 10.27 PM 1000 outer-skin honeycomb panels, aerospace industry [2, 51].

10.6
Summary and Outlook

This chapter on PM-MMC shows the large variation in material systems and powder manufacturing processes, which gives a variety of custom-made powders suitable for each application. Accordingly, the application possibilities are manifold. This compilation can give only a first idea of the potential of powder metallurgically manufactured composite materials and does not claim completeness. However, from the literature further information can be obtained, and at present the Internet and its enquiry possibilities offers various starting points to obtain data about PM-MMC. This applies both to the manufacturing methods and routes as well as to the materials currently used. Ever more company brochures are available on-line and permit fast information access.

References

1 http://mmc-assess.tu-wien.ac.at.
2 http://www.plansee.com.
3 Rohatgi, P. K., et al., Solidification, structures, and properties of cast metal ceramic particle composites, *Int. Mater. Rev.*, *31*(3), (1986), 115–119.
4 Carreño-Morelli, E., et al., Processing and characterisation of aluminum-based MMCs produced by gas pressure infiltration, *Mater. Sci. Eng. A*, *251*, (1998), 48–57.
5 Taheri-Nassaj, E., et al., Fabrication of an AlN particulate aluminum matrix composite by a melt stirring method, *Scr. Metall. Mater.*, *32*(12) (1995), 1923–1929.
6 Degischer, H. P., et al., Design rules for selective reinforcement of Mg-castings by MMC inserts, in *Magnesium Alloys and Their Applications*, K.U. Kainer, Ed., 2000, pp.207–214.
7 Cantor, B., Optimizing microstructure in spray-formed and squeeze cast metal-matrix composites, *J. Microsc.*, *169*(2), (1993), 97–108.
8 Bytnar, J. H., et al., Macro-segregation diagram for dry blending particulate metal matrix composites, *Int. J. Powder Metall.*, *31*(1), (1995), 37–49.
9 Taya, M., Arsenault, R. J., *Metal Matrix Composites*, Pergamon Press, New York, 1990.
10 Everett, R. K., Arsenault, R. J., *Metal Matrix Composites: Mechanisms and Properties*, Academic Press, London, 1991.
11 Chawla, K. K., Interfaces, in *Composite Materials – Science and Processing*, Springer Verlag, Berlin, 1987, pp.79–86.
12 Harris, J. R., et al., A comparison of different models for mechanical alloying, *Acta Mater.*, *49*, (2001), 3991–4003.

13 Gomashchi, M. R., Vikhrov, A., Squeeze Casting: an overview, *J. Mater. Proc. Technol.*, *101*, (2000), 1–9.
14 Kaczmar, J. W., et al., The production and application of metal matrix composite material, *J. Mater. Proc. Technol.*, *106*, (2000), 58–67.
15 Dieringa, H., Kainer, K. U. Partikel, Fasern und Kurzfasern zur Verstärkung von metallischen Werkstoffen, dieser Band, 2002.
16 Hegeler, H., et al., Herstellung von faserverstärkten Leichtmetallen unter Benutzung von faserkeramischen Formkörpern, in *Metallische Verbundwerkstoffe*, K.U. Kainer, Ed., DGM Verlagsgesellschaft, Oberursel, 1994, 101–116.
17 Buschmann, R., Keramische Formkörper zur Herstellung von MMCs, dieser Band, 2002.
18 Chadwick, G. A., Squeeze casting of metal matrix composites using short fiber performs, *Mater. Sci. Eng. A*, *135*, (1991), 23–18.
19 Mortensen, A., et al., Infiltration of fibrous preforms by a pure metal: Part V. Influece of preform compressibility, *Metall. Mater. Trans. A*, *30*, (1999), 471–482.
20 LePetitcorps, Y., Processing of titanium matrix composites: discontinuous of continuous reinforcements? CIMTEC 2002, Florence, Italy, 2002, im Druck.
21 Schatt, W., Wieters, K.-P., *Pulvermetallurgie*, VDI-Verlag, 1994.
22 He, Y., et al. Micro-crystalline Fe-Cr-Ni-Al-Y_2O_3 ODS alloy coatings produced by high frequency electric-spark deposition, *Mater. Sci. Eng. A*, *334*, (2002), 179–186.
23 Berkowitz, A. E, Walter, J. L. Spark erosion: A method for producing rapidly quenched fine powders, *J. Mater. Res.*, *2*(2), (1987), 277–288.
24 Müller, B., Ferkel, H., Properties of nanocrystalline Ni/Al_2O_3 composites, *Z. Metallkd.*, *90*(11), (1999), 868–871.
25 Naser, J., et al.. Laser-induced synthesis of nanoscaled tin oxide based powder mixtures, *Lasers Eng.*, *9*, (1999), 195–203.
26 Naser, J., Ferkel, H., Laser-induced synthesis of Al_2O_3/Cu-nanoparticle mixtures, *NanoStruct. Mater.*, *12*, (1999), 451–454.
27 Naser, J., et al., Dispersion hardening of metals by nanoscaled ceramic powders, *Mater. Sci. Eng. A*, *234–236*, (1997), 467–469.
28 Konitzer, D. G., et al., Rapidly solidified prealloyed powders by laser spin atomisation, *Metall. Trans. B*, *15*, (1984), 149–153.
29 Riehemann, W., Mordike, B. L., Production of ultra-fine powder by laser atomisation, *Lasers Eng.*, *1*, (1992), 223–231.
30 Naser, J., et al., Grain stabilisation of copper with nanoscaled Al_2O_3-powder, *Mater. Sci. Eng. A*, *234–236*, (1997), 470–473.
31 Ferkel, H., et al., Electrodeposition of particle-strengthened nickel film, *Mater. Sci. Eng. A*, *234–235*, (1997), 474–476.
32 Ying, D. Y., Zhang, D. L., Processing of Cu-Al_2O_3 metal matrix nanocomposite materialsby using high energy ball milling, *Mater. Sci. Eng. A*, *286* (2000), 152–156.
33 Bergmann, H. W., et al., Die Erzeugung von Metallpulvern durch Verdüsung ihrer Schmelzen mit flußsigen Gasen, *Steel Metals Mag.*, *28*(10), (1988), 985–1003.
34 Dunkley, J. J. Metallpulverherstellung durch Wasserzerstäubung, *Metall*, *32*(12), (1978), 1282–1285.
35 Kainer, K. U., Mordike, B. L., Oil atomisation – A method for the production of rapidly solidified iron-carbon alloys, *Mod. Dev. Powder Metall.*, *18–21*, (1988), 323–337.
36 Yule, A. J., Dunkley, J. J., *Atomization of Melts*, Oxford University Press, Oxford, 1994.
37 Brewin, P.R., et al., Production of high alloy powders by water atomisation, *Powder Metall.*, *29*(4), (1986), 281–285.
38 Gerling, R., et al., Specifications of the novel plasma melting inert gas atomization facility PIGA 1/300 for the production of intermetallic titanium-based alloy powders, *GKSS-Bericht*, *92*/E/4 (1992).
39 Hort, N., Kainer, K. U., Crucible free gas atomisation of special metals and alloys, in *International Conference on Powder Metallurgy PM'98*, (1998).
40 Naka, S., et al., Oxide-dispersed titanium alloys Ti-Y prepared with the rotating

electrode process, *J. Mater. Sci.*, 22, (1987), 887–895.
41 Seshadri, R., et al., Production of titanium alloy powders, in *Proceedings of a Symposium on Powder Metal Alloys*, Bombay, India (1980), pp. 23–26.
42 Salmang, H., Scholze, H., *Keramik, Teil 2: Keramische Werkstoffe*, 6 Auflage, Springer Verlag, Berlin, 1983.
43 Parvizi-Majidi, A., Fibers and whiskers, in *Materials Science and Technology, Structure and Properties of Composites*, T.-W. Chou, Ed., Vol.13, VCH, Weinheim, 1993, pp.25–88.
44 Chawla, K. K., Metal Matrix Composites, in *Materials Science and Technology, Structure and Properties of Composites*, T.-W. Chou, Ed., Vol.13, VCH, Weinheim, 1993, pp.121–182.
45 Peebles, L. H., *Carbon Fibers – Formation, Structure and Properties*, CRC Press, London, 1995.
46 Beczkowiak, J., et al., *Karbidische Werkstoffe für HVOF Anwendung*, 1995.
47 Beczkowiak, J., et al., *Neue boridhaltige Werkstoffe für das Hochgeschwindigkeits Flammspritzen: Pulver- und Schichteigenschaften*, 1995.
48 Beczkowiak, J., Schwier, G., *Pulverförmige Zusätze für das thermische Spritzen*, 1995.
49 Luyckx, S., et al., Fine grained WC-VC-Co hardmetal, *Powder Metall.*, 39, (1996), 210–212.
50 Seegopaul, P., McCandlish, L. E., Nanodyne advances ultrafine WC-Co powders, *MPR*, 51(4), (1996), 16–20.
51 Plansee, *Dispersionsverfestigte Hochtemperaturwerkstoffe, Datenblatt Werkstoffeigenschaften und Anwendungen*, 1998.
52 Shakesheff, J. A., DERA takes aim at MMC advances, *MPR*, 53(5), (1998), 28–30.
53 Arnold, V., et al., Pulvermetallurgie – Eine Verfahrenstechnik zur Herstellung von Verbundwerkstoffen und Verbundbauteilen, *VDI-Berichte*, 734, (1989), 215–239.
54 Arnold, V., Hummert, K., Properties and applications of dispersion strengthened aluminum alloys, in *Proceedings of an International Conference on New Materials by Mechanical Alloying Techniques*, DGM Informationsgesellschaft Verlag, Oberursel, 1998, pp. 263–280.
55 Sauer, C., et al., Effects of dispersion addition on the properties of Ci-Ti P/M alloys, in *Proceedings of 1993 Powder Metallurgy World Congress*, 1993, pp.143–149.
56 Lotze, G., Stephanie, G., Eigenschaften von über das Hochenergie-Mahlen hergestellten dispersionsgehärteten Cu-Werkstoffen, in *Verbundwerkstoffe und Werkstoffverbunde* (1995), G. Ziegler, Ed., DGM Informationsgesellschaft Verlag, Oberursel, pp. 533–536.
57 Troxell, J. D., GlidCop dispersion strengthened copper: Potential applications in fusion power generators, in *Proceedings of IEEE 13th Symposium on Fusion Engineering (1989)*, Vol. 2, pp. 761–769.
58 Mordike, B. L., et al., Effect of tungsten content on the properties and structure of cold extruded Cu-W composite materials, *Powder Metall. Int.*, 23, (1991), 91–95.
59 Moon, I. H., et al., Full densification of loosely packed W-Cu composite powders, *Powder Metall.*, 41, (1998), 51–57.
60 Sepulved, J., Valenzuela, L., Brush Wellman advances Cu/W technology, *MPR*, 53(6), (1998), 24–27.
61 Bormann, R., Mechanical Alloying: Fundamental Mechanisms and Applications, in *Materials by Powder Technology – PTM'93*, F.Aldinger, Ed., DGM Informationsgesellschaft Verlag, Oberursel, 1993, pp. 247–258.
62 Takacs, L., Self-sustaining reactions induced by ball milling, *Prog. Mater. Sci.*, 47, (2002), 255–414.
63 Nityanand, N., et al., An analysis of radial segregation for different sized spherical solids in rotary cylinders, *Met. Trans. B*, 17, (1986), 247–257.
64 Lewandowski, J. J., et al., Microstructural effects on the fracture micromechanisms in 7xxx Al P/M-SiC particulate metal matrix composites, in *Processing and Properties for Powder Metallurgy Composites*, P.Kumar et al., Eds., The Metallurgical Society, 1988, pp.117–137.
65 Velasco, F., et al., TiCN-high speed steel composites: sinterability and properties, *Composites Part A*, 33, (2002), 819–827.
66 Kim, Y.-J., et al. Processing and mechanical properites of Ti-6Al-4V/TiC

in situ composite fabricated by gas-solid reaction, *Mater. Sci. Eng. A, 333,* (2002), 343–350.

67 Abenojar, J., et al., Reinforcing 316L stainless steel with intermetallic and carbide particles, *Mater. Sci. Eng. A, 335,* (2002), 1–5.

68 Oliveira, M. M., Bolton, J. D., High speed steel matrix composites with TiC and TiN, in *Proceedings of the International Conference on Powder Metallurgy,* World Congress, Paris, 1994, pp. 459–462.

69 Lindqvist, J.-O., Pressing and sintering of MMC wearparts, in *Proceedings of the International Conference on Powder Metallurgy,* World Congress, Paris, 1994, pp. 467–470.

70 Kainer, K. U., et al., Production techniques, microstructures and mechanical properties of SiC particle reinforced magnesium alloys, in *Proceedings of the International Conference on Powder Metallurgy-Aerospace Materials,* 1991, pp. 38.1–15.

71 Loh, N.H., et al., Production of metal matrix composite part by powder injection molding, *J. Mater. Proc. Technol., 108,* (2001), 398–407.

72 Clyne, T. W., Whithers, P. J., *An Introduction to Metal Matrix Composites,* Cambridge University Press, Cambridge, UK, 1993.

73 Feng, C. F., Froyen, L., In-situ P/M Al($ZrB_2 + Al_2O_3$) MMCs: Processing, microstructure and mechanical characterisation, *Acta Mater., 47,* (1999), 4571–4583.

74 He, D. H., Manory, R., A novel electrical contact material with improved self-lubrication for railway current collectors, *Wear, 249,* (2001), 626–636.

75 Upadhyaya, A., et al., Advances in sintering of hard metals, *Mater. Des., 22,* (2001), 499–506.

76 Borrego, A., et al., Influence of extrusion temperature on the microstructure and the texture of 6061 Al – 15 vol.-% SiC_w PM composites, *Composite Sci. Technol., 62,* (2002), 731–742.

77 Kainer, K. U., Tertel, A., Extrusion of short fiber reinforced magnesium composites, in *Proceedings of the 12th Riso International Symposium on Materials Science,* 1991, pp. 435–440.

78 Saito, T., et al., Thermomechanical properties of P/M _ titanium metal matrix composites, *Mater. Sci. Eng. A, 243,* (1998), 273–278.

79 Kennedy, A. R., Wyatt, S. M., Characterising particle-matrix interfacial bonding in particulate Al-TiC MMCs produced by different methods, *Composites A, 32,* (2001), 555–559.

80 Erich, D. L., Metal matrix composites: Problems, applications and potential in the P/M industry, *Int. J. Powder Metall., 23*(1), (1987), 45–54.

81 Mordike, B. L., et al., Powder metallurgical preparation of composite materials, *Trans. PMAI, 17,* (1990), 7–17.

82 Kamiya, A., et al., Alumina short fiber reinforced titanium matrix composite fabrication by hot-press method, *J. Mater. Sci. Lett., 20,* (2001), 1189–1191.

83 Ogel, B., Gurbuz, R., Microstructural characterisation and tensile properties of hot pressed Al-SiC composites prepared from Al and Cu powders, *Mater. Sci. Eng. A, 301,* (2001), 213–220.

84 Tjong, S. C., Ma, Z. Y., Microstructural and mechanical characteristics of in situ metal matrix composites, *Mater. Sci. Eng., 29,* (2000), 49–113.

85 Korb, G., et al., Thermophysical properties and microstructure of short carbon fiber reinforced Cu-matrix composites made by electroless copper coating or powder metallurgical route respectively, in *Proceedings of the International Symposium on Electronic Manufacturing Technology IEMT,* Potsdam, 1998, pp. 98–103.

86 Radhakrishna, B. V., et al., Processing map for hot working of powder metallurgy 2124 Al-20 vol pct SiC_p metal matrix composites, *Metall. Trans. A, 23,* (1992), 2223–2230.

87 Abkowitz, S., Advanced powder metallurgy technology for manufacture of titanium alloy and titanium matrix composites to near net shape, in *Proceedings of the International Conference on P/M in Aerospace and Defense Technologies,* Seattle, Vol. 1, 1989, pp. 193–201.

88 Ray, A. K., et al., Fabrication of TiN reinforced aluminum metal matrix composites through a powder metallurgical route, *Mater. Sci. Eng. A,* 2002, in press.

89 Froes, F. H., Suryanarayana, C., Powder processing of titanium alloys, *Rev. Particulate Mater.*, *1*, (1993), 223–275.
90 Abkowitz, S., et al., The commercial application of low-cost titanium composites, *JOM*, *8*, (1995), 40–41.
91 Jüngling, T., et al., Manufacturing of TiAl-MMC by the elemental powder route, in *Proceedings of the International Conference on Powder Metallurgy*, World Congress, Paris, 1994, pp. 491–494.
92 Buchgraber, W., et al., Carbon fiber reinforced copper matrix composites: Production techniques and functional properties, in *EUROMAT (1999)*, p. 1111.
93 Godfrey, T. M. T., et al., Microstructure and tensile properties of mechanically alloyed Ti-6Al-4V with boron additions, *Mater. Sci. Eng. A*, *282*, (2000), 240–250.
94 Saito, T., et al. Development of low cost titanium matrix composite, in *Recent Advances in* Titanium Metal Matrix Composites, F. H. Froes and J. Storer, Eds., 1995, pp. 33–44.
95 Blawert, C. Noble and non-ferrous metal matrix composite materials, this book, 2002.
96 Hort, N., et al., Magnesium matrix composites, in *Magnesium and Its Alloys*, B. L. Mordike, H.Friedrich, Eds., Springer Verlag, Heidelberg, 2002.
97 Srivatsan, T. S., et al., Processing of discontinuously reinforced metal matrix composites by rapid solidification, *Prog. Mater. Sci.*, 39, (1995), 317–409.
98 Gorsse, S., et al., In situ preparation of titanium base composites reinforced with TiB single crystals using a powder metallurgy technique, *Composites Part A*, 29, (1998), 1229–1234.
99 Gorsse, S., et al., Investigation of Young's modulus of TiB needles in situ produced in titanium matrix composites, *Mater. Sci. Eng. A*, (2002), in press.
100 Abkowitz, S., et al., Particulate reinforced titanium alloy composites economically formed by combined cold and hot isostatic pressing, *Industrial Heating*, 9, (1993), 32–37.
101 http://www.dynamettechnologie.com.
102 Edelstahl WittenKrefeld GmbH, company brochure, 2001.
103 Berns, H., Wewers, B., Development of an abrasion resistant steel composite with in situ TiC particles, *Wear*, *251*, (2001), 1386–1395.
104 http://www.sinter-metal.com.
105 http://www.ospreymetals.com
106 Beane, A., et al., Cold forming extends the reach of P/M, *MPR*, *52*(2), (1997), 26–31.
107 Special Metals Corporation. company brochure: Inconel alloy MA 758, 1993.
108 Special Metals Corporation company brochure: Inconel alloy MA 754, 1999.
109 Special Metals Corporation company brochure: Inconel alloy MA 956, 1999.
110 Materials Innovations Inc., Cold forming P/M.
111 Kahl, W., Leupp, J., Spray Deposition of high performance aluminum alloys via the OSPREY Process, in *Advanced Materials and Processes*, H. E. Exner, V. Schumacher, Eds.,DGM Informationsgesellschaft, Oberursel, 1990, pp. 261–266
112 Kainer, K.U., Verstärkte Leichtmetalle – Potential and Anwendungsmöglichkeiten. *VDI Berichte*, *965*(1), (1992), 159–169.
113 Lashmore, D. S., Christie, P., Cost effective, cold forming powder metallurgy process fabricates dense parts without sintering, *Mater. Technol.*, *11*(4), (1996), 131–144.
114 Lawley, A., Apelian, D., Spray forming of metal matrix composites, in *Proceedings of the 2nd International Conference on Spray Forming*, J.V. Wood, Ed., Woodhead Publishing Ltd., Abington, UK, pp. 267–280.
115 N. N., Soft magnetic composites offer new PM opportunities, *MPR*, *51*(1), (1996), 24–28.
116 Persson, M., *SMC Powders Open New Magnetic Applications*, 1997.
117 Suryanarayana, C., Mechanical Alloying and milling, *Prog. Mater. Sci.*, 46, (2001), 1–184.
118 Suryanarayana, C., et al., The science and technology of mechanical alloying, *Mater. Sci. Eng. A*, *304–306*, (2001), 151–158.
119 Verghese, L., Westam composite nickel powders, *Powder Metall.*, *41*, (1998), 16–17.

11
Spray Forming – An Alternative Manufacturing Technique for MMC Aluminum Alloys

P. Krug, G. Sinha

11.1
Introduction

Nowadays aluminum-based materials are applied in many technical areas. Due to its low specific weight, high corrosion resistance and outstanding technological properties, aluminum (as pure metal) already competes in many applications with the conventional material steel. The various economic advantages, especially the advantages of high performance aluminum, make a convincing case for its use.

At the present time Peak Werkstoff GmbH of Velbert, Germany is leading in the area of spray forming of high performance aluminum alloys. The processed final products (brand name "DISPAL") from the primary material are used in the automotive industry and in general mechanical engineering. Further market segments, for example the aerospace industry, are also available for alloys from Velbert.

Spray compacted aluminum alloys show a number of outstanding properties (see also Fig. 11.1 to 11.5):

- high strength alloys with yield stress of 700 MPa
- high strength in combination with good fracture toughness
  (strength up to 520 MPa with fracture toughness up to 120 MPa m$^{1/2}$)
- high Young's modulus up to 100 GPa
- outstanding hot resistance up to application temperatures of 400 °C
- excellent wear resistance
- low thermal coefficient of expansion, adjustable via Si content
- high corrosion resistance in different media
- good deformability on conventional machines
- trouble free recycling

The properties mentioned can only be realized by powder metallurgical (P/M) techniques, which belong to the rapid solidification category. After DIN the powder metallurgy is assigned to the technology of melting and casting, either the conventional P/M technique or the advanced powder metallurgy.

Metal Matrix Composites. Custom-made Materials for Automotive and Aerospace Engineering.
Edited by Karl U. Kainer.
Copyright © 2006 WILEY-VCH Verlag GmbH & Co. KGaA, Weinheim
ISBN: 3-527-31360-5

Yield stress of 7xxx alloys, quenched (-W) and aged (-T6); longitudinal, 23°C

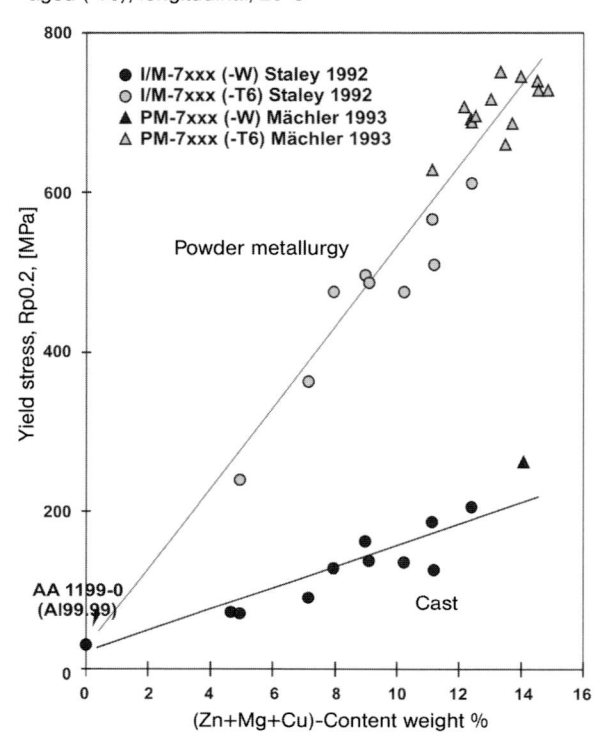

Fig. 11.1 Strength of AlZnMgCu alloys, processed at different cooling rates [1, 2].

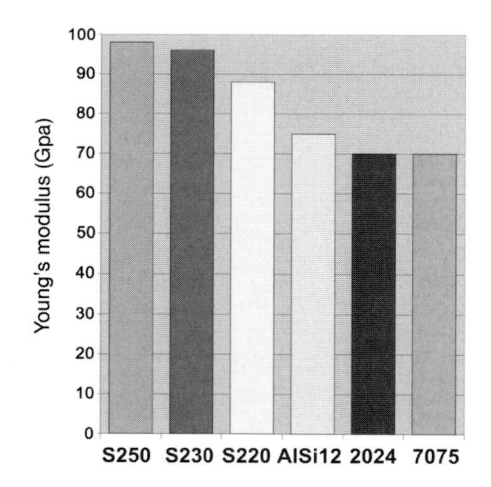

Spray Compacted AlCuMgAgX - Alloys

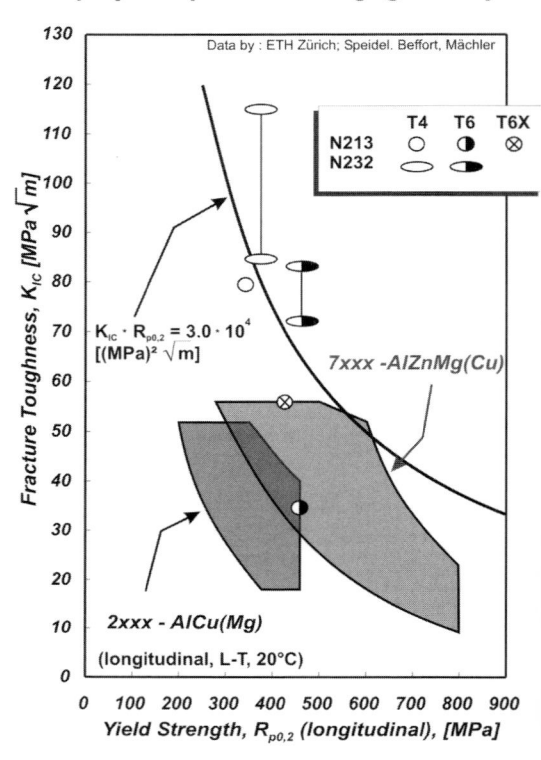

Fig. 11.2 Comparison of fracture toughness for conventional and spray formed alloys [3].

Fig. 11.3 Young's modulus of different alloys. A higher Young's modulus is obtained with a higher silicon content.

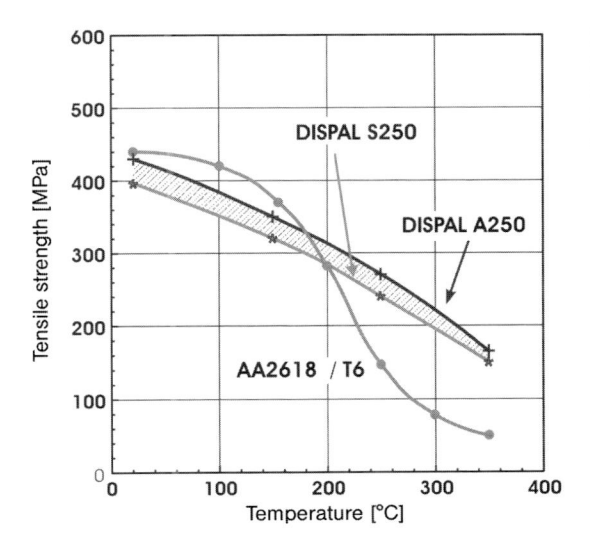

Fig. 11.4 High temperature strength of spray formed materials in comparison with that of conventional alloys AA2618.

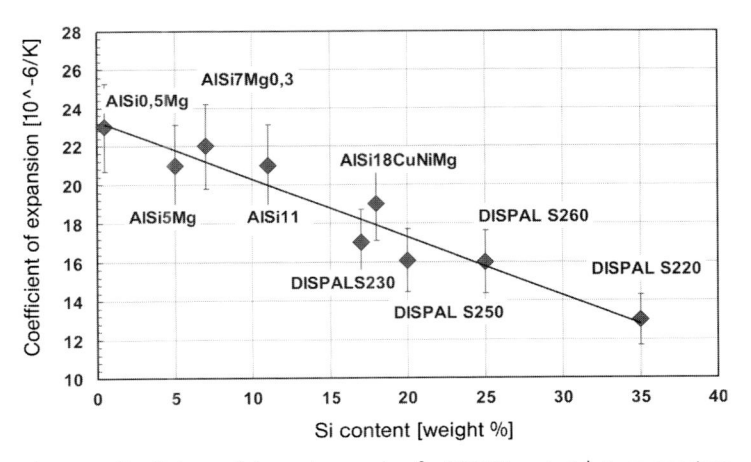

Fig. 11.5 Coefficients of thermal expansion for DISPAL materials in comparison with those of other aluminum alloys.

The aim of the conventional P/M technology is to manufacture parts or assembled units as close to the final geometry as possible. The new powder metallurgical techniques have the aim to develop innovative materials, capable of replacing conventional materials in a widely diversified application area. In fact tailored materials can be created which are recommended for special applications.

11.2
Spray Forming

Regarding the economic and technical advantages, spray forming, as a consequent development of conventional powder metallurgy, is very important. The fundamental technical background is an extreme quench speed (Fig. 11.6) during the solidification of high alloyed and high overheated alloys in the range of 10^3 to 10^4 K s^{-1}.

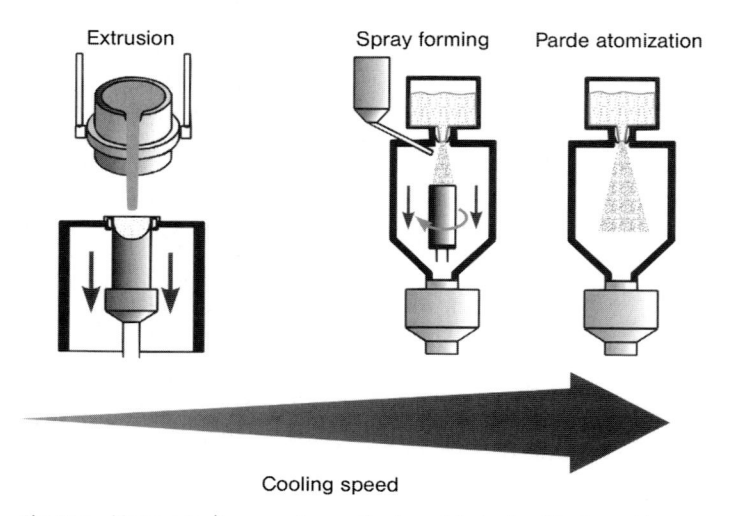

Fig. 11.6 Various melt processing methods and their classification with respect to quenching rates.

The resulting structure at this high quench rate shows a monomodal distribution of very fine primary silicon crystals and intermetallic precipitations (Fig. 11.7).

The manufacture of high strength aluminum alloys is achieved by the alternation of hardening and solidification mechanisms of traditional precipitation hardening with dispersion strengthening. Furthermore, the transformation from brittle, hypereutectic aluminum cast alloys into ductile wrought alloys succeeds by rapid solidification. All available aluminum alloys within the possible alloy spectra show at least one property, or one property combination, which is not represented by conventional casting techniques. These properties are, as mentioned already: high strength, wear resistance, hot strength and high Young's modulus.

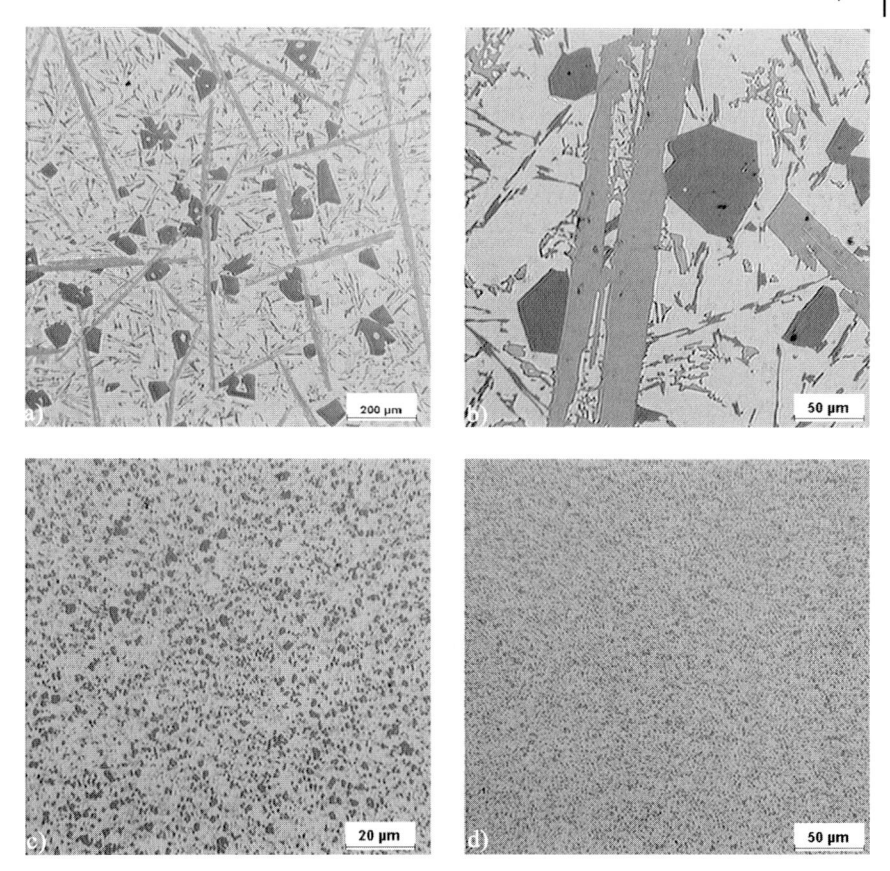

Fig. 11.7 Microstructure of alloy AlSi20Fe5Ni2, "as cast" and "as sprayed". (a) as cast 50×; (b) as cast 200×; (c) spray formed 200×; (d) spray formed 500×.

11.3 Techniques

11.3.1 Rapid Solidification (RS) Technique

The so-called RS technique is applied with gas atomization. The melt beam of a highly overheated metal melt is atomized by a cooling gas (nitrogen, air). The resulting molten droplets are cooled down very quickly by the cooling gas; the oversaturation condition within the powder particle is frozen in. Dispersoids of intermetallic composition, which are necessary for the hot strength of the material, are formed at the same time. Pre-compacting of the gas atomized powder is achieved by cold isostatic pressing (CIP). The complete consolidation takes place by extrusion. During this process step, adjacent powder particles are sheared off against each other and rewelded on the newly developed surfaces.

If a later application makes it necessary, further intermediate steps like encapsulating of the CIP billets in aluminum cans, degassing, hot pressing and stripping down can be applied. Gaseous impurities or humidity are thereby removed from the particle surface, leading to minor property increase.

11.3.2
Spray Forming Technique

Spray forming can be considered as a further development of atomization. Spray formed (spray compacted) materials differ from the RS atomized alloys in the extremely low gas content in the form of hydrogen due to the inert gas atmosphere (nitrogen) during the spray process. This extremely low hydrogen content means that the spray formed materials are weldable by fusion weld techniques. The pure atomization technique is identical to the RS technique. Because the semi-finished product is in the form of a compact billet, cold and hot isostatic pressing is not applied at the spray forming stage, which is different to the RS technique. Figures 11.8 and 11.9 show the scheme of a spray forming plant, as is applied at the PEAK Werkstoff GmbH for special alloys and development concepts. From the Tundish the melt is dispensed onto the two atomization units. The melt is atomized by nitrogen and the resulting drop spectra is accelerated onto the sample. The impinged droplets solidify at the surface and little by little a structure grows, forming an almost perfect cylindrical bolt by constant rotation round the vertical axis and a continuous downward withdrawal rate.

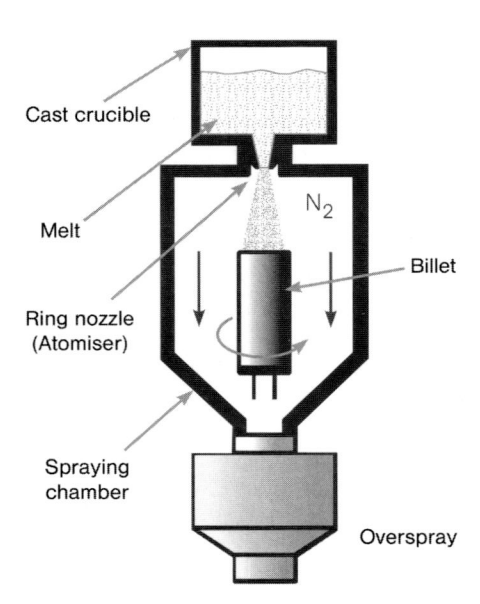

Fig. 11.8 Schematic overview of the spray forming unit.

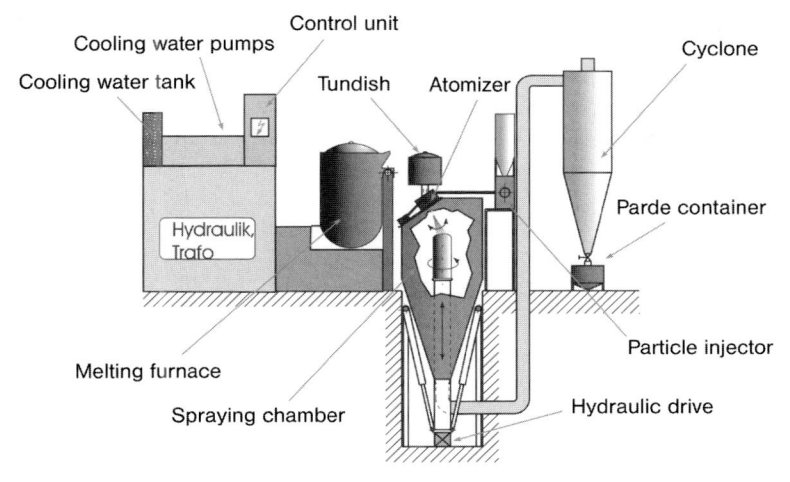

Fig. 11.9 Spraying chamber (schematically shown).

11.3.3
Melting Concept

For large volume production it is recommended to use specially designed spray forming units for a specific alloy. The disadvantage of this is that if there was an unexpected alloy change at the plant this would lead to an extensive service processing time. The manufacturing plant in St. Avold, France consists of a 1.2 t inductive melting furnace, a 2.5 t holding furnace with a flanged fore-hearth and the actual spraying chamber.

The melting furnace is filled from a container by a tilting apparatus. From there the molten metal is transferred via a launder into the holding furnace. The complete unit is situated on load cells to record and control continuously the data related to the metal. By controlled pressure increase in the holding furnace the molten metal is forced into the fore-hearth. The PEAK Werkstoff GmbH runs two serial spray forming units and one F & E spray forming unit.

For spray forming the holding furnace is moved in the direction of the spray chamber and the fore-hearth is docked onto the spray chamber. The pressure of the fore-hearth on the spray chamber can be adjusted by load cells in the hydraulic cylinders, therefore absolute impermeability can be guaranteed and the sealing elements are not mechanically destroyed.

The most important feature of the holding furnace is the pressure control. Using that, it is possible to keep the bath level constant during spraying, independent of the furnace filling level. A constant bath level is a mandatory condition to control the rate of metal flow accurately. This happens by a continuous increase in the overpressure within the holding furnace at a speed which corresponds exactly with the rate of flow. The measured bath level, taken by a float lever in the fore-hearth, is used as the actual value for control. At the end of the spraying process the pressure is reduced only as far as to ensure that the applied foam filter in the fore-hearth is just still wetted by the molten metal. This is important for a long service life of the filter. Furthermore, the cleaning effort of the fore-hearth is thereby considerably reduced.

11.3.4
Atomisation of the Metal Melt

In general there is a difference between a single atomizer and multilevel atomizing. Especially with spray forming plants in production standards, as at PEAK Werkstoff GmbH, the multilevel atomization is limited to a twin atomizer, see Fig. 11.10. The nozzle units, consisting of the atomizer, the primary- and secondary nozzles are identical for each process.

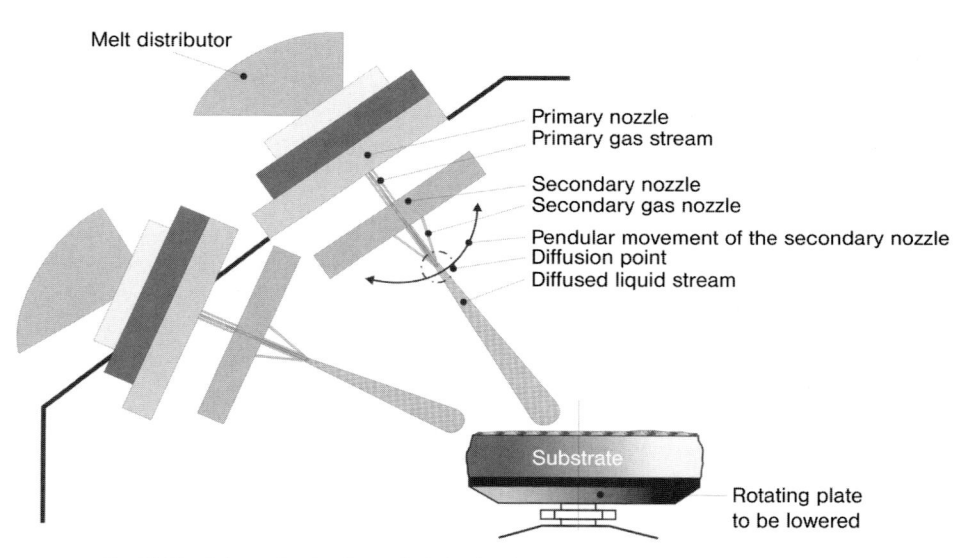

Fig. 11.10 Schematic structure of a twin atomizer.

11.3.5
Nozzle Unit

The atomization unit applied in spray forming is a two-stage arrangement, consisting of two concentric gas ring nozzles, the primary gas nozzle and the secondary gas nozzle. New ceramic funnels are inserted in the primary gas nozzle for each spraying run. The task of these nozzles is to avoid direct melt contact with the steel ring nozzles and their retainers.

Primary and secondary gas nozzles are fixed tightly at the nozzle retainer, at their feeding connection the feed-in of the primary gas, or rather secondary gas, takes place.

11.3.6
Primary Gas Nozzle

The primary gas, which surrounds the discharging concentric melt stream, leads the jet through the secondary gas nozzle up to the atomization point. Furthermore a widening of the melt stream is avoided, so that the formation of enclosed recirculation zones near the nozzle area is prevented. In the case of such "flow dead areas", a pre-diffusion of the melt would take place. Enclosed recirculation zones are avoided by sufficiently high primary gas pressures. The open jet effect developing during the emission from the atomizer, with the formation of a low pressure area between primary and secondary gas nozzle, is mainly compensated for by the primary gas stream. A constant gas pressure of the primary gas stream is an important contribution to good flow behavior.

11.3.7
Secondary Gas Nozzle

The secondary gas nozzle, also called the atomizer nozzle, is responsible for the atomization (diffusion) of the melt stream. The secondary gas, which flow through holes arranged concentrically at a certain angle, destroys the guided melt stream at the atomizing point. This point is a few centimeters below the secondary gas nozzle.

During the impact of the secondary gas stream on the melt stream, an impulse transmission occurs. A spray cone builds up, whose particle spectrum consists of particles varying in size from 1 to 400 µm. The shape of the spray cone formation is dependent on different spraying parameters, for example the nozzle geometry, the pressure and speed relation and the gas / metal ratio.

Besides diffusing the melt stream, the secondary gas has a further function to accelerate the diffused droplets onto a substrate, which is situated in the center of the spraying chamber, and to cool down the droplets before impact. Figure 11.11

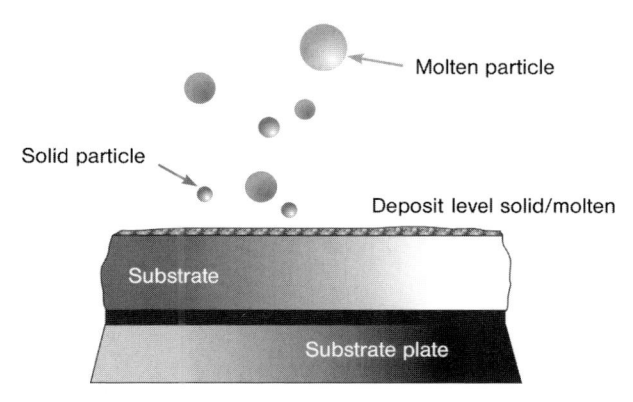

Fig. 11.11 Model of particle deposition.

shows schematically the structure of a spray formed billet, the structure of the deposit layer is also shown.

The particles, which experience a sufficiently high impulse, (i.e. which have a certain mass at a relevant speed) spray directly onto the substrate. Particles, for which the impulse is too low, follow the gas flow and are sprayed far out. This powder is called overspray powder and is sucked off the spraying chamber. Due to the particle size spectrum of the atomized melt stream, the solidification state of the particles is different at the impact.

Investigations have shown that 80% of the atomized droplets are solidified before impacting. Due to the low mass after diffusion, energy, in the form of heat, has been taken from these particles, corresponding to the flight time. These particles impact as solids on the deposit layer. The remaining 20% of the diffused droplets impact in the form of a semi-solid or liquid phase on the deposit layer. These droplets cannot release their latent heat and cool down quickly enough; therefore they do not solidify completely during the flight. This part is of great importance for the actual spray forming process. The impacting and solidified particles on the deposit surface are compacted by particles in the liquid or semi-solid state.

In this process the annealing effect in the deposit level is of great importance for different alloying systems. To reach an appropriate billet structure, the liquid part of the top billet area should not exceed a certain thickness.

The substrate is set in rotation after the start of the process to build up a cylindrical body, the so-called spraying billet. The cooling path of the diffused particle (spraying path), i.e. the path which the particle follows before hitting the deposit surface, is kept constant over the spraying time. To realize that, the spraying plate is constantly lowered.

Billets of diameter from 150 to 400 mm with a maximum length of 2500 mm can be manufactured nowadays, dependent on the unit design. A spray formed billet in its usual dimension has a weight of approximately 450 kg, see Fig. 11.12. The densities of spray formed billets obtained are above 97% of the theoretical density of the operated alloy. The residual compression of the material takes place by a subsequent deformation.

The total annual production of currently available spray forming plants, dependent on the peripheral devices, is 3500 tons.

Spray forming was developed in the 1960s by the company Osprey Metals Ltd. of Neath, Wales (UK). Fundamental research on spray forming was carried out on different metal systems. Nowadays spray forming plants exist Europe-wide for iron alloys, copper alloys and aluminum alloys. The main focus of the Osprey Patent is not only the rigid nozzle unit to build up spray formed bodies, but also the combination of static and scanning nozzles.

In the spray forming of billets, as manufactured at the PEAK Werkstoff GmbH, the static nozzle unit is aligned to the deposit edge, the scanned nozzle unit moves around the area of the deposit radius. Induced by different peripheral speeds by the deposit radius, the scanner nozzle additionally scans a speed profile. Continuous supply of a well defined amount of atomised material per billet rotation is a basic condition.

Fig. 11.12 Spray formed billets of PEAK Werkstoff GmbH.

11.3.8
Plant Safety

Handling aluminum powder requires special precautions with regard to explosion protection. The resulting overspray powder during spray forming is categorized as fine powders with an average particle size of less than 50 μm and needs special consideration with respect to dust explosions.

A dust explosion of overspray particles can take place, if the following three conditions are met:

- presence of a critical powder–air mixture, as from 60 g m^{-3}
- minimum ignition energy from 25 to 50 mJ
- presence of oxygen at a level of 7–9 %

The smaller the particle size of the overspray powder, the lower the minimum ignition energy to start an explosion. To avoid a dust explosion at least one of these factors must be eliminated. Thus oxygen is not present in the spraying chamber; the atomization of the melt takes place using nitrogen. The need for the absence of oxygen was taken into account during the conception of the total plant to ensure explosion protection. Thus a high safety standard can be guaranteed.

The complete spraying chamber is gas-proof, including the spraying chamber door. Therefore a reaction of the overspray powder with oxygen in the atmosphere is avoided. In case of any leaks, the nitrogen overpressure in the spraying chamber provides an escape of nitrogen rather than the ingress of oxygen into the spraying chamber. A closed circuit cooling within the double wall spraying chamber with noncorrosive cooling oil prevents thermal deformation and thus possible leaks. An explosion panel in one of the spraying chamber walls acts as a rated break point, which dissipates released energy in the case of a dust explosion. The operational safety of the plant is controlled by numerous sensors and measurement devices over the entire spraying process. Upon falling below, or exceeding, fixed critical values, corresponding messages are given and if necessary the complete plant operates automatically in an emergency stop loop.

11.3.9
Re-injection of Overspray Powder

The resulting overspray powder during spray forming per spraying run is classed as "waste product", which limits the degree of efficiency of the total process to a maximum of 65%. The technique of re-injection of overspray powders is based on the idea of bringing the powder back into the spraying process and thus increasing the efficiency of the plant. For economic and quality reasons, the idea of remelting the overspray powder is ruled out. The only possibility to use the overspray powder again is to inject it into the melt stream.

Before injection the overspray powder is classified; only a specific size of powders can be brought into the melt stream.

For safety and quality reasons the total classifying process, as with the spraying process, runs in a protective gas atmosphere. Because of transport problems of the powder due to the occurrence of overpressure in the spraying chamber, special injectors for the injection of the overspray powders have been developed.

11.3.10
Functional Ways of Injecting for Re-injection

Injectors for the re-injection of overspray powders work complying with the laws of the pneumatic transport, see Fig. 11.13. To avoid powder deposit the transported powder is always in abeyance in the transport pipes. The speed of the transport gas is therefore higher than the speed of the powder particles. Powder deposits in pipes would lead to temporary intermittent transport cycles of each powder package and consistent melting of the injected powder in the melt stream could no longer be guaranteed; nonmelted powder nests would arise within the spray formed billet.

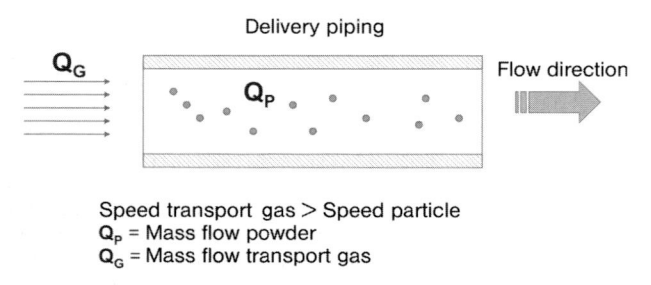

Speed transport gas > Speed particle
Q_P = Mass flow powder
Q_G = Mass flow transport gas

Fig. 11.13 Principle of pneumatic transporting.

The sieved overspray powder is filled into a stainless steel container and set on the particle injectors. The particle injectors are equipped with a differential dossier carriage to always supply (inject) the same amount of powder into the transport pipes over the duration of the spraying process. The powder injected is directly brought into the primary gas flow over the transport pipes. The powder particles hit the atomized melt stream at the diffusion point and after melting they are accelerated onto the spraying plate.

Metallographic investigations of spray formed billets show no difference to common sprayed billets. The injected overspray is melted up to 100%.

11.4 Materials

11.4.1 Spray Forming Products for Automotive Applications

Hypereutectic aluminum silicon alloys with very high silicon contents, which cannot be handled by conventional casting techniques, show excellent tribological properties. Quasi-MMC materials are produced by the primary crystallization of silicon. The high cooling velocity, which can be found during spray forming, enables an even contribution of fine primary silicon crystals in the size range 1–5 µm. In normal castings the largest portion of silicon occurs in the eutectic and does not contribute to tribological behavior. Spray formed, hypereutectic Al–Si alloys, which precipitate all silicon in the form of primary crystals behave in a completely converse way [5].

Cylinder liners of PM aluminum are one example of using a spray formed component in series production. DaimlerChrysler AG and PEAK Werkstoff GmbH have developed spray formed Al liners for application in V-engines as well as in in-line engines of the DaimlerChrysler AG in a joint project (Fig. 11.14).

The spray formed billets are thereby extruded to thick walled tubes in one hot forming process, the indirect extrusion of seamless tubes is carried out using a mandrel. The extruded tubes are hot swaged to thin walled tubes in a subsequent

Fig. 11.14 High pressure die cast crankcase with cylinder liners of the spray formed alloy AlSi25Cu4Mg [6].

deformation process. There they reach their final dimensions of inner and outer diameter. After cutting to length and chamfering, the surface is activated by grit blasting with corundum, which guarantees an excellent metallurgical bonding with the cast alloy of the crankcase during high pressure die casting.

Spray forming technology is the only serial technique which allows, *in situ*, a high amount of primary silicon in the alloy. Alloys with contents up to 35 wt% silicon can otherwise only be obtained by expensive conventional P/M methods.

11.4.2
Spray Forming MMC Materials

Spray formed aluminum materials can be further optimized by inducing a second phase. By injecting hard materials, particle reinforced aluminum alloys can be manufactured. Due to their high stiffness and strength parameters compared to nonreinforced materials these are of great interest for lightweight construction, even at slightly higher density, see Fig. 11.15. Furthermore the wear resistance can be increased significantly by the addition of hard phases, even harder than silicon. Especially when ductility is demanded in combination with high strength and wear resistance, silicon-containing spray formed aluminum alloys do not reach the desired properties; the ductility is significantly reduced due to the high silicon content. Al–Cu–Mg alloys could be used but they exhibit a poor wear resistance. The desired properties can be achieved by addition of wear resistant particles.

There are a large number of particles possible, like for example silicon carbide (SiC), silicon dioxide (SiO_2), titanium carbide (TiC), aluminum oxide (Al_2O_3) or boron carbide (B_4C). SiC was used in pre-investigations due to its ready availability, especially since a cheap and reliable alternative technique to produce SiC reinforced aluminum alloys was sought. This technique is supposed to replace the more expensive powder metallurgical manufacturing route of the pre-material of

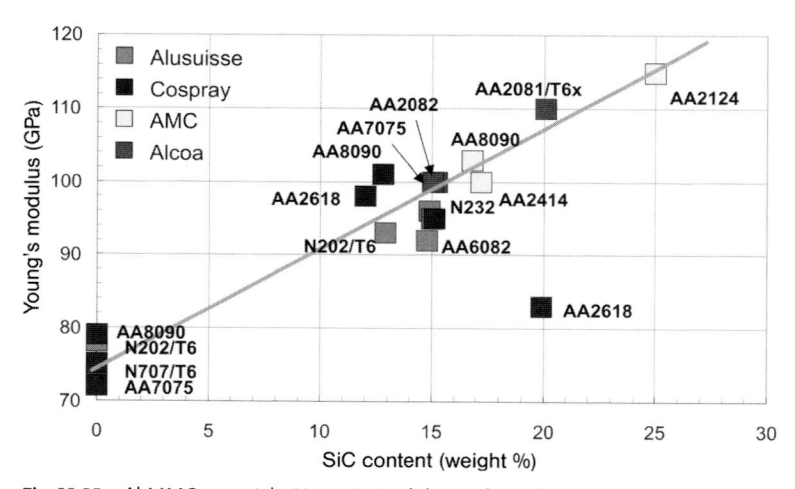

Fig. 11.15 Al-MMC materials, Young's Modulus in dependence on the SiC content.

turbine guiding vanes. Tests have been carried out on two different grain sizes so far. Figure 11.16 shows scanning electron microscopy (SEM) images of both SiC powders.

However, injecting of such highly abrasive hard material cannot be realised with common particle injectors. Due to the relatively high transport speeds damage of all powder touched parts would take place. Furthermore, powders with a very small diameter (<40 µm) are difficult to transport due to their affinity to agglomerate and thus block nozzles and transport pipes. This causes an uneven entry of particles, which is expressed by a heterogeneous contribution in the structure. As an alternative powder pumps can be applied to transport abrasives as well as small particles.

The PEAK Werkstoff GmbH has developed, in cooperation with the Swiss company DACS, such a turn key, production proof powder transport pump especially for SiC powder. The principle of powder transport pumps is based on the "transport of powder packages", that means only small powder clumps are transported over small distances. As Fig. 11.17 shows, the amount of hard material to be transported is controlled by a defined, time-controlled pressure decrease and not by a constant gas velocity. This powder transport is realized by a small powder cavity, which is evacuated by a vacuum pump. The produced underpressure is able to

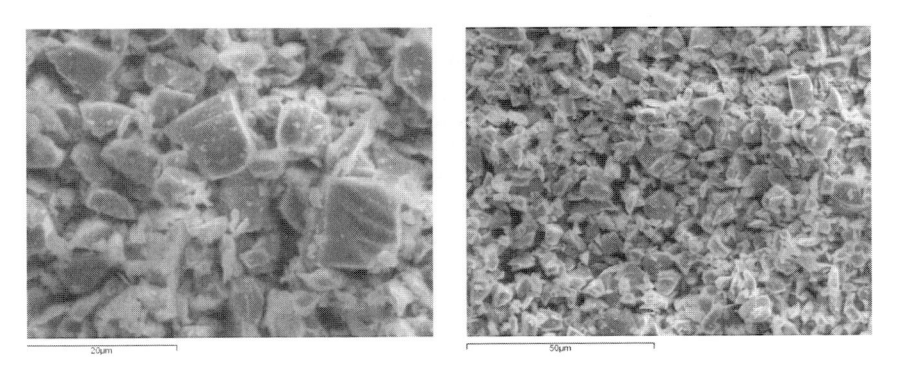

Fig. 11.16 Silicon carbide, SiC F500 with d_{50}=15 µm and SiC F1000 with d_{50}= 5 µm.

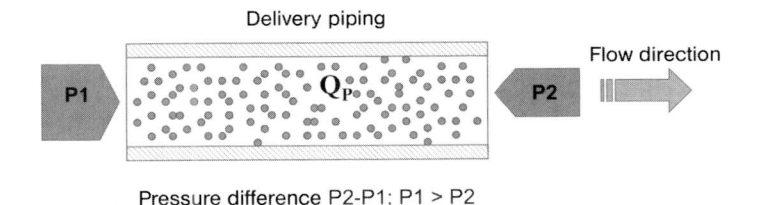

Powder transport is determined by the pressure!
Speed transport gas = Speed particle

Fig. 11.17 Sketch of the principle of powder package transporting.

suck a powder clump of approximately 1 g into the cavity. A membrane protects the vacuum pump against pollution from the transported particles. Afterwards a pressure impulse is given to the powder cavity and a valve of the transport pipe is opened at the same time. This procedure is repeated several times per second, depending on the desired transport rate. The skilful overlaying of several alternating cavities effects a quasi-continuous powder transport, see Fig. 11.18.

By applying this transport principle a homogeneous injection of hard material can be achieved. Figure 11.19(a) shows a structure, which is processed with common hard material injectors. The inhomogeneous contribution of SiC particles is clearly seen (by image analysis systems marked for better recognition). A further phenomenon is also clearly seen. The original drop size can be determined by decorating their original surface. Again that means that the impulse of the pneumatic transported particle is not sufficient to inject the molten droplets and to achieve mixing before compacting. It should also be mentioned, that particle size, spraying parameter (especially gas/metal ratio) and of course the desired volume concentration of hard materials have an influence on the resulting structure. It should be possible to adjust these parameters so that a homogenous structure can be obtained, see also Fig. 11.19(b).

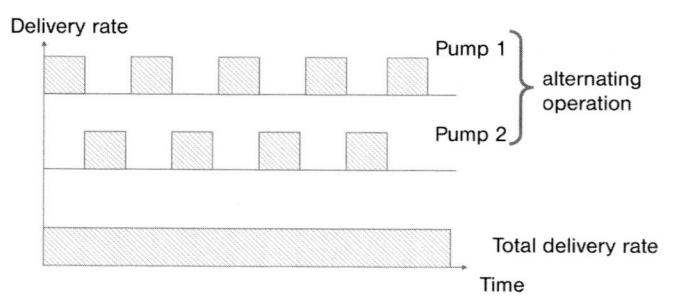

Fig. 11.18 Alternating principle of continuous transporting.

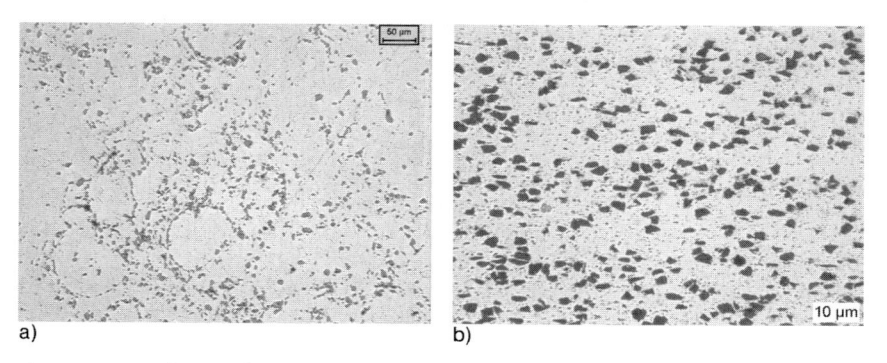

Fig. 11.19 Distribution of SiC particles in the microstructure: (a) inhomogeneous, former drop sizes decorated by particle; (b) homogeneous particle contribution by using a powder pump.

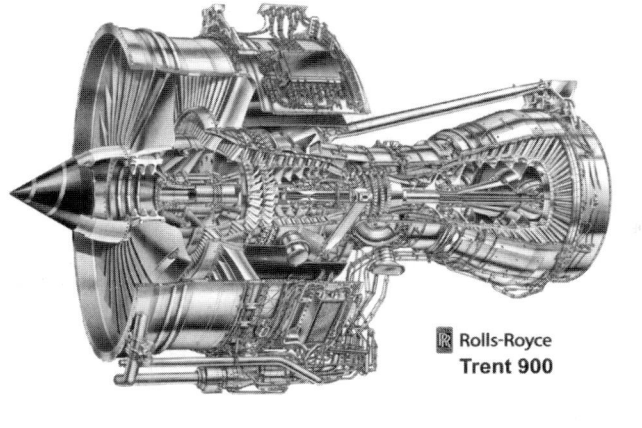

Fig. 11.20 Application of Al-MMC material as turbine guiding vane.

Spray forming in combination with powder transport provides a production method for aluminium-based MMCs at low maintenance cost and therefore suitable for serial production. Further tests with different parameters, powder sizes and types are necessary to ensure uniform properties batch by batch. In cooperation with Thyssen Automotive and Rolls Royce, within a BMBF funded project, the possibility has been validated using a demonstrator part (turbine guiding vane).

Acknowledgement

We would like to thank BMBF for kind support of the funded project "Entwicklung geschmiedeter Leitschaufeln aus partikelverstärktem Aluminum-Matrix-Verbundwerkstoff (Al-MMC) für zivile Hochleistungstriebwerke", Number 03N3090B.

References

1 R. Mächler, Diss. ETH Nr. 10332, 1993, p. 72.
2 J.T. Stanley; *Properties Related to Fracture Toughness*; ASTM STP605, 1976, pp. 71–102.
3 O. Beffort, Diss. ETH Nr. 10289, 1993, p. 102.
4 DIN 8580, Fertigungsverfahren, Einteilung, 1976.
5 W. Kurz, D. J. Fisher, *Fundamentals of Solidification*, Trans Tech Publication Ltd., Switzerland, 1989, p. 110.
6 P. Stocker, F. Rückert, K. Hummert, *MTZ 58*, No. 9, 1997.

12
Noble and Nonferrous Metal Matrix Composite Materials

C. Blawert

12.1
Introduction

Composite materials based on noble metals such as silver (Ag), gold (Au), platinum (Pt) and nonferrous metals such as copper (Cu) and nickel (Ni) are predominantly used for electro-technical applications. In some cases certain properties e.g. thermal expansion, thermal and electrical conductivity, or property combinations, like good conductivity at high strength must be obtained and cannot be achieved by a single material. In other cases pure metals and alloys may reach their application limits because of high thermal and mechanical loads. In these cases composite materials are often used to extend application areas and limits.

The predominant application of composite materials with a noble and/or nonferrous metal matrix is in the area of electrical contact and bimetal materials. Further applications, particularly of layer composite materials are in jewellery and as wear and corrosion protection layers.

12.2
Layer Composite Materials

For layer composite materials a large variety of deposition or coating processes and substrate/layer combinations exists. Even restricting ourselves to noble and nonferrous metals as substrate and/or coating materials, a detailed description of the various processes and coating is beyond the scope of the present chapter. For decades copper or its alloys have been combined, by plating or pressure welding, with steel or nickel alloys to give corrosion resistant construction materials and with aluminum or aluminum alloys to give conducting materials. Copper and nickel are used for coin plating [20]. The production of slide bearings is done by composite casting; rolled gold doublé is produced by pressure welding gold alloys onto less noble substrate materials (silver, copper, brass, bronze, nickel); steel is coated with tin, aluminum or zinc by melt immersion processes; nickel, copper

Metal Matrix Composites. Custom-made Materials for Automotive and Aerospace Engineering.
Edited by Karl U. Kainer.
Copyright © 2006 WILEY-VCH Verlag GmbH & Co. KGaA, Weinheim
ISBN: 3-527-31360-5

and chromium are often galvanically deposited and with plasma processes nearly unlimited layer combinations can be deposited. An overview of the different deposition processes and layer types is given by Pursche [40]. In the following are considered mainly layer composite materials, in which a second phase (usually nonmetallic) is embedded in the layer, thus forming a composite for itself.

12.2.1
Contact- and Thermo-bimetals

The most important properties of contact materials are the contact resistance, contact wear (abrasion, migration of material, burn-off), weldability, reliability and lifetime of electrical contacts. They can be influenced by suitable selection of the contact material. For applications in low voltage technology, pure metals (silver, gold, platinum) and/or their alloys are often sufficient as contact materials. For applications in high voltage technology, particle composite materials (silver/nickel and silver/cadmium oxide) are used besides certain alloys (for example hard silver) [1]. The contact material is then bonded by a suitable joining process to a suitable contact holder material (usually copper and/or copper alloys), see Fig. 12.1. The bonding is often obtained by weld plating, soft or hard brazing.

Thermo-bimetals are layer composite materials in which an active component, with a high thermal expansion coefficient, is joined to a passive component, with a low expansion coefficient (with or without intermediate layers). Additional surface coatings can be applied for corrosion protection. The intermediate layers (copper or nickel) are used, in order to reduce the resistance at higher currents. As a passive component Invar alloy (iron with 36% nickel) is often used, because of its very low expansion coefficient of 1.2×10^{-6} K^{-1} at 20°C. Alloys with higher nickel contents are used for more constant expansion coefficient at higher temperature. The Invar effect is not only found in the iron–nickel alloys but also in further systems, such as Fe–Pt, Co–Cr–Fe, etc. However, for economical reasons $FeNi_{31}Co_7$, $FeCo_{26}Ni_{20}Cr_8$ and $CoFe_{34}Cr_9$ alloys are predominantly used. In many cases the active component is an austenitic steel with nickel contents of around 25 wt% and

Fig. 12.1 Typical layer composites – contact bimetals (left to right Ag/Cu/CuNi$_{44}$, Ag/CuZn$_{28}$ and AgCdO/AgCd/Cu [1].

chromium contents of 3–10 wt%, which can be replaced alternatively by Mn. Manganese alloys with copper and nickel contents of approximately 10–20 wt% each are another alterative. The material selection is based on the thermal expansion behavior, on the melting temperature (>1000 °C) and Young's modulus (>100 GPa), so that, in principle, only the alloys mentioned above are used (Fig. 12.2). The indi-

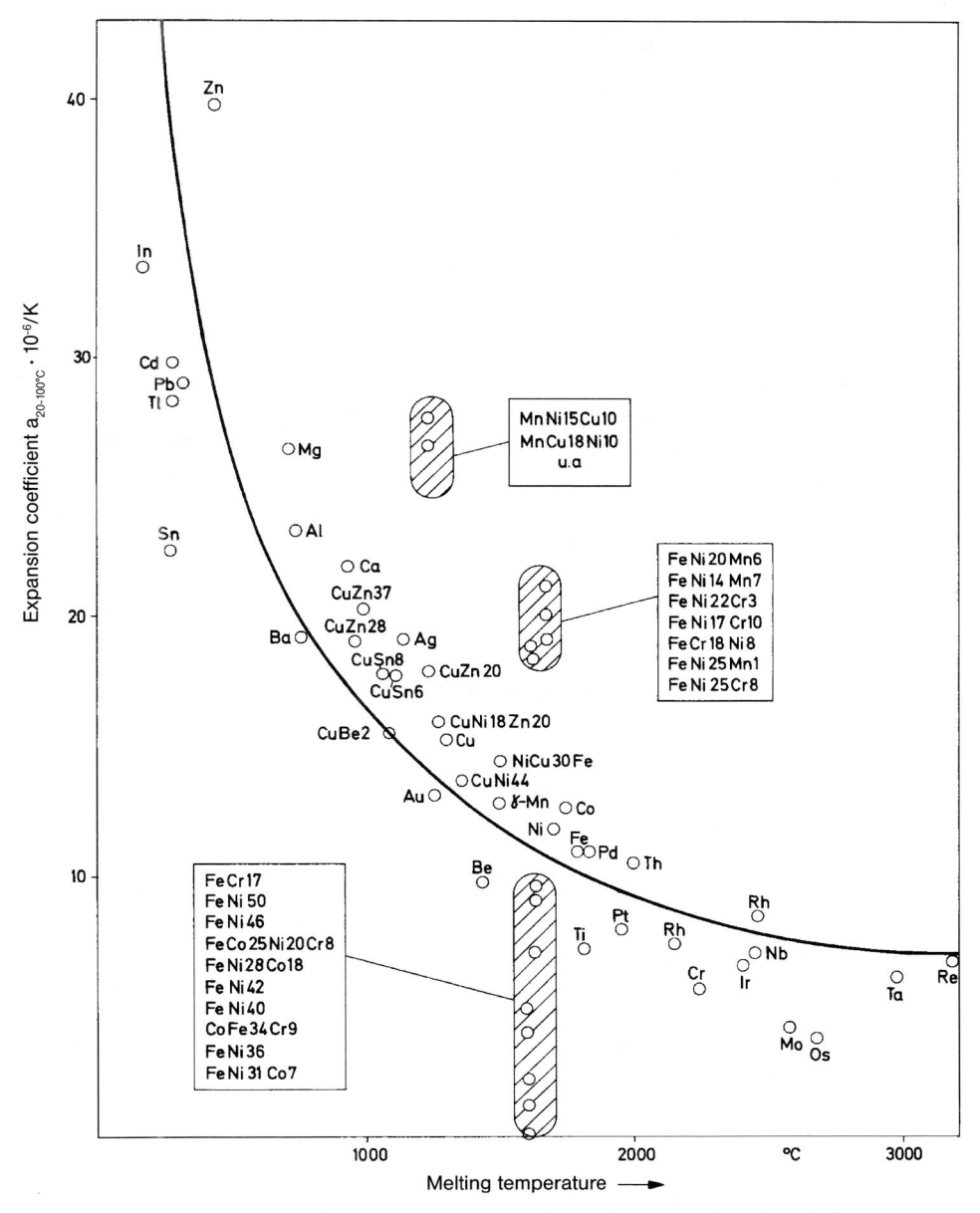

Fig. 12.2 Relationship between melting temperature and coefficient of expansion of metals and alloys [1].

vidual components are combined to the thermo-bimetal by cold-rolling or roll welding [1].

Further applications in electric engineering are magnetostrictive bimetals made from nickel/iron–nickel layer composites, spring materials for switching and contact springs. For the latter good spring characteristics and good conductivity are obtained by plating the conductor, e.g. CuCrZr with a high-strength alloy such as Cu-Ni$_{20}$Mn$_{20}$ [2]. Other applications are cladded wires. The materials for the cladded wires are selected, depending on the application, and range from copper cores with noble metal-, steel- and nickel-cladding and vice versa.

12.2.2
Wear-protection Layers with Embedded Ceramic Particles

These dispersion layers are often deposited galvanically from electrolytes, which contain a certain quantity of highly dispersive powders. The dispersoids are held in suspension by mechanical stirring, air injection or ultrasound. When applying electric current (voltage) the metal deposits at the cathode (work piece, to be coated) and the dispersoids are embedded into the metal matrix simultaneously. As disperse phases, wear-resistant components such as boride, carbides, nitrides etc. and self-lubricating components such as graphite can be inserted. Depending on the selection of the matrix, embedding components and the composition of the layer, certain properties of the layer such as hardness, strength, wear, corrosion behavior, electrical conductivity etc. can be adjusted and controlled.

Nickel-based dispersion layers are often used, because nickel behaves inertly with respect to most solid additives and layers can be relatively easy deposited. Nickel layers with nonmetallic hard phases (SiC, BN, WC, Al$_2$O$_3$ etc.) are used for wear applications and because of their good heat resistance. Nickel–chromium multilayer systems provide increased corrosion resistance and self-lubricating characteristics are obtained (good running-in, dry running properties and low coefficients of friction) by embedding MoS$_2$, graphite or boron nitride into the layer. Ni–SiC dispersive layers are applied in engine construction for the coating of cylinder contact surfaces (aluminum cylinders and/or liners) [42, 44]. Dispersion layers can be also used in order to transfer selected forces by friction. Here Ni–diamond layers are used, e.g. for coated rotors for open end spinning (textile machines), or for locking devices of round tables of numeric control processing machines [43].

Dispersion layers based on copper are often used with self-lubricating properties. Copper corundum dispersion layers are the classical example for good mechanical properties and increased recrystallization temperatures. Dispersion reinforced silver layers have a higher hardness and wear resistance, improved anti-friction characteristics and a large resistance to spark erosion compared to solid-free silver layers. The solid reinforcements are mainly corundum (Al$_2$O$_3$) and/or other oxides, but solid lubricants are also used e.g. for prevention of welding of silvered contacts [4].

While the integration of microscale particles is state of the art, and industrially relevant applications have been developed (e.g. Ni–SiC, NiP–diamond, Ni–PTFE,

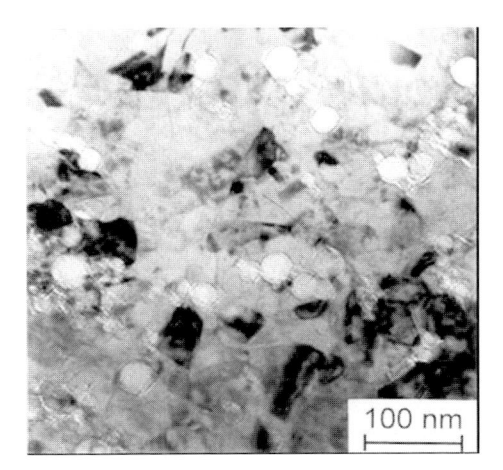

Fig. 12.3 n-Al$_2$O$_3$-particles in a nickel layer [22].

Ni–Co–Cr$_2$O$_3$) [10], the integration of nanoparticles is still under intensive research (Fig. 12.3). A necessary requirement is optimal particle dispersion in electrolytes as well as a uniform particle distribution in the layer. The following nanoparticles have been deposited successfully in nickel layers so far: Al$_2$O$_3$ [11, 21, 22, 25, 23], SiO$_2$ [25], TiO$_2$ [25], SiC [11], diamond [11]. Lead dispersion layers have also been deposited with SiO$_2$ and TiO$_2$ embedded [25].

Spray coatings reinforced with hard phases are state of the art for the improvement of the wear characteristics [26]. However, the choice of reinforcement components is limited due to thermal decomposition and/or reactivity with the matrix metal. For example, in thermal spray processes the decomposition of SiC particles cannot be prevented completely. A possible alternative is under investigation to lower the reactivity of Ni alloys by saturating silicon and carbon in solid solution [9].

Furthermore, multifunctional layers and layer combinations can be manufactured with the help of the modern plasma and vacuum technologies. A recent example of the efficiency and applicability of these technologies is given in the following. Silver, lead, MoS$_2$ and WS$_2$ are candidates for solid lubricants in vacuum applications. In order to increase the wear resistance, and lower the coefficient of friction, additional hard phases can be embedded. By simultaneous deposition of silver (by magnetron sputtering) and TiC (by pulsed laser deposition) such composite layers can be easily manufactured [41].

12.3
Particle Reinforced Composites

To the particle reinforced composites belong the large group of dispersion-hardened materials from which a special form or rather manufacturing process was already introduced in the above section on dispersion layers. In order to obtain optimal properties, the second phase (oxides, carbides, nitrides, silicate, boride, insoluble metals or intermetallic components) should be fine and homogeneously distributed, since the mechanical properties are influenced by the particle size and the distance between particles. Additional requirements are a low, or even better no solubility in the matrix, a high melting temperature and high hardness, so that the properties, once adjusted, remain even at higher temperatures.

In other particle composite materials, the hardening can play a subordinate role and the second phase takes over another function, as for example, the reduction of the burn-out with silver–nickel contact materials, or the change in the wear resistance of sliding bearing materials (e.g. by embedded solid lubricants or hard phases).

The particle reinforced composites are often processed by powder metallurgical routes (see also Chapter 10), e.g. by mixing and compacting powder mixtures in different ways. By mechanical alloying already dispersed powders are produced due to intensive milling of metallic powders and dispersoids. Composite powders e.g. $AgSnO_2$ can also be precipitated from solutions by chemical methods [14], or powders can be coated using CVD processes (fluidization process) [30]. Frequently used processes for compaction are extrusion, hot-pressing, hot isostatic pressing, powder forging and powder rolling. Some material combinations can be processed to compact materials in one single step; others are only pre-pressed and sintered afterwards, with or without pressure, to a compact composite. Nowadays spray forming is often used to avoid difficult and intricate powder handling (see also Chapter 11). To the metallic powder produced in the spraying process, particles

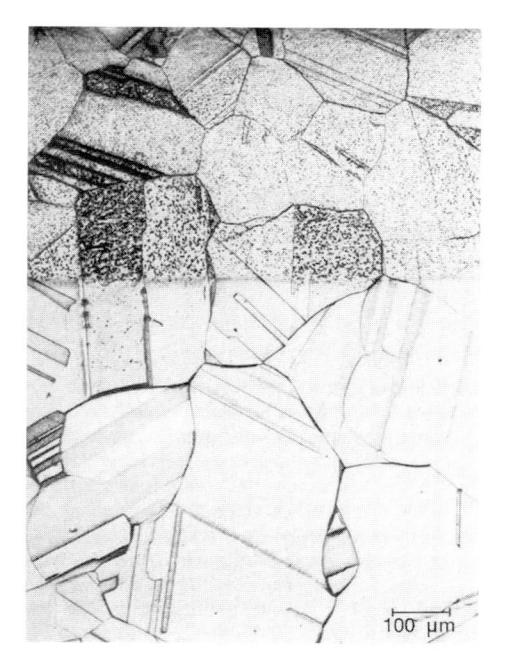

Fig. 12.4 Oxidation front in an AgCd2 alloy [1].

are added directly and deposited together in a semi-solid condition on a substrate. Usually the material will be further compacted in an additional processing step (e.g. extrusion), in order to remove or reduce porosity. Particle reinforced composites can also be produced by melt stirring and by melt infiltration of preforms.

Noble metals with high oxygen solubility, which contain certain alloying elements with high affinity to oxygen, offer another elegant way to manufacture particle reinforced composites. The diffused-in oxygen can react with the alloying elements forming fine dispersive oxides in the noble metal matrix. This process is called inner oxidation and is often used for silver alloys with small additions of cadmium, aluminum, magnesium or indium to obtain a dispersion hardening (Fig. 12.4). By the use of atomic oxygen, produced in a glow discharge of oxygen gas, the inner oxidation can be accelerated. Further possibilities to enhance oxygen up-take exist by sintering porous metal powder preforms in an oxygen atmosphere [1].

Nonferrous and noble metal matrix composites are available with a large number of matrix/particle combinations. Thus the range of possible applications is rather large:

- zirconium oxide dispersion-hardened platinum for devices of glass production [1,16]
- oxide-dispersion-hardened nickel alloys (Ni–Y_2O_3) for high heat resistance, oxidation and corrosion resistance [28]
- aluminum oxide dispersion-hardened copper for conductor materials with increased strength and good long-term strength behavior (resistance welding electrodes, commutation segments, bases for high-power transmitter tubes) [1]
- inner oxidized silver magnesium and silver manganese alloys for conducting spring materials [1]
- electrical contact materials (copper–graphite, silver–graphite, tungsten–copper, molybdenum–copper, silver–nickel, silver–tungsten, silver–molybdenum, silver–cadmium oxide and, for higher joining security silver with copper, tin or zinc oxide) [1, 2, 15, 28, 35, 36]
- heat sinks, heat conductors and carrier plates made of WCu or MoCu composite materials for electronic components [28, 29]
- dispersion-hardened electrode materials made of copper with embedded carbides and oxides [18, 20]
- dispersion-hardened copper (materials C15715, C15725 and C15760 after UNS-list, USA) by internal *in situ* oxidation of aluminum alloy components [18]
- spray formed copper composites for sliding elements, welding electrodes, wear resistance components or materials with optimised chipping behavior (Cu–C) [19, 20].

One of the most important new developments in the last decades is the contact material Ag–SnO_2 [39]. Ag–SnO_2 composites should increasingly replace the Ag–CdO composites. To adjust the higher stability of SnO_2 compared to CdO (SnO_2 evaporates at a substantially higher temperature) and to avoid enrichment of the oxide (isolating surface layer) at the switching surface, additives, such as WO_3, Bi_2O_3, MoO_3, In_2O_3 and CuO (0.05–3 wt%) are added [14, 17].

The development of SiC particle reinforced copper composite materials produced by powder metallurgical routes is under investigation and their use for applications in power devices (e.g. base plate of IGBT modules) are being evaluated. By addition of titanium into the copper matrix property improvements (adhesion) occur, but the carbide formation leads to Si impurities in the matrix, and a reduction in the heat conductivity might be the unintended result. Nevertheless, materials with 40 vol% SiC are now being tested for certain applications [7].

The production and application of particle reinforced active solder for metal-ceramic composites are also subject to investigation. Particles of SiC, TiC, Si_3N_4, Al_2O_3, ZrO_2, TiO_2, SiO_2, C in the size range 1–20 μm are used in active solder of $AgCuTi_3$ [13]. By use of the reinforcement the property transitions between metal/solder layer/ceramic can be adjusted and will be more flexible.

Great improvements in thermal stability and mechanical properties can be obtained by adding nanoparticles into copper (Fig. 12.5). By addition of 3 vol% n-Al_2O_3 the hardness can be increased from approx. 60 HV10 to 130 HV10 and the yield strength from 105 MPa to over 300 MPa. A grain growth is clearly reduced up to temperatures of 900 °C [24].

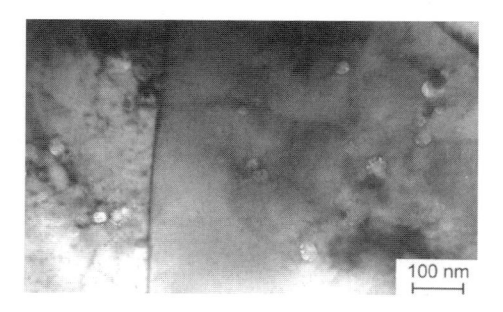

Fig. 12.5 TEM image of a Cu/n-Al_2O_3-composite after 8 h milling and extrusion of feedstock powders [24].

12.4
Infiltration Composites

Infiltration composite materials consist of a porous frame of a higher melting metal, in which a lower melting metal is absorbed. The infiltration process can take place by dipping the frame into a melt of the infiltration material (dipping infiltration) or the infiltration material and the porous substrate are heated together above the melting temperature of the infiltration material. The high-melting component is often tungsten, molybdenum or tungsten carbide, which is usually infiltrated with lead, copper or silver [2].

Typical applications for tungsten–copper, and also tungsten–silver and tungsten carbide silver infiltration materials, are highly loaded contact materials (Fig. 12.6) [28, 35, 36], electrodes for welding machines and for processing of work pieces by spark erosion [1]. In slide bearings, lead is infiltrated into a framework of tin bronze [1]. Tungsten–copper and MoCu composites are also used as heat sinks for electronic components [28].

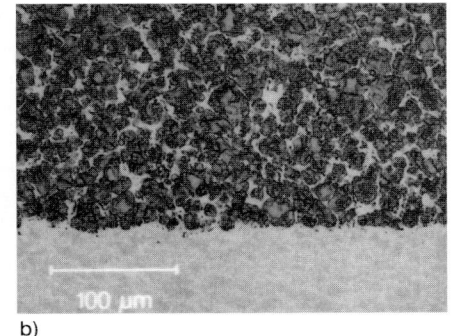

Fig. 12.6 Cuwodur® -contact parts (a) and Cuwodur® 75H – structure (b): gap free transition to the copper substrate [35].

Recent work considers the infiltration of ceramic foams (AlN) and/or SiC preforms by reactive infiltration and centrifugal casting. For higher loaded sliding bearings the porous ceramic framework is infiltrated with aluminum bronze (G-CuAl$_{10}$Ni) and tin bronze (G-CuSn$_{10}$). The materials produced by pressure-free reactive infiltration (SiC with CuAl$_{10}$Ni) show no wear in tribological investigations, even after several test runs, and perform better than the materials produced by centrifugal casting. However, the latter is still better than a nonreinforced material [5]. Due to the smaller ceramic portion, foams can be infiltrated substantially better and more completely by centrifugal casting compared to SiC preforms. Altogether, centrifugal casting can be considered as an attractive process for the production of circular composites [6].

12.5
Fiber Reinforced Composites

Similar to particle reinforced composites, a large variety of combinations of matrix metals and reinforcement components exist. Many of the particle reinforcement materials are available in the form of short or long fibers (see also Chapter 1). The selection of a suitable combination has to consider the chemical and mechanical compatibility, the required mechanical and physical properties, and not less important, the availability and the cost [1].

For production of the composites a large variety of technologies exists [1]:

- melt infiltration under pressure or in vacuum
- powder metallurgical route by sintering and/or pressure sintering
- extrusion, forging and rolling (for short fibers)
- foil plating process
- thermal spray processes (flame and plasma spraying)
- galvanic (in particular for nickel and copper as matrix) deposition

- pressure sintering and pressure welding of metal coated fibers
- joint deformation of cladded wires
- directed solidification of eutectic alloys

For some applications in power trains, rocket technology and thermal energy conversion, nickel and cobalt matrix alloys are used with high melting metal fibers of tungsten, tungsten–rhenium, tungsten–thorium oxide, molybdenum, niobium and tantalum alloys [1]. Copper reinforced with tungsten fibers is considered as a possible coating material for combustion chambers of rocket engines [34]. Ni- and Co-based eutectic composites for turbine blades are often manufactured by directed solidification [38].

However, the predominant number of applications is found again in the field of electro-technology. The most well-known examples are superconductors, where the superconducting material is embedded into a copper matrix as multifilaments. The copper matrix transfers the current in the case of normal conduction (flux jumps) and prevents a burning through of the conductor. The alloy Nb_3Sn has gained commercial significance among metallic materials as a superconductor. By inserting niobium wires into an encased tube of a copper–tin alloy, followed by a set of cold and hot deformation steps, and a final annealing, the superconducting Nb_3Sn phase is formed in the copper matrix (Fig. 12.7).

The properties of contact materials can be further increased by using fiber reinforcements instead of particle reinforcements. A typical example is a silver–nickel composite (Fig. 12.8). By embedding nickel fibers into the silver matrix better burn-out properties are attained compared to the particle reinforced variant, even if the latter was already aligned and elongated by extrusion [1, 3, 36]. Economical production is possible by the joint deformation of cladded wires. Further material combinations for fiber-reinforced metal matrix composites as contact materials are copper/palladium [2, 3] and copper/tungsten [36].

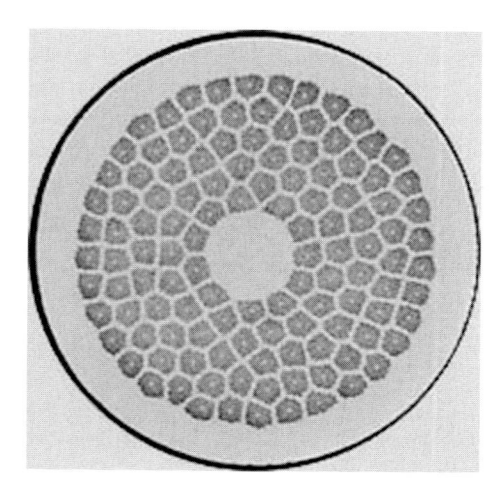

Fig. 12.7 Nonstabilized superconductor type NS 10000 with 120×84 = 10 080 Niob-filaments in a CuSn matrix [31].

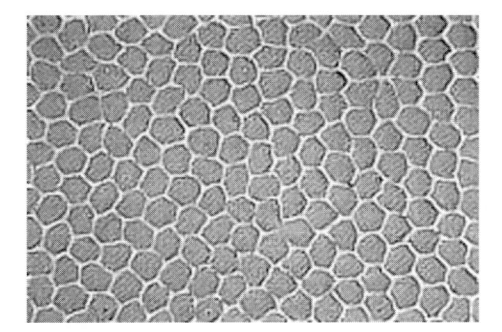

Fig. 12.8 Cross-section of a silver/nickel fiber composite with 60 wt% nickel [3].

Similar improvements are obtained for silver–graphite composite materials with a fibrous structure. To save expensive silver, copper fibers with a graphite core are often embedded into a silver matrix. Oxidation of the copper in the arc is avoided by a self-produced inert gas atmosphere of carbon monoxide (Fig. 12.9) [3].

Spark plug electrodes are exposed to similar loads as electrical contacts. Therefore fiber reinforced matrix composites are also applied for the central electrode. A corrosion resistant nickel alloy is embedded into a matrix of silver or copper, which provides excellent heat conductivity [1].

Under development is the pressure-free melt infiltration of carbon fiber reinforced carbon structures (C/C) with copper titanium alloys. The fibers require a SiC coating for successful pressure-free infiltration (minimization of the contact angle between melt and fiber). Possible areas of application are electrical and/or thermal conductors as well as tribological applications [8].

Copper–carbon composite materials are commercially available for "chip packaging". The carbon fabrics are infiltrated with copper and by the metal volume content the thermal expansion coefficient of the compound can be adjusted to that of silicon, gallium arsenide or aluminum oxide, while a good thermal conductivity is still ensured [32]. The use of copper–carbon short fiber reinforced composites for heat sinks with an expansion coefficient adapted to the electronic components (Si,

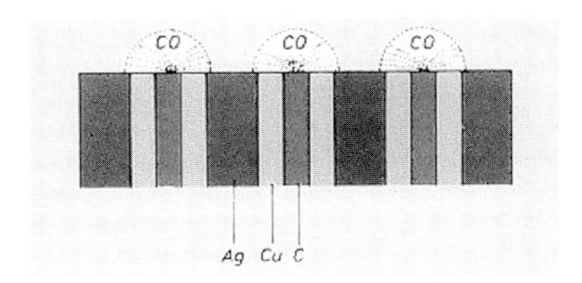

Fig. 12.9 Localisation of the carbon monoxide protection layer around the non-noble metal part as a result of the regular structure of an Ag/Cu/C fiber composite [3].

ceramic) were also the subject of investigations by Kolb et al. Hot-pressed composites with 39–44 vol% PITCH type fibers and an average fiber length of > 60 µm (aspect ratio: 6) are the most suitable combination for application as a heat sink [27]. Potential applications of graphite–copper composite materials (weight reduction compared with Cu/W) are heating and cooling devices for aviation and space application [33, 34]. However, fiber wettability, infiltration and delamination problems are not yet solved, and newer developments aim towards aluminum oxide fibers as a reinforcement component [37].

Another field of investigation is looking into the production and application of C fiber reinforced active plumbs, for connecting metal and ceramics [12, 13]. The fiber reinforcement is used to adjust the properties of the soldering zone over a wide range. The active plumbs are based on silver–copper alloys, which are modified with titanium, zirconium or hafnium, in order to wet ceramics without previous metallization. Usually the solder material is *in situ* reinforced, using laminated foil formed solder and fiber fabrics. The fiber fabrics are coated with copper for a good wettability and infiltration of the fibers.

References

1 *Metallische Verbundwerkstoffe*, Company brochure to celebrate 100 years G.Rau Company, Pforzheim, 1977.
2 *Elektrische Kontakte – Werkstoffe und Technologie*, G.Rau Company, Pforzheim.
3 http://www.rau-pforzheim.com
4 R.S. Sajfullin, *Dispersionsschichten*, VEB Verlag Technik, 1978.
5 A. Dwars, M. Eitschberger, B. Mussler, R. Schicktanz, G. Krauss, in *Verbundwerkstoffe und Werkstoffverbunde*, B. Wielage, G. Leonhardt (Eds.), Wiley-VCH, Weinheim, 2001, pp.101–106.
6 M. Eitschberger, C. Körner, R.F. Singer, in *Verbundwerkstoffe und Werkstoffverbunde*, B. Wielage, G. Leonhardt (Eds.), Wiley-VCH, Weinheim, 2001, pp.114–120.
7 T. Weißgärber, J. Schulz-Harder, A. Meyer, G. Lefranc, O. Stöcker, in *Verbundwerkstoffe und Werkstoffverbunde*, B. Wielage, G. Leonhardt (Eds.), Wiley-VCH, Weinheim, 2001, pp.140–145.
8 J. Schmidt, M. Frieß, W. Krenkel, in *Verbundwerkstoffe und Werkstoffverbunde*, B. Wielage, G. Leonhardt (Eds.), Wiley-VCH, Weinheim, 2001, pp. 322–327.
9 B. Wielage, J. Wilden, T. Schnick, in *Verbundwerkstoffe und Werkstoffverbunde*, B. Wielage, G. Leonhardt (Eds.), Wiley-VCH, Weinheim, 2001, pp. 542–547.
10 J. P. Celis, J. Fransaer, *Galvanotechnik*, 1997, 88(7), 2229–2235.
11 S. Probst, A. Dietz, B. Stindt, M. Söchting, in *Verbundwerkstoffe und Werkstoffverbunde*, B. Wielage, G. Leonhardt (Eds.), Wiley-VCH, Weinheim, 2001, pp. 563–568.
12 B. Wielage, H. Klose, L. Martinez, in *Verbundwerkstoffe und Werkstoffverbunde*, B. Wielage, G. Leonhardt (Eds.), Wiley-VCH, Weinheim, 2001, pp. 611–616.
13 B. Wielage, H. Klose, in *Verbundwerkstoffe und Werkstoffverbunde*, K. Schulte, K.U. Kainer (Eds.), Wiley-VCH, 1999, pp. 716–721
14 F. Hauner, D. Jeannot, in *Verbundwerkstoffe und Werkstoffverbunde*, B. Wielage, G. Leonhardt (Eds.), Wiley-VCH, Weinheim, 2001, pp. 644–649.
15 C. Peuker, in *Verbundwerkstoffe und Werkstoffverbunde*, B. Wielage, G. Leonhardt, Wiley-VCH, Weinheim, 2001, pp. 650–655.
16 H. Gölitzer, M. Oechsle, R. Singer, S. Zeuner, *Verbundwerkstoffe und Werkstoffverbunde*, B. Wielage, G. Leonhardt (Eds.), Wiley-VCH, Weinheim, 2001, pp. 656–661.
17 W. Weise, in *Metall – Forschung und Entwicklung*, Degussa AG, Hanau.

18 M. Türpe, in *Metallische Verbundwerkstoffe*, DGM Seminar 2001, Geesthacht.
19 M. Türpe, *Metall* **1999**, *53*(4), 211–212.
20 M. Türpe, in *Verbundwerkstoffe und Werkstoffverbunde*, G. Ziegler (Ed.), DGM Informationsgesellschaft mbH, 1996, pp. 39–42.
21 B. Müller, H. Ferkel, in *Verbundwerkstoffe und Werkstoffverbunde*, K. Schulte, K.U. Kainer (Eds.), Wiley-VCH, Weinheim, 1999, pp. 658–663.
22 B. Müller, H. Ferkel, *Nanostruct. Mater.* **1998**, *10*, 1285–1288.
23 J. Steinbach, H. Ferkel, Scr. Mater. **2001**, *44*, 1813–1816.
24 H. Ferkel, Nanostruct. Mater. **1999**, *11*(5), 595–602.
25 S. Steinhäuser, B. Wielage, in *Verbundwerkstoffe und Werkstoffverbunde*, K. Schulte, K.U. Kainer (Eds.), Wiley-VCH, Weinheim, 1999, pp. 651–657.
26 S. Steinhäuser, B. Wielage, in *Verbundwerkstoffe und Werkstoffverbunde*, G. Ziegler (Ed.), DGM Informationsgesellschaft mbH, 1996, pp. 315–318.
27 G. Kolb, W. Buchgraber, in *Verbundwerkstoffe und Werkstoffverbunde*, K. Schulte, K.U. Kainer, Wiley-VCH, Weinheim, 1999, pp. 503–508.
28 http://www.plansee.com
29 http://ametekmetals.com
30 http://powdermetinc.com
31 http://www.wieland.de
32 http://enertron-inc.com
33 S. Rawal, *JOM* **2001**, *53* (4), 14–17.
34 *Metals Handbook*, Vol. 2, ASM International, 10th Edn., 1990.
35 Firmeninformation, AMI DODUCO, Pforzheim.
36 D. Stöckel, in *Verbundwerkstoffe*, W. J. Bartz, E. Wippler (Eds.), Lexika-Verlag, 1978.
37 J.S. Shelley, R. LeClaire, J. Nichols, *JOM* **2001**, *53*(4), 18–21.
38 P. R. Sahm, in *Verbundwerkstoffe*, W. J. Bartz, E. Wippler (Eds.), Lexika-Verlag, 1978.
39 J. Beuers, P. Braumann, W. Weise, in *Verbundwerkstoffe und Werkstoffverbunde*, G. Ziegler (Ed.), DGM Informationsgesellschaft mbH, 1996, pp. 319–322.
40 G. Pursche, in *Verbundwerkstoffe und Werkstoffverbunde*, G. Leonhardt (Ed.), DGM Informationsgesellschaft mbH, 1993, pp. 669–681.
41 J.L. Endrino, J.J. Nainaparampil, J. E. Krzanowski, *Surf. Coatings Technol.* **2002**, *157*, 95–101.
42 K. Maier, in *Verbundwerkstoffe und Werkstoffverbunde*, G. Leonhardt (Ed.), DGM Informationsgesellschaft mbH, 1993, pp. 683–690.
43 J. Lukschandel, in *Verbundwerkstoffe und Werkstoffverbunde*, G. Leonhardt (Ed.), DGM Informationsgesellschaft mbH, 1993, pp. 691–697.
44 A.C. Hart, *Nickel Mag.*, Special Issue, August, **1999**.

Subject Index

A

AA2618 279
AA6061 185
– -20p 185
– -20s 185
Acoustic emission 177
Adhesion 17, 26, 27, 31, 35, 36, 38, 39, 55, 63, 100, 115, 118, 119, 120, 123, 133, 148, 150, 157, 162, 181, 200, 201, 205, 211, 218, 220, 221, 227, 231, 232, 236, 237, 238, 302
– thermal sprayed coatings 100, 116, 234
Agglomerates 259
Al metal matrix composites 147
Al–Cu–Mg alloys 290
Al-MMC 81, 293
– brake disks 92
Al_2O_3 68
Al99.85-20s 185
Alloy powders 246
– systems 6, 45
Alloys 215
– over-eutectic 215, 220
Aluminum crankcases 90, 221, 239
Aluminum cylinder head 90
Aluminum diesel piston 77, 88
Aluminum matrix composites 96
Aluminum oxide (Al_2O_3) 290
Aluminum-based MMCs 264, 272
Aluminum-matrix composite materials 95
Alusil® 90, 91, 98, 99, 101, 108, 222–224
AlZnMgCu alloys 278
Arc spraying 118, 126, 127
Armouring 133ff.
Atomisation 246, 284, 282
– advantages result 246
Atomization of metal melt 284
Automotive applications, products 289

B

B_2O_3 68
Base materials 119, 129

Blasting 115
Bonding of materials 111
Boring of Al cylinder surfaces 161
Boron carbide (B4C) 290
Bowl edge, fiber reinforced 96
Brake components 92
Brake drums, particle-reinforced 154
Buffering 133
Build-up welding
– autogenous 137
– metal active gas welding (MAG) 140
– metal inert gas welding (MIG) 140
– open arc (OA) 137, 138
– plasma hot wire 143, 144
– plasma MIG 141
– plasma transferred arc welding (PTA) 142
– resistance electro slag (RES) 139, 140
– under powder (UP) 138

C

C/Mg–Al composites 199, 201, 203, 205, 207, 209
Capillary effect 31
Carbon fibers 5, 35, 65, 66, 76, 211, 248, 274
– PAN 65
– pitch 65
– production of 65
Casting process 98, 101, 104, 105, 106
Ceramic fibers, oxidic 70
Ceramic particle, embedded 239, 241
Ceramic reinforcements 174
Characterization, composites 180, 212
Chemical deposition 246
CIP 261
Cladding 132ff.
–, autogenous- 137
–, metal inert gas- 140f.
–, open arc- 137f.
–, plasma hot wire- 143f.

Metal Matrix Composites. Custom-made Materials for Automotive and Aerospace Engineering. Edited by Karl U. Kainer.
Copyright © 2006 WILEY-VCH Verlag GmbH & Co. KGaA, Weinheim
ISBN: 3-527-31360-5

Subject Index

–, plasma mig- 141
–, plasma powder transferred arc- 142f.
–, resistance electro slag- 139f.
–, underpowder- 138
Classification 4
Coating materials 127, 130, 133, 134, 230, 231, 295
Coating techniques, thermal 111, 218
Coefficients of thermal expansion 279
Cold gas spraying 112, 118, 125
Cold isostatic pressing (CIP) 260, 281
Combustion bowl 88
– fiber reinforced 88
Combustion engine 231
Comparison of 232
Comparison of different MMCs 186
Compo-casting 8
Composites 215
– aluminum matrix 49, 50, 56, 90, 95, 96, 101, 174, 223, 241, 265
– characterization 180
– fatigue behavior 87, 94, 174, 180, 183, 184
– fiber
– infiltration
– in situ 1, 8, 152, 177, 197, 198, 199, 200, 244, 248, 249, 258, 259, 264, 266, 267, 270, 271
– light metal 1, 2, 4, 5, 6, 7, 9, 11, 18, 21, 22, 24, 40, 41, 42, 43, 44, 45, 49, 77, 86, 94, 95, 113, 147, 148, 215, 216, 218, 219, 220, 224, 229, 241
– metal matrix
– noble metal 298, 301, 305
– nonferrous metal matrix 295
– particle
Concepts 52, 95, 96, 99, 100, 108, 282
Consolidation 249, 252, 259, 260, 264, 267, 270, 281
Continuous fibers 5, 41, 52, 61, 73, 81
Cooling lubricants 228, 229, 241
– metal composites 241
Copper-based MMCs 267
Crack 38
Crankcases 90, 95 ff, 147, 152–154, 162, 163, 171, 221, 239
Crankshafts 92
Cutting material selection 147
Cutting parameter 155, 159, 163, 169
Cutting speed 154, 157, 158, 159, 162, 163, 164, 165, 166, 168, 169, 229
CVD-diamond 154
– – thick film 156

Cyclic deformation behavior 186, 187
Cyclic hardening 186-187
Cyclic softening 187
Cylinder crankcase concepts 95, 96, 155, 161, 162, 163
Cylinder crankcases 103, 104, 106, 108
Cylinder liner reinforcement 90
Cylinder liners 289
Cylinder surface 91

D

DaimlerChrysler AG 289
Damage 176–178, 192, 193
DC plasma spraying 127
Deformation behavior 15, 174, 175, 186, 187, 193
Degassing 282
Deposit layer 286
Design criteria 220
Detonation spraying 122, 123, 124
Diamond tools 262
Diesel piston 88
Dilution 132ff.
Direct squeeze casting 11
DISPAL 277
Dispersion strengthening 280
Drilling 101, 108, 141, 152, 154, 155, 165–168, 171, 218, 220, 221–223, 229, 232, 234, 239

E

Edge angle 27
Effective geometry factor 23
Embedded ceramic particle 239, 241
Expanding system 228
Expansion coefficient, thermal 5, 20, 24, 25, 45, 46, 47, 50, 55, 174, 296, 305
Extrusion 260, 261
– profiles, particle reinforced 152, 166, 171

F

Fatigue 173, 178, 183, 184
– Basquin law 181, 189, 193
– crack growth rate 182
– cyclic deformation curve 179, 186–188
– cyclic stress–strain curve 180
– Manson-Coffin law 181, 189, 193
– stress-strain hysteresis loop 179
Fatigue life 189–191, 194
– – behavior 188
Fatigue properties 173, 256
Fiber 248, 261
– application 304

– damage   261
– long   248
– production   303
– reinforced composites   303
– reinforcement   153
– short   248
Fiber composites   13
Fiber reinforcement   13, 16, 102, 186, 193, 194, 205, 248, 306
Filler wires   127, 133, 138
Fine bore processing
Fine processing   221
Finish bore processing   163
Flakes
Flame spraying   100, 112, 118, 120, 121, 122, 123, 124, 127, 130
Forging   260
Form accuracy
Fracture   38
Fracture toughness   277, 278

G

Gas atomization   281
Gas pressure infiltration   8

H

Hard metal alloys   134
Hard particle in melting
Heat dissipation   35
Heat treatment   185
HF plasma spraying   128, 129
High stiffness   290
High strength   277, 280
High velocity oxyfuel spraying
HIP   261
Honing   222, 227, 228, 232, 236
– comparison with   232
– processing technologies   232
– thin coatings   144, 220, 236
– thin coatings   236
Honing tests   234, 236, 238
Hot gas corrosion   113
Hot isostatic pressing (HIP)   260
Hot strength   280
Hot-pressing   260
Hybrid preforms 85, 91
Hydride preforms

I

Importance, material scientific
In situ composite material   258
Indirect squeeze casting   11
Infiltration   216

– composites   302
– process   32
Injectors   128, 288, 289, 291, 292
Inner interfaces
Interface   26, 197
– improvement   197
– influence   26
– optimization
Interfaces and layers   120
Interlayers   197, 198
– improvement   197
– microstructural characterisation   198
Intermetallic precipitations   280
Inverse mixture rule: Reuss-model (IMR)   23
Iron-based MMCs   269, 270

L

Laser alloying   100, 103, 218, 220
Layer adhesion   100
Layer composite materials   295
– contact- and thermo-bimetals 296
– wear-protection layers 298
Layer composites   296, 298
Light metal composites   42, 215, 229
Linear mixture rule 13
– – – Voigt-model (ROM)   22
Lokasil®   90, 91, 95, 96, 101, 103, 104, 106–108, 153, 155
– -cylinder crankcases   107
Long fiber reinforcement   13
Lubrication concept   158

M

Machinability   154
Machining   221
– problems   147
– technology   147
Magnesium-based MMCs   263
Manufacture of MMCs   249
Manufacturing techniques, alternative   73, 75
Materials
– for light metal composites   42, 215, 229
– metallic
Matrix alloy systems   6
MCrAlY   113, 128
Mechanical alloying   130, 245, 248, 249, 253, 254, 255, 257, 258, 259, 264, 266, 271
Mechanical behavior   173, 176
– stress-strain curve   176
Mechanical tests, fatigue behavior   177, 186
Melt stirring   8

Melting concept   283
Melting metallurgical processes   7
Melting metallurgy   8
Metal composites   42, 48, 215, 229, 241
Metal injection molding   260
Metal matrix composite   103, 152, 198, 243
Metal melt, atomization
Metal-ceramic-coatings   281
Metallographic testing method   131
Metallic powder   247, 300
Metallic powders   246
Metallurgical reactions   126
Micromechanical model (shear lay model)   16, 20
Microstructure   292
– of alloy 281
Microtomography   177
Milling   166, 221
Minimum pressure   32
Minimum quantity lubrication   157
Mixing   45, 132, 135, 137 ff., 249, 253, 254, 258, 259, 260, 262, 263, 264, 267, 268, 270, 292, 300
– parameters   259
– process   259
MMC   198, 263
– aluminum alloys   277
– aluminum-based   264
– atomic dimension   198
– copper-based   267
– iron-based   269
– magnesium-based   263
– materials   290
– materials, spray forming
– nickel-based   271
– titanium-based   266
Monofilament   4, 61, 62
– chemical vapor deposition (CVD)   62
Moulded paddings   130
Multifilament   61
– fibers   61, 65, 70, 71, 72

N

Nanostructured materials   130
Nickel-based MMCs   271
NiCrBSi   113, 121, 135
Noble metal matrix composite   295, 301
Nonferrous metal matrix composite   295
Nozzle unit   284, 286

O

ODS alloys   244
Osprey Metals Ltd.   286
Over eutectic alloys   215, 220
Over spray powder, re-injection
Over-eutectic AlSi alloy 149
Overspray powder   287–289
Oxide ceramic fibers 68, 70
Oxide dispersion strengthened   244

P

Particles   4, 56, 57
– Al_2O_3 57
– AlN   57
– application   301
– B_4C 57
– BN   57
– ceramic   237, 239, 241
– composites
– dispersion-hardening   299
– injectors   289, 291
– inner oxidation   301
– processing   300
– reinforced composites   299
– reinforced extruded profiles   166
– reinforced extrusion profiles   152, 166, 171
– reinforcement   8, 22, 56, 86, 148, 184, 194, 236, 302
– SiC   57
– TiB2   57
– TiC   57
Peak Werkstoff GmbH   277, 282–284, 286, 287, 289, 291
Peripheral zone   159
Permeability   34
Pickling   116
Plant safety   287
Plasma coatings   218, 220, 230, 232, 234, 238, 239, 241
– adhesion
– metallic
Plasma spraying   100, 112, 118, 127, 128, 129, 130, 132, 230, 231, 232, 303
Plastic flame spraying   121
PM-MMC   244, 249
– manufacture   249
Pneumatic transport   288
Polycrystalline diamond (PCD)   154, 156
Powder flame spraying   120, 121, 122
– metallurgical processes   8
Powder metallurgical techniques   277
– metallurgically   243
– metallurgy   1, 6, 8, 51, 220, 244, 245, 249, 255, 263, 264, 266, 271, 277, 280
– transport pumps   291
Pre-boring operation   161, 162
Preform   77, 78, 104

Subject Index

- infiltration   107
- manufacture   103
- manufacture   78, 102, 103
Pressure casting   11
Primary gas nozzle   284, 285
Primary silicon   280, 289, 290
Process engineering   7
Process parameters   153
Processing   221
Processing technologies, comparison with honing   236
Products for automotive applications   289

Q
Quality assurance   131
Quasi-MMC   289
Quasi-monolithic concepts   100, 108
Quench rate   280

R
Rapid solidification (RS) technique   280, 281
Re-injection   288
Re-injection of over spray powders   288
Reactive components in melts   218
Recirculation   52
Recycling   52, 102, 277
Reinforcement   4, 248
- continuous   248
- discontinuous   248
- of light metals   77
- particle   248
Reinforcement components, ceramic   77, 248, 266
Reinforcement of cylinder surfaces   43, 152, 153, 163, 165, 171
Reinforcements
- by light metals   5, 20, 58
- by long fibers
- by particle
- by short fibers
Remelting   113, 121
Residual stress, thermal
Resistance heating   143
Reuss model   23
Rod flame spraying   121, 122
Rolling   260
Rota plasma system   130
Rough honing   236

S
Schapery   24
Secondary gas nozzle   284, 285
Selection criteria   219

Service life behavior
Short fiber performs   78, 94, 244
Short fiber reinforcement   16, 185, 193, 194
Short fibers   73
SiC   292
- multifilament fibers   70, 71
- particle reinforced brake drums   152, 153, 154, 160, 171
Silicon carbide (SiC)   290
Silicon dioxide (SiO2)   290
Sintering   8, 69, 103, 105, 114, 133, 136, 216, 220, 249, 254, 260, 261, 266, 267, 301, 303, 304
Sintering 216, 260
SiO_2 68
Solidification   280
Spark erosion   246
Spray coatings   116, 299
Spraying   111ff.
-, arc-   126
-, cold gas-   125
-, cold gas flame-   130
-, DC plasma-   127
-, detonation-   122ff.
-, flame-   120
-, HF plasma-   128
-, high velocity flame- (HVOF)   123ff., 130
-, plastic flame-   121
-, powder flame-   120
-, singe wire arc-   127
-, suspension plasma-   129
-, vacuum plasma-   128ff.
-, wire/rod flame-   121f.
Spray forming   218, 220, 246, 247, 249, 262, 264, 267, 277, 278, 280, 282, 283, 284, 286, 287, 288, 289, 290, 292, 293, 300
Spray forming   218, 246, 262, 277, 280, 282, 283, 284, 286, 288, 289, 290, 293
Spraying techniques, thermal   103, 111, 132, 133, 144, 249, 252
Squeeze casting   7, 11
Stiffness   177, 192
Stress–strain curve   15
Stress–strain hysteresis loops   194
Structure   40
Substrate materials   115, 137, 259
Surface layer influence   47, 115, 120, 124, 147, 149, 151, 152
Surface processing   234

T
Tailored materials   279
Thermal coating processes   111
Thermal coefficient of expansion   277

Thermal expansion coefficient 24, 25
Thermal residual stresses 174
Thermal sprayed coatings 100
Thermal spraying 103, 111, 132, 133, 144, 249, 252
Thermo bi-metal
Titanium carbide (TiC) 290
Titanium-based MMCs 266
Turbine guiding vane 291, 293
Turning 100, 148, 149, 150, 151, 152, 154, 155, 156, 157, 158, 159, 160, 163, 171, 221, 222, 232
Twin atomizer 284

U
Ultrasonic precleaning 115

V
Voigt model 23

W
Wear protection coatings 252
Wear resistance 277, 280, 290
Web openings, fiber reinforced
Wettability 9, 26, 27, 29, 32, 33, 35, 71, 74, 114, 306
Wetting 31
Whisker 4, 43, 74, 75, 147, 248
Wire flame spraying 121, 127

Y
Young's modulus 22, 174, 175, 277, 278, 280, 290